GLOBAL ENGINEERING, MANUFACTURING AND ENTERPRISE NETWORKS

IFIP - The International Federation for Information Processing

IFIP was founded in 1960 under the auspices of UNESCO, following the First World Computer Congress held in Paris the previous year. An umbrella organization for societies working in information processing, IFIP's aim is two-fold: to support information processing within its member countries and to encourage technology transfer to developing nations. As its mission statement clearly states,

IFIP's mission is to be the leading, truly international, apolitical organization which encourages and assists in the development, exploitation and application of information technology for the benefit of all people.

IFIP is a non-profitmaking organization, run almost solely by 2500 volunteers. It operates through a number of technical committees, which organize events and publications. IFIP's events range from an international congress to local seminars, but the most important are:

- The IFIP World Computer Congress, held every second year;
- open conferences;
- working conferences.

The flagship event is the IFIP World Computer Congress, at which both invited and contributed papers are presented. Contributed papers are rigorously refereed and the rejection rate is high.

As with the Congress, participation in the open conferences is open to all and papers may be invited or submitted. Again, submitted papers are stringently refereed.

The working conferences are structured differently. They are usually run by a working group and attendance is small and by invitation only. Their purpose is to create an atmosphere conducive to innovation and development. Refereeing is less rigorous and papers are subjected to extensive group discussion.

Publications arising from IFIP events vary. The papers presented at the IFIP World Computer Congress and at open conferences are published as conference proceedings, while the results of the working conferences are often published as collections of selected and edited papers.

Any national society whose primary activity is in information may apply to become a full member of IFIP, although full membership is restricted to one society per country. Full members are entitled to vote at the annual General Assembly, National societies preferring a less committed involvement may apply for associate or corresponding membership. Associate members enjoy the same benefits as full members, but without voting rights. Corresponding members are not represented in IFIP bodies. Affiliated membership is open to non-national societies, and individual and honorary membership schemes are also offered.

GLOBAL ENGINEERING, MANUFACTURING AND ENTERPRISE NETWORKS

IFIP TC5 WG5.3/5.7/5.12 Fourth International Working Conference on the Design of Information Infrastructure Systems for Manufacturing (DIISM 2000)
November 15–17, 2000, Melbourne, Victoria, Australia

Edited by

John P.T. Mo
Laszlo Nemes
Division of Manufacturing Science and Technology
Commonwealth Scientific and Industrial Research Organisation (CSIRO)
Australia

KLUWER ACADEMIC PUBLISHERS
BOSTON / DORDRECHT / LONDON

Distributors for North, Central and South America:
Kluwer Academic Publishers
101 Philip Drive
Assinippi Park
Norwell, Massachusetts 02061 USA
Telephone (781) 871-6600
Fax (781) 871-6528
E-Mail <kluwer@wkap.com>

Distributors for all other countries:
Kluwer Academic Publishers Group
Distribution Centre
Post Office Box 322
3300 AH Dordrecht, THE NETHERLANDS
Telephone 31 78 6392 392
Fax 31 78 6546 474
E-Mail <services@wkap.nl>

Electronic Services <http://www.wkap.nl>

Library of Congress Cataloging-in-Publication Data

IFIP TC5 WG5.3/5.7 International Working Conference on the Design of Information Systems for Manufacturing (4th : 2000 : Melbourne, Vic.)
Global engineering, manufacturing and enterprise networks : IFIP TC5 WG5.3/5.7/5.12 Fourth International Working Conference on the Design of Information Infrastructure Systems for Manufacturing (DIISM 2000), November 15-17, 2000, Melbourne, Victoria, Australia / edited by John P.T. Mo, Laszlo Nemes.
p. cm. — (International Federation for Information Processing ; 63)
Includes bibliographical references.
ISBN 0-7923-7358-8
1. Production engineering—Congresses. I. Mo, John P.T. II. Nemes, L. III. Title. IV. International Federation for Information Processing (Series); 63.

TS5 .I35 2000
670'.285—dc21 2001029436

Printed on acid-free paper.

Printed in the United States of America.

Contents

Preface

The availability of effective global communication facilities in the last decade has changed the business goals of many manufacturing enterprises. They need to remain competitive by developing products and processes which are specific to individual requirements, completely packaged and manufactured globally. Networks of enterprises are formed to operate across time and space with world-wide distributed functions such as manufacturing, sales, customer support, engineering, quality assurance, supply chain management and so on. Research and technology development need to address architectures, methodologies, models and tools supporting intra- and inter-enterprise operation and management. Throughout the life cycle of products and enterprises there is the requirement to transform information sourced from globally distributed offices and partners into knowledge for decision and action.

Building on the success of previous DIISM conferences (Tokyo 1993, Eindhoven 1996, Fort Worth 1998), the fourth International Conference on Design of Information Infrastructure Systems for Manufacturing (DIISM 2000) aims to:

- Establish and manage the dynamics of virtual enterprises, define the information system requirements and develop solutions;
- Develop and deploy information management in multi-cultural systems with universal applicability of the proposed architecture and solutions;
- Develop enterprise integration architectures, methodologies and information infrastructure support for reconfigurable enterprises;
- Explore information transformation into knowledge for decision and action by machine and skilful people;

These objectives reflect changes of the business processes due to advancements of information and communication technologies (ICT) in the last couple of years.

DIISM 2000 has attracted a large number of contributors. All proposed abstracts were considered by the Organising Committee before accepted for presentation of the full paper. Each full paper was reviewed in a confidential reviewing process by three of the forty members of the International Program Committee (IPC). For a paper to be accepted, the paper must be accepted by at least 2 reviewers. We would like to thank the Chairman of IPC, Prof. John Mills, Dr. Jan Goossenaerts and members of IPC who worked hard to ensure the quality of the papers.

A total of 56 papers were accepted for presentation and are included in this book. These papers describe the state-of-the-art development in information infrastructure systems with strong emphasis on the applications to manufacturing. We are very pleased to see a wide variety of application areas being tackled by authors in DIISM 2000.

This book is divided into nine parts. Part 1 features the keynote of the conference on corporate knowledge issues by Mr. Ron Beckett, a senior executive in industry. Vast amount of information can be generated by computers nowadays but these data would not be useful unless they are organised and retained in the company. Rapid changes in virtual enterprises make it difficult for companies to maintain a consistent knowledge base for supporting their businesses.

Part 2 explores the nature of virtual enterprises and their operating characteristics. This provides an understanding of how business is changing by the application of ICT. To understand the virtual enterprises from a theoretical point of view, Part 3 introduces methodologies on the modelling and analysis of virtual enterprises.

Supply chain management involves the logistics and management of supplies with developed products and services. It is a subset of virtual enterprises issues and is discussed in Part 4. Part 5 explores the latest e-Commerce and e-Service business activities and discusses some of the potential application models.

Virtual enterprises are formed when there is a business opportunity. In manufacturing environment, product development and knowledge management is critical to the success of a company. Part 6 discusses various aspects of product development and life cycle issues.

In Part 7, we elaborate further on the issues of knowledge management. Contributions in the topics of knowledge models, ontology, enterprise knowledge creation, capturing and software systems are useful references.

Part 8 is devoted to contributions with new developments in information technologies for supporting information infrastructure systems in manufacturing. Papers in Part 9 describe systems and work in integrating manufacturing activities at a more traditional CIM level.

Dr John P.T. Mo
Project Manager
Global Manufacturing
CSIRO Manufacturing Science and Technology

Dr Laszlo Nemes, FTSE
R&D Manager
Manufacturing Systems and Automation
CSIRO Manufacturing Science and Technology

Foreword by the Conference Chairman

Very few countries need the more advanced information infrastructure for manufacturing than Australia. The country's population is concentrated in five major cities 800-4000 km apart. The remaining 39% of its people are living in small towns scattered across an area roughly equivalent to that of the USA. Manufacturing industries want to serve these markets effectively in timely manner. As customers change their expectations, producers need to adapt themselves quickly since suppliers have to accommodate their products to the wishes of the purchasers.

In the present world of complex supply chains, Australian component suppliers are linked to many overseas companies. To enhance the operational agility and global networking capabilities of Australian manufacturing firms they need tools and methodologies for enterprise management, cooperative planning, concurrent engineering, adaptable production automation, decision support, and interaction protocols for operating as "virtual" organisations. Fast and reliable communication is essential. Complex application programs are required to communicate data, information and knowledge both ways for the benefit of increased businesses.

Globalisation of manufacturing will require many companies, particularly Australian SMEs, to focus on providing one or more specific core capabilities within global alliances. Australian industry must be able to participate in these virtual enterprises in response to the continual dynamics of global market opportunities. Of critical interest for them will be design of the extended enterprise, network information management, computer supported collaborative work, scheduling and coordination, and the robustness, stability, and optimisation performance of the supply networks.

More than ever, operating conditions will be characterised by frequent change in product, process, market, or supply and distribution networks. Success therefore also demands well-coordinated agility in all internal aspects of an enterprise. For such agility, aspects of critical importance are techniques for rapid product development, adaptable production systems, and the related information and decision support systems. These issues identify many important research areas.

Understanding the behaviour and performance of highly dynamic distributed enterprise networks and characterising the influences of key factors. This will provide fundamental insights into the most appropriate architecture on which to base such enterprises, as well as the decision support requirements for the management of complete supply network life cycles. Performance metrics are needed for, and stability of, distributed system control in dynamic environments. Coordination protocols, control mechanisms, risk assessment and algorithms are needed for conflict resolution in a global dynamic environment.

Studying the theoretical foundation for robust agent-based manufacturing systems. We need characterisation of manufacturing domains receptive to an agent-based approach. We have to identify potential advantages and develop migration strategies from legacy systems to distributed manufacturing.

Industry demands corporate knowledge management systems. One focus here is the integration of information along the supply chain. This is crucial for cooperative decision making and inter-enterprise resource management to provide high quality global customer support, service and problem resolution. A complementary requirement is to underpin the full range of decision support needs associated with supply network life cycles.

I am sure many of you can add topics to this list. This conference has highlighted some of these problems and it is an important step in sharing knowledge in many very important areas. As the outcome of this conference, this proceeding represents a major milestone in research and application for information infrastructure for manufacturing.

Dr Laszlo Nemes, FTSE
R&D Manager
Manufacturing Systems and Automation
CSIRO Manufacturing Science and Technology

Foreword by the Chairman of International Program Committee

The 4th International Conference on the Design of Information Infrastructure Systems for Manufacturing - or DIISM 2000 for short – continues the tradition started in Tokyo in 1994 and continuing every two years in Eindhoven, The Netherlands, Fort Worth, USA and now Melbourne, Australia. The intent of the originators of the conference series was to evaluate and demonstrate *current* activities, progress and results in the development of infrastructures supporting the creation and management of information in manufacturing in the large.

The aim of *this* conference was to extend the understanding and agreement about design methods, models, modelling languages, reference frameworks, novel implementation approaches (e.g. agents), services and architectures for information infrastructure systems for global and multicultural manufacturing. This is a rapidly changing domain, with new developments being announced almost weekly.

The papers in this volume, reviewed by the International Program Committee and revised by the authors address the issues in the following areas:

- Establishment and management of the dynamics of inter and intra-enterprises for the support of global and multi-cultural manufacturing and engineering;
- Modelling and co-ordination of information system requirements and development of solutions for virtual and extended enterprises;
- Current and potential enterprise integration architectures, models, implementations, methodologies and information infrastructure support for extended and virtual enterprises; and
- Knowledge in Manufacturing and how information is related to and transformed into knowledge;

The papers in this volume address issues in areas ranging from highly theoretical topics such as semiosis, knowledge and its representation, and intelligent and autonomous agents to practical experiences with introducing and implementing existing technologies such as CORBA, XML and expert systems for production operations. Information infrastructures for virtual and extended enterprises were discussed in some detail, but no conclusions about standard models or methodologies were reached. A number of papers cover progress in international projects such as Globemen, and HUMACS/PSIM, facilitating interchange of ideas within these projects and with others outside of the projects. The excellent quality of the papers is a reflection of the hard work and dedication of the IPC to whom I am much in debt. The overall field continues to be dynamic and the conference achieved not only the goals of the current organizers, but the intent of the originators of the series.

The next DIISM conference is already in the planning stages under the outstanding leadership of Dr Arai of Osaka University.

John J. Mills, Ph.D.
Chairman, the International Program Committee

PART ONE

Keynote

Accessing Corporate Memory – Some Knowledge Structure Concepts

Ronald C Beckett
Hawker de Havilland, Australia
Beckett.ronald@hdh.com.au

Abstract A model of a "Corporate Memory" that influences an organisation in carrying out its purpose, and is a repository of information and knowledge beneficial to the future operation of the organisation is used to suggest ways of deriving value from different components of that "Corporate Memory". The model is combined with a generic representation of key organisation functions (in this case, representative of a manufacturing organisation) to provide a basis for knowledge "maps" that reflect what an organisation knows and where that knowledge is located.

BACKGROUND

If, as many argue, intellectual assets are more important than tangible assets in effectively achieving the purpose of an organisation in the 21st century, serious effort must be put into using these intellectual assets efficiently, in the same way that efficient use of tangible assets is the norm today. But what are these assets, how do we retain them, and how do we know if they are being used efficiently.

Intellectual assets are developed using knowledge and learning processes, and throughout this paper, different views of knowledge and knowledge transfer will be explored. There is reference to the value of knowledge. Under some circumstances people in a particular community will take a philosophical view of knowledge and learning, placing value on the improved understanding of their environment provided. In other circumstances, a pragmatic view will be taken, placing value on the use of knowledge to produce beneficial outcomes, and that is the view taken in this paper. Historically, Universities would be associated with the former view, and Business with the latter, but today it is necessary to strike a balance between the two. If a University does not commercialise its knowledge to some extent, or if a Business does not understand how its environment is changing, then both face extinction.

Thus, for a particular organisation, there will be internal and external aspects to the knowledge important to it that will contribute to a distinctive "corporate memory". And it is suggested here that the balance is changing, and will continue to change, with external knowledge becoming more important. For example, in a "virtual organisation", the bulk of the knowledge accessed will be outside of the notional "organisation". So in discussing "Corporate Memory" in this paper, whilst internal and external elements are featured, one needs to think flexibly about where the boundaries might be.

If a representation of corporate memory is to be of value, then a number of issues have to be addressed:

- What aspects of corporate memory might be of most value (is it the company policy manual, or is it something else)?
- How is this memory made visible and accessed (particularly if significant components of it are outside of the organisation)?

The approach taken to considering these issues is to use a model as a framework for discussion, but consider ways that knowledge sets may be represented and structured to simply provide visibility of different facets of knowledge.

AN INTELLECTUAL ASSETS PERSPECTIVE

Historically the perceived value of a company has been driven by its financial and capital assets. But today, some of the worlds largest companies have a market value many times the value of their capital base. The additional value is considered to be related to "customer assets" and "intellectual assets", with an emphasis on the latter. Some [14] seek ways to characterise and value these intangibles. Some component parts of these assets are individual competencies, internal structures (eg unique practices and systems) and external structures (eg networks of contacts).

Efforts to identify and enhance core competencies are seen as the source of a company's sustainable competitive advantage, and considerable efforts to codify expert knowledge to make it easier to both retain by a company and to share internally to enhance operations [15]. Legal protection of ownership by patent, copyright, trademark or whatever else makes sense is a focus for many companies. Such approaches have an underlying assumption that drawing all these resources within a company will offer an advantage.

But other approaches pull together a network of companies that between them have all of the resources to tackle a particular opportunity or task, and this involves sharing intellectual property. Some, such as Australian company Moreton Bay Ventures go further, and adopt an open source approach, making new knowledge freely available on the condition that subsequent enhancements and applications will be available back to the company at no cost. Then Moreton Bay Ventures develops higher level applications using the enhanced knowledge.

These quite different strategies; *formalise knowledge to retain control of it, or formalise knowledge to share it and stimulate its growth, have a common objective – obtain leverage from what the company "knows".* Just what an organisation "knows" and how to characterise it will be discussed in subsequent sections of this paper.

OBTAINING LEVERAGE FROM A REPRESENTATION OF CORPORATE MEMORY

What does an organisation "know", and how can this provide leverage. What an organisation "knows" can be characterised using the notion of a "Corporate Memory" that influences the organisation in carrying out its purpose, and is a repository of information and knowledge beneficial to the future operation of that organisation. It will be reflected in the repertoire of practices and routines that are the norm for the organisation, and will have both tacit (vested in people) and explicit (documented and codified) components [12]. It will have both internal (e.g. company computer systems) and external (e.g. a network of contacts) aspects.

In Europe, some information technology researchers are exploring possible codified corporate memory attributes. In broad terms, they have identified the following:

- Different kinds of interfaces that might suggest decisions to the user, or explain results, or critique input decisions
- An administration function that inserts rules, finds redundancies and contradictions
- A data base that has case-specific information, general information on external rules and data attributes, and an ontological or meta-information layer that controls the evolution of the information repository

Another group trying to establish a knowledge reference model is focussing on several key design objectives:

- Ease of use, building on experience with book referencing, library science and such-like existing analogues
- Semantic precision, with information relationships and descriptors

- Freedom from buzzwords
- Portability of content, such that the system is not dependant on a specific technology solution
- Adaptability to continuous change and growth that can benefit from the cumulative judgements of multiple experts.

Bearing all of this in mind, a high level systems engineering style model was evolved over a period of a year or so with contributions and critique from some colleagues. The corporate memory model is made up of a mixture of knowledge sets (some of which may be outside of the business) that could be treated like sub-systems of a total system, with information or knowledge flows between them. The model (Figure 1) has 8 sub-tier knowledge sets:

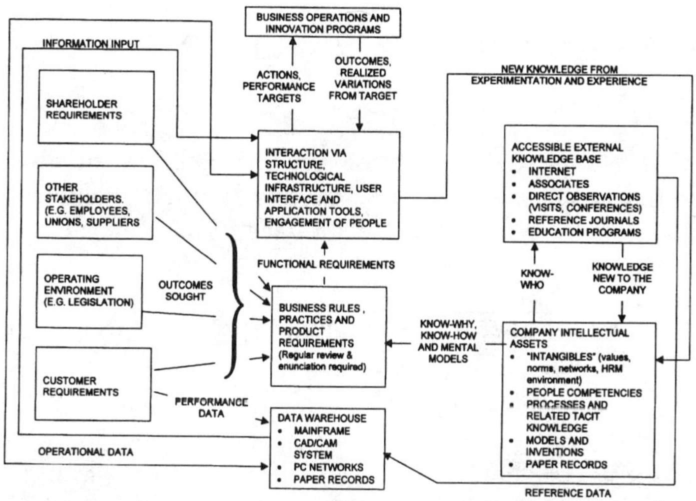

Figure 1 Model of Corporate Knowledge Attributes

- Various kinds of external contacts (generally a “know – who “ knowledge set)
- An internal know how knowledge set (commonly thought of as “intellectual assets”)
- Owner influences and rules
- Employee/ Community influences and rules (eg through union intervention)
- Customer influences and rules
- Company data warehousing of different sorts
- Operational/Business rule sets and routines

- Operations implementation strategies that determine how the knowledge flows will interact with the firms primary business.

POTENTIAL SOURCES OF LEVERAGE

From examination of the sub-tier knowledge sets described above, different kinds of leverage a particular business might develop by focussing on a particular sub-tier set can be envisaged, as shown in the table below. It might be noted however, that individual sets are part of a total system, and to realise and sustain leverage, multiple knowledge elements are involved. For example, a good Franchise Operation commonly has a Data Warehouse and Business Rule-set as integral parts of the Operational Implementation

DESIRABLE KNOWLEDGE REPRESENTATION ATTRIBUTES

So we have a model, and we can see ways of potentially obtaining leverage from our knowledge assets, but how can the "knowledge" within each sub-tier set be represented so it can be shared and enhanced (Table 1)?

Table 1 Examples of sub-tier knowledge set

SUB-TIER KNOWLEDGE SET	EXAMPLE OF POTENTIAL LEVERAGE
COMPANY INTELLECTUAL ASSETS	• STRENGTH IN CORE COMPETENCIES • PATENTS • COPYRIGHT MATERIAL
EXTERNAL KNOWLEDGE BASE	• EARLY TREND & OPPORTUNITY SPOTTING, DEAL BROKERING • POWERFUL "VIRTUAL" ORGANISATION
DATA WAREHOUSE	• RE-USE OF KNOWLEDGE (EG EXPERT ASSISTANTS) • TARGETED MARKETING (EG READERS DIGEST APPROACH)
OPERATIONS/BUSINESS RULES	• MAKE OR CHANGE INDUSTRY NORMS
OPERATIONAL IMPLEMENTATION	• AGILE SYSTEM • FRANCHISE OF WELL DEVELOPED, PROFITABLE BUSINESS SYSTEM

Lundvall and Johnson [10], noted four kinds of knowledge:

- Know-what, which is knowledge about facts. Know-what is information that can be broken down into bits and easily codified
- Know-why, which is knowledge about principles and laws – it reduces the frequency of errors in technological trials

- Know-how, which is skills, the capability to undertake a given task successfully
- Know-who, which is information about who knows what and who knows how to do what

They also suggested that in most organisations, these types of knowledge must cover at least three distinct domains: technical competencies and capabilities, organization capabilities and "system" capabilities in terms of interactive links.

It was noted that know-what and know-why are closest to the traditional concepts of science that can be readily transmitted as information. Know-how and know-who are not so easily transmitted, requiring personal contact, observation opportunities and social interaction with an extended network.

In considering ways to exchange knowledge in a recent international research program, knowledge to be transferred was classified as available in Documentary form (reports, e-mail), or in Procedural form (models, processes) or as Background knowledge (personal or organisational tacit knowledge). In the same project, management roles of co-ordinator, collaborator and communicator were formalised to facilitate operations in a "virtual" project environment. In broad terms, they dealt with technical, organisational, and systems knowledge domains respectively, and were thus consistent with the above observations [10].

With ready access to the Internet and other sophisticated data search possibilities, Berreby [3] notes that the situation can lead to a paradox: extra details can obscure patterns and make it harder to get useful facts; and quotes the view of a colleague concerned with knowledge management – that the emphasis is no longer on information processing, storage and analysis; but on representation. He discusses the use of a range of sensory perceptions besides words and numbers: colour, texture, sound; to be able to rapidly assimilate "represented" information through metaphor and analogy. A related view is that the expanding supply of codified knowledge is increasing the demand for skills relating to the recognition of patterns in data and selecting relevant data for scrutiny [11].

In dealing with large volumes of information in the past, people have developed special "maps" (eg a street directory) and "Indexes" (eg library indexing systems). Such devices enable information to be organised in a hierarchial way that enables big-picture visibility and top-down searching, or using supplementary information, detailed visibility and bottom-up searching. They also have a relatively stable structures and a small number of attributes (eg the standard symbols for roads, railway lines etc on a map) that provide semantic precision, but which combined together provide a large amount of information at a glance.

Some of the notions discussed above are put together in Table 2 to suggest a knowledge classification and representation approach. The notion of "maps" will be considered again later in this paper.

Table 2 Knowledge classification and representation

KNOW-LEDGE TYPE	CLASSIFICAT-ION APPROACH	REPRESENTATION APPROACH	EXAMPLE
KNOW-WHAT	Documentary (eg reports)	Analogue map Plus associated digital data	• Street directory using standardised symbols • Document framework with standard software (eg Microsoft Word)
KNOW-WHY	Procedural (eg models)	Analogue map Plus associated digital data	• Street directory using standardised symbols • Model framework with exchange standards (eg STEP)
KNOW-WHO KNOW-HOW	Background (eg experience unique to an individual or organisation)	Shared personal information within a classification structure Plus Personal contact processes	• "Who's–Who" indexes • White pages / yellow pages telephone directories • Video-conferencing

The approaches to knowledge representation discussed so far require a stable framework within which the representation resides. The model of corporate memory presented earlier is considered to be stable, and is useful in understanding *where* knowledge may be found, and how different knowledge flows lead to action, but it does not address the issue of different knowledge domains. It can be argued that at some level of abstraction, all organisations are the same, for example, as noted earlier, all will have technological, organisational and systems aspects to their operations. All manufacturing operations buy things, work on them, and package and sell things. All organisations have some kind of Human Resources management system.

DIFFERENT VIEWS OF ORGANISATIONAL KNOWLEDGE

Here are some illustrative examples of models used within some parts of an Aerospace company, Hawker de Havilland, over a number of years to provide a systems framework for corporate knowledge.

The first was used in the early 1990's during a period of intensive Business Process Re-engineering [2]. It was derived from work done in an associated company to try and develop a coherent computer systems approach between disparate divisions of a recently agglomerated business. The business process was defined as a number of functional systems linked by information flows. The systems were:

- *Operations research:* containing executive information systems, some longer term business opportunity research and planning functions
- *Engineering design:* containing product development and definition sub-systems and compliance demonstration functions
- *Manufacturing engineering:* containing process development and build definition sub-systems and change management functions
- *Materials management:* containing provisioning and purchasing sub-systems and inventory management functions
- *Shop floor management:* containing daily planning and scheduling sub-systems and various reporting functions
- *Customer order support:* containing estimating and pricing sub-systems and order administration / technical support functions
- *Quality management:* containing quality system definition and maintenance sub-systems and continuous improvement functions
- *Personnel and administration:* containing payroll and job recording sub-systems and other personnel functions
- *Financial control and accounting:* containing financial and cost accounting sub-systems and various reporting functions
- *Production scheduling and control:* containing production strategic planning sub-systems and overall manufacturing control functions
- *Project management:* containing programme and configuration control sub-systems and various liaison and reporting functions

Each functional system and sub-system had a brief description, and a data dictionary provided a consistent terminology for information that flowed within and between functional systems.

This model was used for a number of purposes:

1. As a framework to collect information about a myriad of mainframe, personal computer and manual systems used by individuals within the organisation to carry out their daily tasks
2. As a checklist to confirm that all activities had been assigned to someone after a substantial re-organisation
3. As a benchmarking framework to compare resource utilisation in different Operating Divisions of the organisation

A second model, currently in use in some parts of the business was developed by a colleague, Ross Penfold, building on his experience with such modelling in another industry. This model defines, at a very general level, systems that it is considered any business must have as:

- *Leadership and management systems:* including vision, mission, strategic planning and strategic change elements
- *Customer and commercial systems:* including marketing, financial management, information technology and procurement elements
- *Human resources systems:* including employee and labour relations , people and organisational development, performance measurement and remuneration system elements
- *Technology systems:* including production processes, process capability development and process control elements
- *Innovation systems:* including research and development, and continuous improvement methodologies
- *Quality systems:* including quality assurance systems and quality control processes elements
- *Asset management systems:* including maintenance, OH&S assurance and environmental management elements

As with the previous model there are sub-systems that are probably company or industry unique. There are three views available of each system; a daily operations view, a tactical planning and organising view, and a longer term strategic view. The model can be accessed from the company intranet system, with hyperlinks to existing documents, procedures, computer programs or web-based information. At this stage, the full representation is only available for some systems.

It is noted that how useful particular knowledge is depends on environment and context. If this changes with time, then maintaining an up-to-date knowledge system can be a significant task. For example, a Company may retain copies of some particular legislation to keep up to date, but if the rate of change is too high, or the company does not have some-one who can interpret the significance of changes, this information will not be useful. The company may choose to take expert advice from a consultant instead, moving from an internal knowledge set to an external one. Examples like this highlight the need for a stable reference architecture that is kept relevant by some simple process.

Another issue relates to the opportunity to convert new knowledge to action within a particular organisation. Acquiring new knowledge that cannot be used may not be regarded as value adding. A number of potential cultural barriers are observed [13]:

- *Ignorance:* those who have the knowledge don't realise others may find it useful, those who could use it don't know it exists
- *No absorptive capacity:* lack of resources or study time to make adaptation of an idea useful

- *Lack of pre-existing relationships:* no personal dialogue that builds confidence and shares views
- *Lack of motivation:* No clear reason for making a change perceived.

O'Dell and Grayson [13] have also observed a number of systemic barriers that tend to dominate some companies

- *The silo company:* focussed units with no incentive to share information
- *The NIH company:* that values local knowledge creation over knowledge sharing
- *The Babel company:* with far-flung employees that lack a common purpose
- *The By-the-book company:* that considers documented knowledge is the only valid form
- *The Bolt-on company:* that considers adding knowledge transfer responsibilities to a duty statement is all that is needed.

It is suggested that different forms may exist within one organisation.

SOME PARTICULAR ISSUES IN THE "VIRTUAL ORGANISATION"

In considering a conventional organisation using the model of corporate memory described earlier, there were four internal knowledge sets that influenced the working of the organisation, and four external ones.

It is suggested here that in a "Virtual Enterprise", there is only one internal knowledge set: that related to a negotiated set of business rules, with an implementation strategy knowledge set being evolved as the work of the enterprise proceeds. This may be illustrated by drawing on a case study of a Film-making Enterprise [8], where an Enterprise is established for the project, then completely disbanded when the project is finished. The project starts with an idea by one or a few individuals plus some shareholders who determine artistic and financial "rules". The project accesses knowledge through clusters of industry specialists and resources (eg around Hollywood) who help develop an implementation strategy, then do the work. Knowledge transfer is primarily through socialisation processes occurring in parallel with film production, and some elements of "corporate memory" " are held at an Industry level, not at the level of an individual firm.

Today, the expression "Virtual Enterprise" tends to be associated with electronically connected, remote participants (but as the example above illustrates, this need not necessarily be the case). Where a team is separated in time and/or place of work, it may be considered that "rules" and "intellectual assets" are being represented in terms of information exchange standards and software. Considerable effort is being put into the development of these tools, which facilitate transfer of documentary and procedural knowledge. But it seems that tools to facilitate exchange of background knowledge are less developed.

Coleman [4] has recently presented some views on electronic collaboration and the evolution of "community". Several definitions of electronic collaboration are offered, but the one selected to support discussion here is "intentional group processes plus software to support them", where collaboration is seen as many-to-many and goal oriented, whereas communication is seen as one-to-one and unstructured. A kind of scorecard is presented to help people think about the readiness of their organisation to pursue full scale electronic collaboration, with the following factors being rated out of 10:

- Technology (provides everything needed to collaborate) – weighting factor = 1
- Culture (trust, common goals, acceptance of risk-taking and sharing) – weighting factor = 2
- Economics (is it economically critical to collaborate)–weighting factor=3
- Politics (management believes it is important) – weighting factor = 4

It is noted that technology is not the main driver. Coleman has also observed that the weighting factors may be different in different countries, depending on the national disposition towards teamwork. The factors shown above are for North America. The evolution of "community" is seen as developing from network applications such as e-mail and groupware in the early 1990's through knowledge management in the late 1990's. Whilst real-time collaboration tools (audio, visual and data) are rapidly evolving, there is considerable turbulence in the product range available.

There are suggestions that a virtual enterprise be treated like a project, with a finite life. This is effectively the approach supported by the GERAM architecture referred to earlier. Many conventional firms are becoming more project oriented as product life cycles decrease and as more work is outsourced, but there are special issues associated with distributed project management [5].

DISCUSSION

This paper set out to explore some knowledge structure concepts for accessing corporate memory. A representation of that memory, characterised as a number of interlinked generic knowledge sets has been presented. It has been noted that all knowledge sets will have both explicit and tacit knowledge components. In further discussing approaches to structure, it appears that a multiplicity of views, each with sub-tier components is needed. A particular organisation's operating environment and infrastructure may help or hinder knowledge transfer. The diversity represented in these different perspectives is possibly what makes the concept of knowledge management complex and fuzzy when it comes to implementation of a program of some sort.

Allee [1] observes this fuzziness and contends knowledge is "too complex and fluid to be designed, processed and managed from an old thinking perspective". She observes twelve qualities of knowledge:

- "Knowledge is "messy"" (interconnections between aspects of knowledge and the contextual significance associated with it make it hard to compartmentalize it)
- "Knowledge is self-organizing" (it has a life of its own, it is created and killed off as purposes and values change)
- "Knowledge seeks community" (as illustrated by the explosive growth of the Internet)
- "Knowledge travels on language" (without language and the jargon associated with a particular field, we cannot communicate what we know)
- "Knowledge is slippery" (too much formality, e.g. in codification, can lead to the unwanted side effect of stifling creativity and new knowledge)
- "Knowledge likes looseness" (we can waste resources trying to control knowledge processes too tightly – the survival rate of diverse, decentralized systems is higher)
- "Knowledge experiments" (the on-going conversation about knowledge is more important than the right answer in opening up new options to explore)
- "Knowledge does not grow forever" (unlearning, letting go old ways, contributes to the vitality and evolution of knowledge)
- "Knowledge is a social phenomenon" (only people together can make knowledge happen. Knowledge managers cannot manage knowledge itself, only some processes for acquiring it and using it)
- "Knowledge grows organically" (it is a waste of time to create rules about knowledge, it is better to remove barriers to self-organization)
- "Knowledge is multi-modal" (it must be supported at multiple levels and in various ways that support a systems approach, reflection, experimentation)

- "Knowledge is multi-dimensional" (privileged positioning of explicit knowledge, communication and sharing of tacit knowledge and enhancing knowledge competencies all lead to more effective ways of creating, adapting and acquiring knowledge)

Particular organisations have tried to deal with this complexity by focussing on a subset thought important to the organisation, but this has not always been successful. Lucier and Torsilieri [9] noted that some initiatives taken in the name of knowledge management (eg putting a company manual on-line) may not lead to improvement unless there is associated action to beneficially change some current practices. Similarly, Davenport [7] observes that just building a framework, without content that will make business sense and stimulate its use, may be wasteful.

In this paper, the intention is to establish a multi-tier framework that is relatively simple. Earlier discussion had suggested primary links between particular corporate memory knowledge sets and the way they might provide leverage to deliver value to a particular firm, eg franchising might build on well developed implementation strategies. This kind of view might be used to make business sense of the content of a corporate memory representation. Taking this approach, for each knowledge set, documentary, procedural and background knowledge elements would be identified for each domain relevant to a particular business. And for each domain an application area would be identified (eg strategic, tactical or operational). This would yield the kind of framework element illustrated below, and could provide the basis of Knowledge "maps" of an organisation (Figure 2).

For example, if the 8 knowledge sets of the corporate memory model described earlier were combined with the 7 business systems identified by Hawker de Havilland, then there would be 56 framework elements of the type represented in the diagram above. Current knowledge access status could be readily appreciated by scanning a 8 X 7 matrix representation where each entry contained a symbol (eg tick / cross) representing status (eg substantial knowledge accessible; or some knowledge accessible; or little knowledge accessible or status unknown). This would enable scanning of the matrix to see where gaps exist, for comparison with the Enterprise's strategic needs, as there might be some framework elements less important than others in particular circumstances (eg individual employee career development may not be a key issue in a virtual enterprise). A similar approach could be taken within each framework element. Whilst most businesses may require a number of sub-tier domains (eg project management as a subset of Leadership and Management systems) to be identified to provide adequate visibility, (substantially increasing the number of elements), this relatively simple pictorial overview is still considered to be useful .

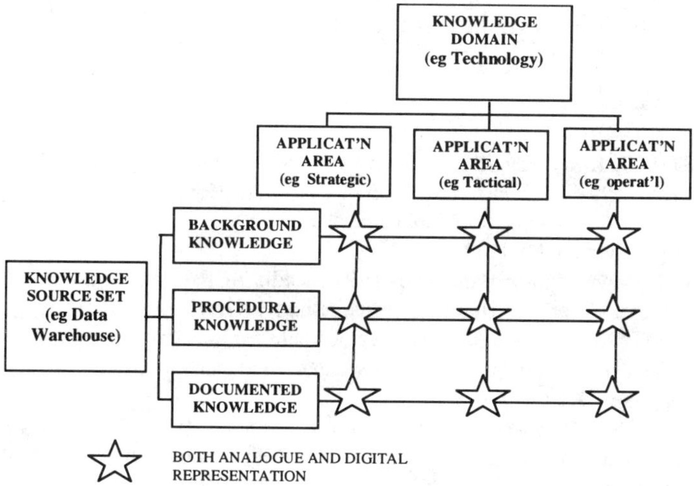

Figure 2 Knowledge maps of an organisation

It was noted in the figure above that each knowledge type / knowledge application area combination should have both an analogue and a digital representation. An example of this would be a "mindmap" created utilising *MindManager* software that can flexibly represent knowledge artifacts, with access to more detail via sub-tier maps and hyperlinks (refer www.mindjet.com for further details).

The notion of knowledge "maps" described here is consistent with the views of Davenport [6] who felt that knowing where knowledge can be simply accessed is important to the success of a knowledge program.

CONCLUSIONS

A model of "corporate memory "has been used to indicate how different internal and external facets of that memory might deliver value to a particular organisation. It is suggested that a "Virtual Enterprise" has a larger number of external components to its "corporate memory", and has some special knowledge transfer needs.

The model plus its characteristic components (documentary, procedural and background knowledge) has been used in conjunction with generic knowledge domain and knowledge application views to suggest a stable framework that assists in "mapping" organisational knowledge.

A number of success factors and potential barriers to be considered in any knowledge program have been noted. These indicate that whilst technology and culture aspects are important to the successful operation of a knowledge program, sound economics and a supportive Enterprise political climate are critical in the establishment phase.

REFERENCES

[1] Allee V (1997) Knowledge management: a life of its own. HR Monthly, May

[2] Beckett R.C (1999) "A Characterisation of Corporate Memory as a Knowledge System" Proceedings of Knowledge Management conference – KNOW'99, University of Technology, Sydney, September 26/27

[3] Berreby D (1996) Finding the knowledge needle in the data haystack. Strategy and Business, Booz-Allen and Hamilton Issue 5, pp 84 –92.

[4] Coleman D (2000.1) Accelerating the value chain for the virtual organisation. Presented at a conference on "Project and Knowledge Management in the Virtual Corporate World", Sydney, August 16-17

[5] Coleman D (2000.2) Successful Practices in Distributed Project Management. Presented at a conference on "Project and Knowledge Management in the Virtual Corporate World", Sydney, August 16-17

[6] Davenport T H (1996) Some principles of knowledge management. Strategy and Business, Booz-Allen and Hamilton, Issue 2 pp 34 – 40.

[7] Davenport T H (1997) Known evils – Some pitfalls in knowledge management. C.I.O, July

[8] DeFillippi, RJ and Arthur M.B (1998) Paradox in project-based enterprise: The case of film making. California Management Review, Berkley, Winter issue, Vol 40, Issue 2, pp 125-139

[9] Lucier C E and Torsilieri J D (1997) Why knowledge management programs fail: a CEO's guide to managed learning. Strategy and Business, Booz-Allen and Hamilton, New York. Issue 9, pp14 – 28

[10] Lundvall, B.A. and Johnson, B. (1994) "The Learning Economy" Journal of Industry studies Vol 11, No 2 pp23-42

[11] Marceau J., Manley K. and Sicklan, D. (1997) "The High Road or the Low Road - Alternatives for Australia's future" Australian Business Foundation, Sydney, August

[12] Nonaka I and Takeuchi H (1995) The knowledge creating company. Oxford University Press

[13] O'Dell C and Grayson C J Jr (1998) If only we knew what we know – the transfer of internal knowledge and best practice. The Free Press, New York

[14] Sveiby, KE (1997) The New Organisational Wealth: Managing and Measuring Knowledge –Based Assets. Berrett-Koehler

[15] Whyte A (1997) Knowledge – its creation and management as a key business driver at Morgan and Banks. Knowledge Management Conference, Sydney, Australia. 22 – 23 September

PART TWO

Virtual Enterprises

Virtual Enterprise Architecture and its Supporting Methods/Tools for Managing Supply Chain System Life Cycle

Yoichi Kamio, Eiichi Yamamoto
Toyo Engineering Corporation, Japan
Email: kamio@ims.toyo-eng.co.jp

Kazuo Morita
Mitsui Engineering & Shipbuilding Co., Ltd, Japan

Yoshiro Fukuda, Yasuyuki Nishioka
Hosei University, Japan

Keywords Virtual Enterprise, Application Service Provider, Supply Chain Management, Supply Chain Planning, Inter-Enterprise Management

Abstract In global and competitive business environment, agile virtual enterprise set-up to meet customer needs has become more important. Many companies make a lot of efforts to reengineer their business processes and for example reform existing rigid and inefficient inter-enterprise data interchange to open and efficient supply chain by utilizing advanced information technologies. This paper discusses set-up of virtual enterprise that deliver services of life cycle management of supply chain system.

1 INTRODUCTION

Engineering firms have been accustomed since their origin to organizing a project group involving enterprises concerned, in order to implement Engineering/Procurement/Construction (EPC) for client's one-of-a-kind product, which is often a plant or factory. The engineering industry has long been holding work execution conformation that could serve as the origin of today's Virtual Enterprise (VE). Engineering firms have been concentrating their efforts on project management based on such work execution conformation.

As telecommunication and information processing technologies developed, this work execution conformation has become accepted by other one-of-a-kind industries, such as shipbuilding, heavy machinery, and building. This work methodology has recently been attracting attention of general machinery industries, as Supply Chain Management (SCM). In the Globeman 21 Project [1][2][3][4] we defined this work execution methodology as Virtual Enterprise (VE) and studied its features.

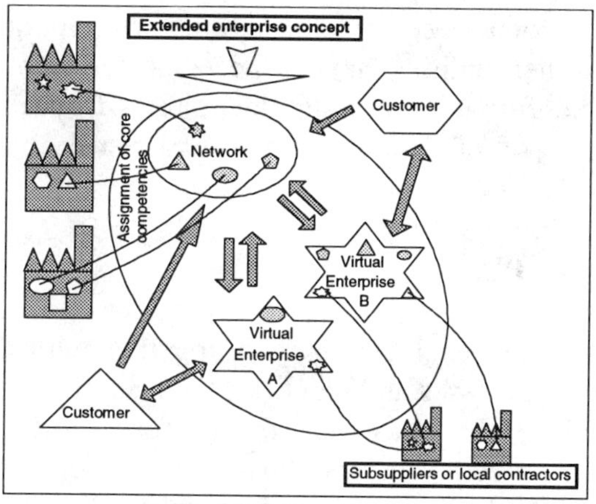

Fig.1 VE Concept

The VE is initiated with consultation for clarifying specifications based on client's specific requirements. As to design, construction, operation, and specific activities, competent enterprise groups are combined to implement the project, as shown in Fig. 1. When an enterprise has completed its role in the project, the enterprise goes out of the VE, leaving technical data about the object. Upon completing the project, the VE is disorganized, while related data is accumulated through the network in the real enterprises that have participated in the project, for use for renewal and maintenance of the project plant or factory. In order to realize such an enterprise entity, information infrastructure and management are needed.

This paper discusses the basic structure of inter-enterprise management as well as infrastructure for constructing the required VE.

2 LIFE CYCLE MANAGEMENT FOR VE

A long-life product, such as plant and factory, serves for 30 to 40 years from the beginning of the project to scrapping, through construction and operation. The project EPC, implemented from the project planning stage to the start of plant operation, takes only 2-3 years. While conventional engineering work covered the 2-3 years, a new work-type engineering firm should consider Virtual Enterprise (VE) to offer services during the whole product life cycle. The product life cycle covers: sales, consulting, contracting, EPC, delivery, and operation support and maintenance.

Each of these phases requires project management, technical support, special technologies, labour, etc. Since it is almost impossible for any enterprise entity to offer, control, and implement all technologies for all the phases, it has become necessary to organize a temporary team, involving suppliers and other entities having necessary technologies. This report analyses work conformation at individual phases from the viewpoint of VE and on the basis of engineering firm's experience, and summarizes the requirements.

2.1 Consulting Phase

In this phase, the engineering firm, together with the client, analyzes existing business processes and identifies the specifications required for "to-be" systems. This service must be performed for a short period, although it is normally paid. The work requires client's basic information to facilitate the business process analysis. Since confidentiality is crucial, the engineering firm excludes suppliers from the VE team and gathers necessary information without involving suppliers. In other words, from the viewpoint of VE, the client and the engineering firm implement the work, under a **1:1** relationship. It sometimes occurs that the client performs the work in-house using engineering firm's tools and methodology because the client does not want to disclose its own information. In this phase, therefore, an environment appropriate for prompt work support should be built, and temporary lending out of methods and tools may be considered.

The roles of the engineering firm in this phase are (1) providing analysis tools and methodology, (2) providing a concept of the new system, and (3) providing methodology for modelling specifications. In this phase, the engineering firm is not requested to provide management services. The output of this phase is identified product (system) scope, purpose, structure, and components, and serves as the input to the EPC phase.

2.2 EPC Phase

The Engineering/Procurement/Construction (EPC) phase starts on the basis of specifications clarified as a result of the consulting phase. In some cases, the prime contractor, or engineering firm, serving for the EPC phase may differ from the prime contractor for the consulting phase. The outcome of the consulting phase should be specifications understandable for third parties.

When the EPC phase starts, the organization of enterprises that compose the VE has not yet been completed. The organization is gradually completed as engineering work progresses. That is, the VE group is composed of a few members around the main contractor that manages the progress of the entire project, and the group grows as the project work develops. The main contractor, or engineering firm, coordinates the client and suppliers; the main contractor works with a 1:N relationship.

In the construction phase, a number of subcontractors participate. At this stage, while information about the VE flows in two ways between VE members, technical information and managerial information should be concentrated at the engineering firm. The roles of the engineering firm at this phase are: (1) splitting work, (2) securing resources, (3) managing the progress, (4) managing the functions of the entire system, (5) managing technical information, and (6) managing information flow inside the project. That is, while an **N:N** relationship is constructed in the VE, all information should be concentrated at the engineering firm.

2.3 Operation Support and Renewal Phase

Previously, engineering firms were not active in the phases after the delivery of product. Engineering firms participated in the Operation Support and Renewal (OSR) by, for example, bidding for revamping etc. Individual suppliers also participated in OSR by supporting operation and maintenance of their equipment. In this paper we define the OSR phase as a stage where client needs can be well understood and client-oriented services can be offered. Based on this understanding we provide "OSR Community Environment" as illustrated in Fig. 2. In this community, the client and suppliers are combined with **N:N** relationship, and a VE is organized of most suitable members to provide services on the basis of the contract with the client (monitoring, diagnosis, training, preventive maintenance, revamping, etc.).

It is necessary to reorganized design information produced in the EPC phase so that the information can be used in the OSR phase. The roles of the engineering firm in this phase are: (1) monitoring the entire system, which is the product, (2) providing tools and methodology required for solving problems, and (3) providing necessary resources.

Fig. 3 summarizes the changes in VE at various phases in accordance with the life cycle.

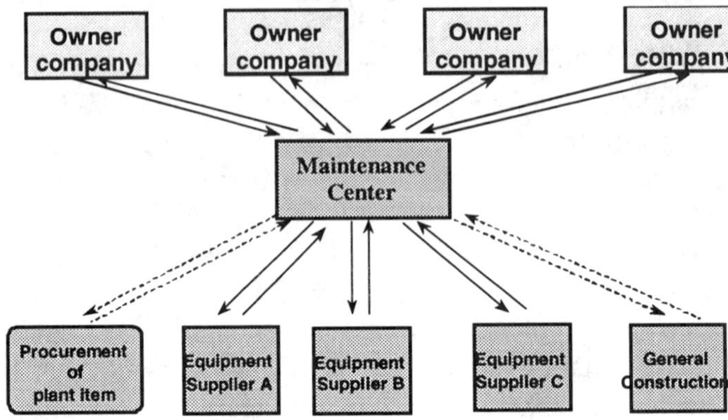

Fig.2 OSR Community Environment

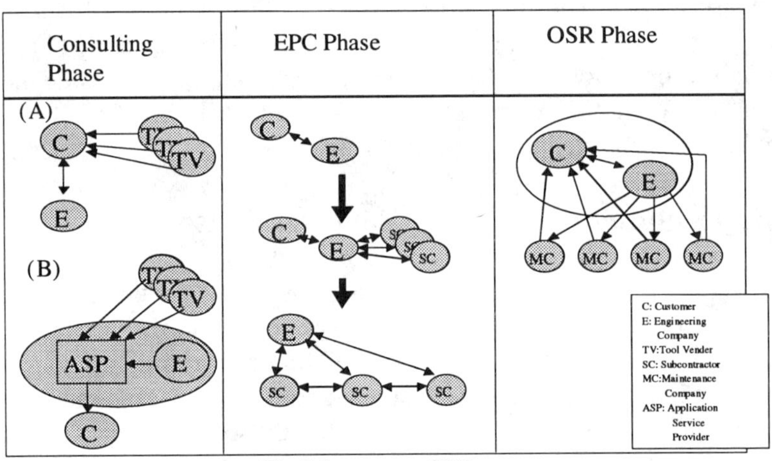

Fig.3 Structure of VE

3 REQUIREMENTS FOR VE LIFE CYCLE

TEC implements engineering activities to create SCM, in addition to conventional engineering firm's activities. With regard to SCM, TEC's activities are focused on the consulting phase. Based on such experience, we have analyzed the VE activities in the entire lifecycle activities, and have revealed the following major requirements. In order to realize VE:

1) VE should assure free, open participation.
2) VE should be able to combine special capabilities.
3) Each VE member should be able to identify his or her roles in the VE.

4) VE should allow information to be exchanged freely.
5) VE should not depend on OS and other specialty.
6) VE should allow VE Technical data and models of the object system to be transferred when the phase changes.
7) VE should be able to guarantee data to be interchangeable during the life of equipment and other hardware.
8) VE should be able to change the basic composition of VE from phase to phase.

Items 1), 2) and 3) relate to organizational problems, 1), 4), 5) and 6) to information infrastructure problems, and 6), 7) and 8) to project lifecycle problems. With regard to project organizational problems, project functions through phases should be broken down and described clearly. For this purpose, the domain of each function and input/output specifications should be defined at least. To satisfy this requirement, techniques for modelling of objects should be utilized. As a tool for modelling, IDEF for example is available for description. For SCM in which the objects are known to a certain extent, modelling techniques with templates such as SCOR are used.

As to information infrastructure, communication functions and information interchange is a problem. This problem can be coped with by the combination of the Internet and browser functions; however, the solution is applicable mainly to documents and screen images prepared in advance. In project execution, real-time information exchange and decision-making through conference are important, and this requires particular infrastructure. In the case of VE, the problem of information exchange is wider, covering documents, drawings, images, applications, etc. Since information exchange is needed throughout the entire product lifecycle, information exchange viewing future prospect is needed. This requires not only exchange of applications that is presently made, but also information exchange intermediate form such as XML, which is independent of applications. Where the applications to be used are known to a certain extent, use of Application Service Provider (ASP) is available. In this case, at least data should be independent.

The problem of lifecycle is important for project management although it has been little aware of. Depending on the lifecycle, resources including organization, application and tool are different. The prime contractor (engineering firm) may be different depending on the lifecycle, so the model must be described in advance. The description can be made with GERAM etc.

Table 1 shows the result of organizing functions needed for supporting the construction and operation of VE and for managing inter-enterprise processes. Fig.4 shows the summary of specific methodology referring to building supply chain system.

Table.1 Methods/Tools for VE and IEM

VE set-up & operation	Open network environment	Internet/Extranet, WEB
	VE & business process Modelling	IDEF, GERAM, ARIS, SCOR
	Application interoperability	Java, CORBA, ASP
	Document exchange	XML, SGML, STEP
	Security management	
Inter-enterprise management	Database management	OODB
	Schedule/Cost/Quality control	PMS tools
	Progress monitoring & control	PMS tools
	Scheduling & coordination	APSTMIZER, PSL

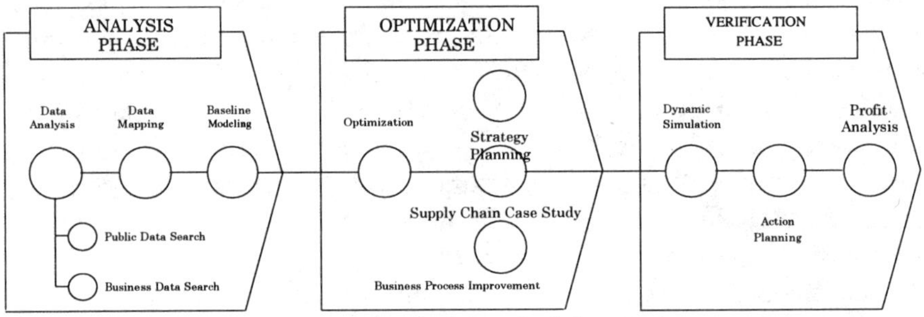

Fig.4 SCP methodology

GLOBEMEN-Japan intends to realize the whole picture of project management as shown in Fig. 5 in accordance with the VE features described in Chapter 2 and the requirements described in Chapter 3, in order to expand the project management in whole.

4 CONCLUSION

This paper summarizes architectures necessary for realizing OKP engineering activities using VE and proposes architecture based on the experience in engineering activities. This architecture will be realised as industrial prototype in GLOBEMEN project [5].

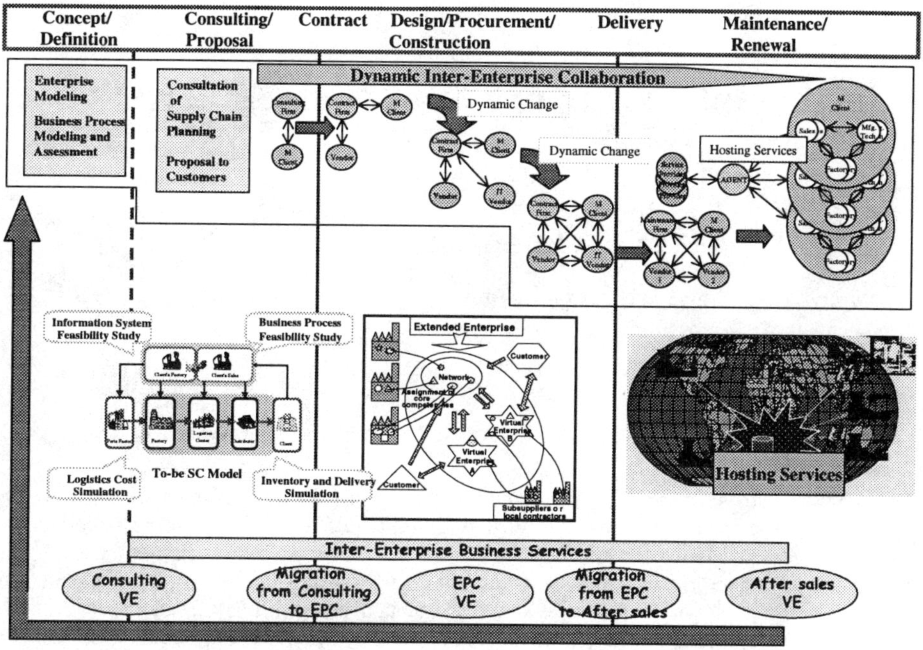

Fig.5 Life Cycle Structure for VE

5 ACKNOWLEGEMENT

This research has been carried out under the GLOBEMEN project, one of the IMS International Programs. The authors would like to acknowledge all the members of GLOBEMEN project for variable support.

6 REFERENCE

[1] Globemen21 project web page http://www.ims.toyo-eng.co.jp/Index2.htm

[2] Bernus P., Mertins K., Schmidt G. (Eds.) Handbook on Architectures of Information Systems, Springer Verlag 1998

[3] GERAM: Generalised Enterprise Reference Architecture and Methodologies, ISO/DIS 15704, mainly developed by IFIP/IFAC "Task Force on Enterprise Integration" and IFIP WG 5.12 "Architectures for Enterprise Integration"

[4] Tolle M, Verterager J., Pedersen J.D. A Methodology for Virtual Enterprise Management – Results from IMS 95001/ESPRIT 26509 Globeman21 project, Proc. Of the 6th International Conference on Concurrent Enterprising, Toulouse, France, 28-30 June 2000, p. 119-127

[5] GLOBEMEN project web page http://globemen.vtt.fi/

Flexible Infrastructure for Virtual Enterprises

A.T.M. Aerts, N.B. Szirbik, J.B.M. Goossenaerts
Eindhoven University of Technology, Eindhoven, The Netherlands
email: A.T.M.Aerts@tue.nl, {N.B.Szirbik, J.B.M.Goossenaerts}@tm.tue.nl

Keywords Virtual Enterprise, Infrastructure, Flexibility, Mobile Agents

Abstract In recent papers we studied the use of a mobile agent framework to support the primary process in a Virtual Enterprise. We argued that the installation of standard software modules, called service bridges or docks, at the participant enterprises provides a suitable infrastructure for the use of mobile agents. The deployment of mobile software agents using such modules has been studied in applications such as networked electronic trading and mediation of negotiations. The emphasis in these applications lies on demonstrating the potential of mobile agents for the support of complex decision problems. As such they focus on the agent - agent interaction. In this paper we study the agent - system interaction. We will discuss a number of change cases and examine their impact on the requirements on the agent - agent and on the agent - system interaction arising from the need for flexibility.

1 INTRODUCTION

A Virtual Enterprise (VE), in our view, is a conglomerate of regular enterprises that collaborate on an ad hoc basis. The term was derived from "Virtual Organisation" [5], which denotes a group of people from different organisational structures. These people usually have a common short-term goal, and they form for a short period a team that can be viewed as an organisational structure, which crosses the boundaries of the long-term organisations they belong to. Similarly, a VE is an ad hoc organisation that responds to unexpected change, for example to one-of-a-kind business opportunities, such as the one-time production of a specific landing gear in the case of some unexpected damage, a batch of mobile phones as part of some promotional activity, or some emerging niche market. Of course, the enterprises that participated in the production of one particular item will know how to find each other when a similar opportunity arises. The result will be a group of enterprises, with complementary capacities, that can collaborate on the production of a set of related products, such as aircraft subsystems, whose complexity transcends the abilities of a single enterprise. Early examples of virtual enterprises include the nineteenth century whaling industry and nowadays the movie industry (for a history of the VE concept see [8]).

Characteristic for the members of a VE nowadays is that they commit only a minor part of the entire production capacity to the VE (see Figure 1). The major part is committed to long-term alliances, e.g., to one or more supply chains in which they are a link.

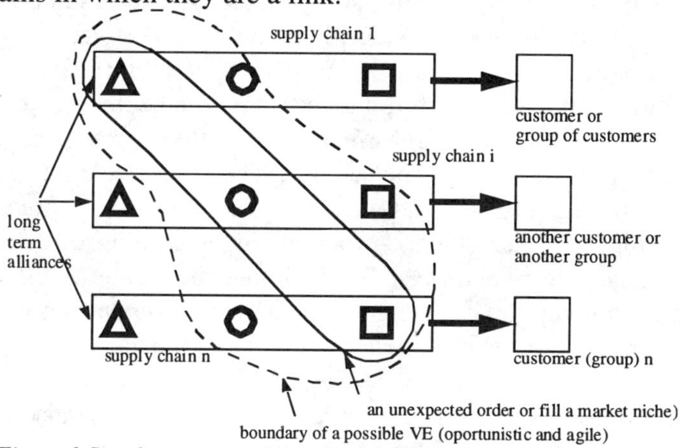

Figure 1 Simultaneous participation of enterprises in an ISC and a VE

1.1 VE and Integrated Supply Chain

Enterprises use the VE strategy to meet unexpected change and unforeseen events (i.e., become agile). One of the beneficial results is that unused capacities or planned over-capacity can be made productive. To cope with momentary unavailability of a particular type of capability, a VE will include several members with similar capabilities (redundancy). This is opposed to the concept of *lean structure*, but it will help the VE itself to be agile [8]. In view of the opportunistic nature of the business, the integration of the ICT systems of the participants will be at the operational level, limited to the exchange of data on availability, prices and products. These characteristics distinguish the VE from a more long-term inter-organisational structure, such as the Integrated Supply Chain (ISC). An ISC is a lean, stable business organisation that caters to a relatively stable market. Usually, such an organisation is designed to cope with foreseeable changes in supply and demand. There are stable procedures for fast global re-planning and local re-scheduling, which make the organisation flexible, but it is not agile, since it will have difficulty coping with unexpected change. The ICT integration in a

Supply Chain is at a higher level, involving also the exchange of forecasts and production schedules, and the capacities are committed for a longer period of time. This level of integration is needed to provide the necessary performance in terms of delivery time and costs.

1.2 VE and e-Business

A VE is also distinct from an e-Business, which typically uses the World Wide Web as an additional communication channel to reach clients. Two new interesting examples of e-Businesses, which among other things address the problem of providing aircraft subsystems, are Aeroxchange [1] and MyAircraft [11]. Both companies define a digital marketplace for aerospace parts and components, where parts can be leased, bought, sold or auctioned.

In addition, services are offered to facilitate the exchange of scheduling and replenishment planning information, so that both customers and suppliers can optimise their supply chains. These services may range from supply chain management, including collaborative demand forecasting and supply planning, to trading and e-procurement, including order promising, and vendor managed inventories. To facilitate the usage of these services system integration support is offered. Both virtual companies are run from a central website that allows all parties involved (customers, suppliers) to selectively communicate with each other. A specific feature of both is that the provider of the software is a separate and highly specialised company, Oracle for Aeroxchange and I2 for MyAircraft.

In this paper we will take the point of view that the VE has been formed. A discussion of the formation process can be found, e.g. elsewhere in these proceedings [15].

2 MOBILE AGENT INFRASTRUCTURE

Enterprises can join or leave the VE at short notice, depending on capability and opportunity. The volatility of virtual enterprises imposes strong requirements on their ICT support. In order for the VE to be agile, the ICT infrastructure must be highly flexible. Typical events that a VE has to be resilient to are:

- An enterprise joins the VE.
- An enterprise leaves the VE.
- The VE expands its catalogue with new products.
- The VE merges with another VE to enter a new market.

2.1 Mobile Agent Support for the VE primary process

One way to provide the required flexibility is to use a mobile agent framework. The deployment of mobile software agents has been studied in applications such as networked electronic trading [4] and mediation of negotiations [13]. In the latter, software agents play the role of a human supervisor, who is in charge of the tracking, monitoring and problem management of a specific product item. This is inspired from real business cases, where rush orders are assigned a special manager to supervise their fulfilment. The Mobile Agent System can be envisaged as a Support System for this human manager. When a customer places an order, e.g. via a Web portal, which is accepted, an agent is made responsible for filling the order. The agent monitors the assembly of the ordered product and sends out agents to provide the necessary components. These agents move to those enterprises in the VE that have the capacity to deliver the required components at the right time. If an enterprise needs components from other enterprises to produce its own component another batch of agents is sent out to supervise the delivery of these components.

The agents are programmed to perform the monitoring task, and also handle exceptions. For instance, it is possible that one of the enterprises in the VE cannot commit, for whatever reason, production capacity that it had previously advertised as available. The agent that comes to claim this capacity finds that it is no longer available and has to find another enterprise in the VE to provide the required capacity. This will of course entail some negotiations, e.g., to free the required capacity at the right time and arrange compensation. The agents take care of routine tasks, including negotiations, and involve the human decision maker in more complicated or unforeseen situations.

2.2 Mobile Agent Framework requirements

A Mobile Agent framework has to satisfy a number of properties to provide a suitable infrastructure [9]. In [2] we argued that specification of agent behaviour by means of rules makes it possible to change to behaviour by changing some of the rules. Given an appropriate rule system this could perhaps even be done by the decision maker the agent has to support.

Deployment of the agent infrastructure should also be easy. A party that wants to join a fairly volatile organisation such as a VE has a low level of commitment and should not be required to invest a lot of time and effort to be able to join. In recent papers we studied the use of a mobile agent framework [7] to support the primary process in a Virtual Enterprise [2]. We argued that the installation of standard software modules, called *docks* and *service bridges* at the participant enterprises provides a suitable infrastructure for the use of mobile agents. At the dock, agents can be created, and sent out, and they can come back there to communicate with the local information system. Since the docks usually are available on several platforms, they shield the agent environment from hardware and OS heterogeneity. The Internet provides the basic connectivity. The service bridge provides the semantic alignment of the local IS to the common shared ontology.

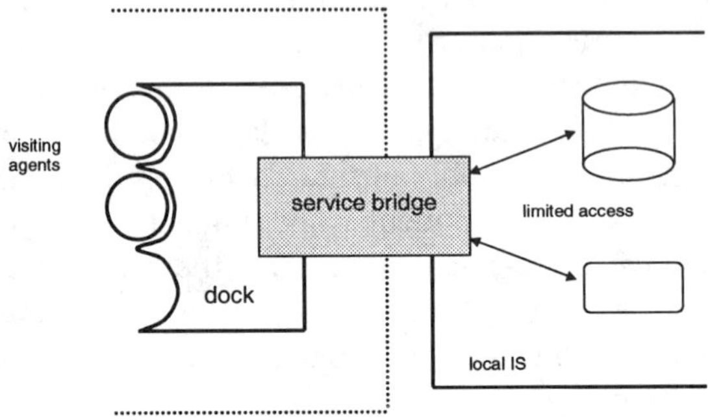

Figure 2 Architecture of the dock and service bridge component

On top of the basic infrastructure the VE support system can be built and consists of the mobile agents that provide the product tracking functionality [13]. In addition, some components, such as the web portal, with a product catalogue and ordering and tracking facilities that serve the VE as a whole are needed, and some components that are specific to an individual enterprise.

2.2.1 VE-level components

In [2] it was argued that in addition to a shared interface to the customers (e.g., a Web-portal), also other services, such as a scheduling service, should be made available at the VE-level, i.e., to the monitoring agent system as a whole. The scheduling service determines whether it is possible to fill the order and allocates the proper enterprises to the order. The scheduler needs data from the enterprises about availability of ready-made components or production capacities, information that may be gathered on demand by special roaming agents. An alternative is to store the information in a central repository and refresh it on a regular basis. When customer orders are incidental, such as those arising from a broken down landing gear, the first approach may be sufficient. But one can also envisage Virtual Enterprises that cater to emerging markets and have to deal with a higher frequency of ordering. In such a case a central body of availability data is more appropriate and moreover, it can serve as a basis for order promising.

The agents should be able to communicate amongst themselves and with the local information systems. The former requires a shared perspective, such as a common ontology, which would detail the representation and semantics of data about, e.g., availabilities, bills of material and routing, and procedures to be followed, e.g., in the case of negotiations. A lot of work on standardization of this kind of information has been done in the context of electronic data interchange (EDI) by large car and aircraft manufacturers, and many others, to organize their supply chains. Also, platform organisations in the domains of, e.g., commerce, health care and transportation have made similar efforts. This implies, that in many areas already the knowledge and experience exists, that is needed to establish a common ontology. Recently, two new developments have taken place in this area. The first is EDI via Internet that will allow also smaller businesses to partake in EDI at a lower cost of investment than would be required to obtain EDI-services from a particular Value Added Network (VAN). Because of this, also participation in a VE becomes quite feasible for these kinds of businesses. The second is the introduction and fast adoption of XML to make the messages self-descriptive. This will add a flexibility of operation, not found in the often proprietary, extensive and rigid EDI-message frameworks. Given the ontology, the agents can communicate among themselves using an agent communication language [6, 10]. A service at the VE level would be to provide this ontology to new VE members.

2.2.2 Enterprise-level components

An enterprise that will join a VE will have to conform to the VE common ontology and map their internal concepts and procedures to the common counterparts. This requires components that are specific to an enterprise since they involve the integration of the information systems of the enterprise with the VE support system. These components reside at the service bridges. In fact, a service bridge (see Figure 2) can be seen as consisting of two parts, connected by a gateway. One part is external to the enterprise and belongs to the general infrastructure of the VE. It is the same at every site and provides the mobile agents with a uniform facility to come and visit the site. This part is, for security reasons, continuously verified for code consistency in order to make sure, that it has not been tampered with by malicious visiting agents. The second part is internal to the enterprise and connected to the external part via a secure gateway. Here the mapping between the VE ontology and the enterprise's internal ontology is executed, the access to the enterprise data and communication with systems or parties in the enterprise is controlled.

3 VE ORGANISATION AND ROLES

3.1 The role of the mobile agents

The role of the agents is to extend the local information systems with mobile functionality to enable enterprises to participate in the VE business process. The advantage is a reduction of the local effort an enterprise has to make to join the VE. Moreover, a change in the business process can be dealt with, as far as support functionality is concerned, at the agent level (e.g., by upgrading the agent rule set).

3.2 The role of the software agent provider

The perceived volatility of the VE poses a curious problem with regard to the provision of the required infrastructure: should the VE develop and maintain its infrastructure (docks, agents, global services) itself? This proposition implies that some enterprises in the VE will have a more than average commitment to the VE. Should one of these leave, then the continuation of the VE may be in danger. Our position is that the development of the agents and the common services, such as scheduling and repositories for data and the ontology is best handled by an independent, trusted third party, that we call SACP (Software Agent Common Provider) [13]. This SA-provider plays a role similar to the Internet-, the VAN- and the ASP-providers. They all offer communication and application services to customers at various levels of performance, security and reliability. They also take care of the development and maintenance of required software. Since they will provide services to more than one VE and solutions can be reused, the costs will be lower and also the threshold in terms of required technology, knowledge, and investments for smaller companies to join will be lower.

An independent third party is also a guarantee for the autonomy of the participating enterprises, especially of the smaller ones. In the context of extended enterprises [3], mobile agents have been considered as a lightweight extension of the dominant company's ERP system to its suppliers. The result was a tight integration of the supplier's information system into that of the extended enterprise. Such a lock-in allowed the dominant company to manipulate the supplier's production and delivery schedules to optimise the performance of the extended enterprise. This effectively blocks a similar collaboration of the supplier with other customers.

We see that in the case of MyAircraft and Aeroxchange some parties have so large a commitment that they have invested in a joint venture to deal with not just incidental, but also the stable provisioning of aircraft parts and services, aimed at supply chain optimisation. These ventures clearly have a more long-term perspective. An infrastructure has been developed with a number of global (in this case even central) services, and with a low threshold (for a number of services a Web browser is sufficient) for other companies to participate. In a VE, one-of-a-kind orders are predominant. To make the investment into VE-level services worthwhile, the confidence has to exist among a number of parties, that more opportunities will present themselves. The possible reuse of the infrastructure for similar products (viz. small changes in the Bill of Material) then is an important requirement.

3.3 The role of the participants

The role of the participants in the VE is to make their production capacities available to the VE. The SACP can assist in building the software needed to transform between the enterprise and the VE data and procedures, by supplying documentation, templates, and even help-desk support. Many back-end systems already offer connectivity support, ranging from JDBC-interfaces and Business connectors to Business connectivity. The final control should stay with the participant to increase the level of confidence in and acceptance of the VE system. The participant should determine what kind of data and IS functions to expose to the VE (see Figure 2).

4 CHANGE CASES

4.1 An enterprise joins the VE

In the previous section we discussed some of the infrastructure needs a VE. The first step for an enterprise that wants to join the VE then is to become familiar with the VE ontology. The VE should therefore offer a mechanism whereby an enterprise can express its intention to join and, upon admission, get access to the necessary information, such as the VE ontology. The VE-Web portal can provide this functionality. As a second step, the prospective participant has to publish its resource types and availabilities to the VE. To achieve this, the prospective participant should install a dock and service bridge and connect it to its back-office systems by implementing the translation methods (semantic alignment) to and from the VE ontology. When the necessary communication channel is established, the enterprise can be included in the list of members of the VE. Agents can now visit the new participant, and the VE-mechanism for updating the availabilities can be used to also include the capacities of the new participant.

4.2 The VE expands its catalogue with new products.

The addition of a new participant by itself only increases the redundancy of the VE. However, a new participant may also possess production capabilities of a type that was not available to the VE before and open up the possibility for new products. A similar situation can arise when an enterprise wants to expand the list of capabilities that it offers to the VE with a capability not yet available to the VE. At this point the VE must be aware of the opportunity to expand its catalogue. This can be done in several ways.

The initiative for this action can be taken by the enterprise that is offering the new capability and wants to make it more productive. This enterprise has to consult the VE-ontology and compare it to its own product catalogue. When the enterprise catalogue contains products that in addition to the new capacity only require production capacities available in the VE, the enterprise can propose to the VE to include these products in its catalogue. Proper procedures to do this have to be set up.

Another trigger for the expansion of the catalogue may come from customers who inquire if the VE can produce a product not in the catalogue. The VE in this case should provide a facility by which a customer can specify a product, complete with the required resources. The customer does not necessarily use the same ontology as the VE does, and so support to produce the correct specification may be needed. Given the product specification, the VE can check to see if the required capacities are available (even if they have not yet been in use), or, if some capabilities are missing, inquire among its participants to see if any may be willing to offer these capabilities. In the case of a positive answer the capacities may be added to the list of VE-resources and the new product may be included in the VE catalogue.

Expansion of the VE catalogue of products has an impact on the ontology. When the addition to the catalogue concerns variants of previous offerings the impact may be minor, just affecting a (part of a) bill of material (BOM) or routing. Flexibility in the agent communication now requires that the BOM can be passed as a parameter (for example, in a self describing format as XML) so that variations in composition pose no problem. When the new product involves new concepts, such as production techniques that are new to the VE also the ontology has to be expanded.

4.3 An enterprise leaves the VE

When an enterprise leaves the VE, the inverse process of a new member joining and a new product being added to the catalogue has to be carried out. The enterprise has to make its withdrawal known to the VE. The VE then has to check whether the withdrawing enterprise has unique production capabilities. If not, then all that has to be done is to remove that enterprise's capacities from the set of available resources. Only the redundancy of the VE is affected in this case. When the enterprise was offering unique capabilities, also those products have to be removed from the catalogue, for which the contribution of the leaving participant is essential.

4.4 The VE merges with another VE to enter a new market

The possibility of a merger between two VE's is a feasible option when there already is considerable over-lap between the activities and capacities of two, e.g. one is specialized in mobile phones, and the other in PDA's, i.e. devices, which have a number of components in common. This implies that some enterprises will be a member of both VE's. The merger scenario is simplest, when both VE's are using the same infrastructure (obtained from the same SACP). In this case, a new VE can be set up by combination of infrastructures. The ontologies must be merged, which probably will require some semantic alignment. Once that is done also the catalogues and lists of resources can be combined. The next step then is the expansion of the catalogue to exploit the enhanced production facilities. In case the two VE's do not use the same infrastructure, the merger will have to be effected by adding the members of one VE to the other. This procedure requires the repeated application of the scenarios discussed in par. 4.1 and 4.2.

We conclude from the use cases above that the VE, in addition to providing flexible support for the production process itself, also will have to provide a number of services, such as an ontology server and capability and product addition and removal services, that will support the dynamics of the composition of the VE. These services rely on sufficiently expressive specification languages and sufficiently widely accepted standards.

5 CONCLUSION

The infrastructure presented here differs from more specialized system integration architectures such as federations [12] and mediators [14], which also provide a unified view of the system to the outside world (user or application). In both architectures, the transformation to the local representations is done centrally, so that the local systems remain unaffected. This, however, requires the local systems to publish a suitable part of their interfaces to the global system, implying a more long-term commitment. Both architectures require synchronous communication with their components and are better suited for local area network environments. Of course, the infrastructure proposed in this paper can be specialized for the support of federation or mediation type architectures.

The agent infrastructure discussed in this paper can also be applied in other situations. In addition to supporting the VE strategy it can be used in e-commerce situations, such as auctions and in supply chains. In the case of auctions, no integration with the back-office systems is required. The agents provide in this case an intelligent interface that allows a customer to specify, e.g. a bidding strategy.

In the case of the integrated supply chain, the agents may provide a more flexible alternative to the EDI-procedures in use at present. When the agents are supplied with error handling functionality to autonomously solve a number of common problems, the system of procedures and messages may be significantly simplified. When suitably extended, the agent framework discussed in this paper therefore may be used in e-market and supply chain situations as well [16].

The mobile agent system will be tested in some small-scale experiments in the PROVE project [2, 13].

6 REFERENCES

[1] Aeroxchange, http://www.aeroxchange.com

[2] Aerts, A.T.M., Szirbik, N.B., Hammer, D.K., Goossenaerts, J.B.M., Wortmann, J.C. (2000) "On the Design of a Mobile Agent Web for supporting Virtual Enterprises ", WET ICE 2000.

[3] Browne J., Sackett P.J., Wortmann J.C., (1995), "Future manufacturing systems - Towards the extended enterprise", Computers in Industry, 25, Elsevier, pp.235-254.

[4] Dasgupta, P., Narasimhan, N., Moser, L.E., Melliar-Smith, P.M., (1999), "Mobile Agents for Networked Electronic Trading", in *IEEE Transactions on Knowledge and Data Engineering*, vol. 11, no. 4, pp.509-525.

[5] Davidow W.H., Malone M.S., (1992), "The Virtual Corporation", Harper and Collins, New York.

[6] The Foundation for Intelligent Physical Agents, "FIPA99 language Specification", http://www.fipa.org

[7] Goossenaerts, J.B.M., Aerts, A.T.M., Hammer, D.K., (1998), "Merging a Knowledge Systematisation Framework with a Mobile Agent Architecture", in *Information Infrastructures for Manufacturing II*, (Mills & Kimura, eds.), Kluwer, The Netherlands.

[8] Goranson, H.T., (1999), "The Agile Virtual Enterprise: case, metrics, tools", Quorum Books, Westport.

[9] Hammer D.K., Aerts A.T.M., Dalmeijer M., (1998), *Mobile Agents Architectures: What are the design Issues?* International Conference on Engineering of Computer Based Systems (ECBS), Jerusalem, Israel, March 1998.

[10] Labrou Y., Finin T., (1997), "A proposal for a New KQML specification", Technical Report CS-97-03, Computer Science and Electrical Engineering Dept., University of Maryland, Baltimore.

[11] MyAircraft, http://www.myaircraft.com

[12] Sheth, A.P., Larson, J.A., (1990),"Federated Database Systems for managing distributed, heterogeneous and autonomous databases", ACM Computing Surveys, 22(3), 183-236.

[13] Szirbik, N.B., Wortmann, J.C., Hammer, D.K., Goossenaerts, J.B.M., Aerts, A.T.M., (2000), "Mediating Negotiations in a Virtual Enterprise via Mobile Agents", AIWoRC'2000: Mobile Technologies and Virtual Enterprises", Buffalo.

[14] Wiederhold, G., (1994), "Interoperation, Mediators and Ontologies", Proc. Int. Symp. On Fifth Generation Computer Systems, Vol. W3, 33-48, ICOT, Tokyo.

[15] Wiendahl, H.-P., Engelbrecht, Hamacher, O., (2000), "From Single Enterprises to Complementary Networks", Proceedings of DIISM2000, the 4th International Conference on Design of Information Infrastructure Systems for Manufacturing 2000, 15-17 November 2000, Melbourne, Australia.

[16] Wortmann J.C., Szirbik N.B., (2000), "ICT Issues among Collaborative Enterprises: from Rigid to Adaptive Agent-Based Technologies", to be published in Intl. Journal of Production Planning and Control.

Assessing Ability to Execute in Virtual Enterprises

Roelof J. van den Berg
Baan Development, The Netherlands
Em:rvdberg@baan.nl

Martin Tølle
Technical University of Denmark
Em: mat@ipt.dk

Keywords Virtual enterprise, ontological convergence, modelling, web assurance services

Abstract The state of the art in ICT enables integration beyond the application per se, but this requires assessment and reconciliation of business processes. This paper describes this challenge to true co-operation in a virtual enterprise, focusing on industries for one-of-a-kind production. It explains why expressive modelling is important in the assessment process, by introducing the underlying mechanisms of inter-subjective maintenance of mental models. Finally, the paper discusses the future of assessment and auditing services in the context of business-to-business e-commerce.

1 INTRODUCTION

Most literature on e-commerce, including commerce in a business-to-business setting, concentrates on trading of standard products. Certainly, supporting the buying and selling of this type of products with advanced information technology can lead to significant gains in efficiency. But most business relationships go beyond mere trading. Their value is based on true co-operation, and synergy through mutual tuning of world views and processes. Prior to transactions in this context it is not automatically clear what each partner has to do nor if his business process execution matches with the contributions from others. Particularly in an arena centred around one-of-a-kind products e.g. power plants, ships or bridges resolving this ambiguity is no trivial endeavour.

This paper highlights some aspects of assessment of ability to execute in virtual enterprises. It will start with a reflection on integration of information systems in section 2, focusing on the difference between deterministic and opportunistic integration. Section 3 and 4 are devoted to the main challenge of the latter type of integration. Section 5 will briefly discuss the context of the assessment, while section 6 contains the conclusions.

2 EVOLUTION OF SYSTEM INTEGRATION

Many predict that Internet will eventually enable one global so-called perfect market, where everybody can deal with everybody else on equal terms. This is based on the idea that price is all that matters. Certainly developments in trade of books, CD's or airplane flights support this. But in large segments of b-to-b commerce it pays to invest in differentiation and uniqueness through better product quality or logistical performance. Thus a consortium can essentially create a qualitative monopoly. In those areas of b-to-b commerce the trend is not so much towards one electronic market, but instead to electronic "balkanization".

In this situation several networks exist. The network is a co-operative alliance of competencies established to jointly exploit business opportunities. One enterprise, a group of enterprises or alternatively a large customer, who can be perceived as the business concept owner, initiates the creation of the network. Each network is relatively closed, allowing each member to exchange intimate business information with the others. The mutual trust between partners is the basis for much more sophisticated co-operation, up to the point that members seem to be part of one and the same "virtual" enterprise.

A virtual enterprise (VE) can be defined as a customer solutions delivery system created by a temporary and reconfigurable aggregation of core competencies. Note that in this view and unlike popular thought, members of successions of VEs always come from one and the same network. The network can be seen as a potential from which different VEs can be established in order to satisfy diverse customer demands. Although it comprises competencies from various partners, the VE appears to the customer as one, unified, and attuned enterprise. Hence its virtual nature. When the customer demand has been fulfilled, the virtual enterprise dissolves into its constituent parts to reassemble into other configurations (other VEs). Competencies and experiences gained in the virtual enterprise are transferred back to the network and its participants. See [4] for a more elaborate discussion on these issues.

The developments concerning VEs mark a new era of integration of information systems. Many difficulties from earlier stages of integration of information systems have been addressed [5], but new challenges have appeared. We will further explain this evolution of enterprise integration with the layered framework, illustrated in figure 1. According to this framework satisfactory integration at a lower level is necessary before integration at a higher level can be achieved. We will briefly discuss each of the layers subsequently.

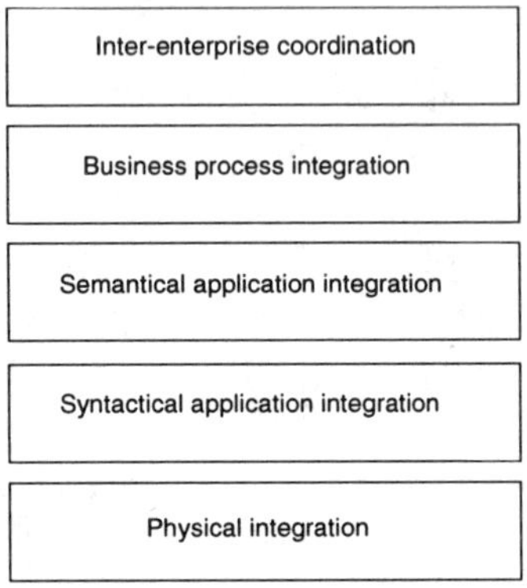

Figure 1 Framework for Enterprise Integration

1. Standards for physical integration

Naturally, physical integration is needed to facilitate co-operating applications and enterprises. Recently wireless integration has rapidly grown in importance. Relevant standards at this level of integration are WAP, MAP, and Ethernet.

2. Syntactical standards for application integration

This concerns the integration of application software systems at the level of "form". Standards in this area are Java, XML, Corba, DCOM.

3. Semantic standards for application integration

Semantic integration should result in application output which is meaningful to other applications. This type of integration abstracts from the technical details of software implementations. Examples of standards at this level are EDIFACT, STEP and BizTalk.

4. Modelling languages for business process integration

Virtual enterprises require a common understanding about the shared business processes. Modelling languages are needed to make these business processes explicit. Examples of standards at this level are IDEF, Petri nets, UML, ER, LOTOS, SDL, VDM, Z, B, •. Time is generally incorporated in these modelling languages, but distribution in space usually is not.

5. Inter-enterprise co-ordination

Dedicated guidelines at the level of inter-enterprise co-ordination are needed, e.g. for partner selection, certification or inter-enterprise best practice definition.

The framework in figure 1 is not specific to the era of "the new economy", but over the years the higher levels of the framework have become more relevant. Until recently integration typically concerned two given systems within two given units, often within one organization. The aim was to share specific data and applications for given purposes in fairly stabile business processes.

Without trying to marginalize that type of integration we could classify it as deterministic. Many aspects were known in advance, e.g. the business context of the integration, the players, their roles and their systems. The solution was dedicated to a specific case. Due to lack of standardization the emphasis in the integration effort was on the technical challenge of making the two information systems operate as one.

Currently most attention is devoted to a type of systems integration which is quite different, one based on much more standardized technology, enabling communication between information systems in enterprises without *a priori* acquaintance. The technical integration challenge for companies nowadays is to create "openness" to systems which are not known and owned by unknown partners, to enable co-operation in business functions which are not automatically streamlined across the VE.

Compared to the first type of integration, the second type is much less directed. It has to be more flexible to a wider variety of business applications. It is more opportunistic. ICT is used to release intellectual potential, through smart forms of co-operation. The two lower layers are becoming more and more standardized and a commodity, leaving more room to creatively exploit the benefits of open technologies. Thus the emphasis has shifted towards the three highest levels. In the next sections we will further discuss aspects of the fourth and fifth level respectively.

3 THE CHALLENGE OF BUSINESS INTEGRATION

Our view on business integration is based on the idea that the world around us is not just a given [3]. We influence it and, at a cognitive level, make it what it is. During our lives we create subjective realities, our world-views. As a productive system our society is based on a separation of concerns, which breeds variety in world-views. A worker will be very much the product of his particular education (e.g. the ideas of a particular professor or "school"), professional culture (e.g. accountancy versus tax law) or organizational (sub) culture (e.g. Dow versus Shell versus BP or sales versus human resources or R&D). We can say that specialization breeds ontological divergence: differentiation in world-views.

The variety in world-views is also reflected in our information systems. In a sense people try to freeze their world-view into the information systems they use. The way a department defines an entity e.g. "customer" or a business process e.g. "purchase order processing" depends on the precise way in which they operate, which in turn reflects their view on their role in the organization and its larger environment. This view is continuously maintained through the department's interaction with its environment. The main reason why large portions of systems development budgets are devoted to systems maintenance is the fact that world-views are not fixed.

The systems we employ become more and more open, but this does not mean their output automatically matches with the mental models of "foreign" users. To the contrary, open systems technology has only increased our possibilities to integrate our systems with those of others, which are based on increasingly different assumptions. The challenge of business integration has increased accordingly, but enterprises have to face it to stay competitive. Thus, especially in an era of open systems, mechanisms have to be available to bring world-views together, achieve ontological convergence and realize the benefits embedded in the state of the art of ICT.

4 FUNCTIONAL CONFORMITY

Concerning integration most literature on e-commerce concentrates on technical conformity: do the companies' respective systems comply with the same ICT-standards, so they can exchange data? We want to stress that in order to be competitive and really exploit the opportunities of enterprise integration another type of conformity should also be addressed: functional conformity. This addresses the question: does a common view exist on the inter-organizational activities and do business processes match (inputs/outputs required, processing sequence, lead times)?

It should be noted that a positive answer to this question does not mean that the ontological divergence itself should be denied and only standard solutions should be provided. As mentioned above clear reasons exist for the divergence in world-views. But in order to rapidly deploy integrations that truly support the operations in the VE it is necessary that the respective members make their views explicit. From there a solution can be set-up for their specific VE.

It is unrealistic to assume that each VE will be supported by a fully dedicated systems implementation. The temporary nature of the VE is one reason. Another is the fact that especially larger organizations will participate in more than one VE at the same time, with a desire to use the same ICT support. It is therefore required that they can support different approaches (for different business partners, regions, products etc.) with the same system simultaneously.

Current standard software solutions are able to accommodate for different approaches along a series of parameters, but they have great difficulty to support alternatives simultaneously within one implementation [2]. One reason is they are not enterprise "aware": the organizational scope of the functionality has not been modeled explicitly. Future systems will have to run in a powerful enterprise modelling environment, which includes an expressive dynamic model of the implemented enterprise systems themselves and of the ICT-infrastructure which is available: the model is the application.

Differentiated business processes have to be described in a form, understandable by humans and interpretable by software. It is especially this interpretation that should be enhanced, e.g. to allow different versions of software to semantically function similarly at different places in an enterprise. A necessary condition is that software can inspect at run time the configuration of versions, functions, processes, enterprise units, and data sets in its environment. On the other hand it should also be possible to access and activate the same enterprise software object through different business model instances.

5 ASSESSMENT ISSUES: PRESENT AND FUTURE

The human interpretation of the business models is an essential part of the assessment of ability to execute. Modelling techniques such as those for process modelling can be used to support this check for functional conformity. Current modelling techniques do not only enable enterprises to define their business processes at a generic "what do we do?" level, but also to clearly differentiate between process variants for different business partners, products, regions, seasons et cetera. Several business modelling tools make it possible to show which variants of business processes are active under specific circum- stances.

While the specification of the business processes has to be done in-house, this does not apply to the assessment itself. After all, the ideal of VEs is driven by the strategic desire to concentrate on one's core business. From this perspective it is not sensible to have a large in-house assessment function.

Assessment should be left to specialized third parties. At this moment certified public accountants (CPAs) are most advanced in this area, especially those in North America [1]. In the past years they launched several web certification services for b-to-c commerce, but still have to launch one dedicated to b-to-b commerce. Other professionals, e.g. EDP-auditors and even public notaries or banks could play a role in certifying enterprises which meet certain thresholds to participate in VEs.

Instead of making a definite choice for one group or the other, we raise the possibility to launch dedicated web assurance agencies, which combine expertise from several professions, e.g. accounting, EDP-auditing, law, payment systems, system integration, electronic marketing and cryptography. These agencies could perform an audit in an inter-disciplinary team. To support recognition of web assurance services, it is important to establish a standardized way of working as much as possible. Among the CPAs mentioned above a tendency exists to develop a firm specific web assurance provision with corresponding seal. A "jungle" of unclear web assurance services and corresponding certifications only leads to confusion in the target market.

More specifically the assessment should be organized as a mix of a public and private function. Public functions are strictly regulated by law to secure standardization of execution and to provide outsiders with a solid basis for interpretation of the results. An example of a public function of CPA's is the annual audit of the financial statements of a company. The generic parts of assessments, resulting in conclusions that go beyond the lifetime of one VE, can be organized as a public function: an enterprise is evaluated in isolation vis-a-vis generally accepted standards. Results of such public assessments could be used to recruit enterprises as part of a network.

Unlike public assessments private ones are reported to the client only. For this reason the precise organization of such an audit is much more left to the discretion of the individual auditing firm. In a web assurance audit a private (part of an) assessment will deal with the specific requirements of the VE that is being set up. The precise nature of the audit will be tuned to the requirements of the network, which requests it, and the exact work that has to be executed in the VE. In a private assessment the enterprise is evaluated reciprocally against the needs of others in the VE. It is not unlikely that application service providers will eventually take the web assurance services under their wing.

6 CONCLUSIONS

Current rhetoric on e-commerce tends to depreciate the complexity of requirements for true co-operation in e-business as opposed to merely trading. Developments in open systems technology certainly have pushed the state of the art in technical integration of systems, but they do not automatically support additional business integration which is required to establish VEs.

This last type of integration cannot be achieved without reconciliation of the world-views that underlie the systems in the VE. Explicit modelling of the assumptions behind each of the partners' systems functionality is needed to support this ontological convergence. Modelling can also help to differentiate functionality of one and the same system along various dimensions. This requires executable modelling environments. The model becomes the application. Eventually dedicated web assurance service providers will assess a partner's ability to execute in a VE. The generic assessment should be organized as a public function similar to a CPA audit of a firm's annual accounts or ISO certification. The subsequent more specific assessment per VE is best organized as a dedicated private function. In the future it is likely that application service providers will cover the assessment function as a natural outcome of their mission to provide the ICT backbones of VEs.

7 ACKNOWLEDGEMENT

This paper benefited from the critical review and valuable remarks from Jens Dahl-Pedersen at the Technical University of Denmark and Arian Zwegers at Baan Development.

8 REFERENCES

[1] Berg, R.J. van den and Lieshout, van, J.M., "Eliminating Hurdles to Trust in E-commerce", in *Global Production Management*, K. Mertens, O. Krause and B. Schallock (eds.), Kluwer, 1999, pp. 522-529.

[2] Berg, R.J. van den, and Zwegers, A.J.R.,"Decoupling Functionality to Facilitate Controlled Growth", in *Studies in Informatics and Control*, Vol. 6, No. 1, March 1997, pp. 57-64.

[3] Berger, Peter L. and Luckmann, Thomas, *The Social Construction of Reality*, Penguin University Books, Middlesex, 1971.

[4] Tølle, M., Vesterager, J., Dahl-Pedersen, J., "A Methodology for Virtual Enterprise Management – Results from IMS 95001/Esprit 26509 Globeman 21 project", Proceedings of the 6th International Conference on Concurrent Enterprising, Toulouse, 28-30 June 2000, pp. 119-127.

[5] Zwegers, A.J.R., *On Systems Architecting*, Ph.D. thesis, Eindhoven University of Technology, 1998.

Project-specific Process Configuration in Virtual Enterprises

C. Rupprecht and T. Rose
Research Institute for Applied Knowledge Processing (FAW), Germany
Em: {Christian.Rupprecht | Thomas.Rose}@faw.uni-ulm.de

E. van Halm and A. Zwegers
Baan Development, Netherlands
Em: {evhalm | azwegers}@baan.nl

Keywords Process modelling, process configuration, project management, virtual enterprises

Abstract Virtual enterprises have great difficulty in defining and adapting their business processes across the members of the virtual enterprise. We introduce an approach that allows (semi-)automatic (re-) configuration of one-of-a-kind business process models fitting the specific requirements and constraints of a project in a virtual enterprise. Our approach is based on the idea of making these requirements and constraints explicit by representing them in a model for each project.

1 INTRODUCTION

Virtual enterprises – temporary networks of enterprises that are formed for one-of-a-kind projects – have great difficulty in defining and adapting their business processes across the members of the virtual enterprise. Current practice is to use a rather general process model and to manually fill in the details from all members from scratch for each project. This practice is time consuming, costly, and error prone.

The business processes of a specific project in a virtual enterprise are influenced by many factors, such as customer requirements, legal constraints, and resources. We introduce an approach that allows (semi-) automatic (re-)configuration of one-of-a-kind business process models fitting the specific requirements and constraints of a project in a virtual enterprise. In this context, semi-automatic means generating proposals for adaptation going along a guided dialog with the user. In our approach, requirements and constraints will be made explicit and represented in a model. We address the definition and adaptation of business processes as a configuration problem.

The configured project-specific models will be transferred to multiple project management applications that control the execution of the defined business processes. Each member of the virtual enterprise will receive its part of the configured models in order to execute its business processes as specified in the model. Software support for our approach will significantly reduce costs and time-to-market and increase quality and responsiveness-to-market in the project industry.

2 PROCESS MODELLING IN VIRTUAL ENTERPRISES

2.1 Project Management

Project management is about controlling cost, scope, and time. Activities during the initial stages of a project are scope management, cost engineering, and scheduling. *Scope management* is the process to define, change and monitor deliverables, the project structure, and the resources required. Scope management is needed before cost engineering and scheduling can start. *Cost engineering* has basically two phases, namely estimation and budgeting. Finally, *scheduling* is the process of determining the project lead-time. An activity network is made and the resource assignment(s) are scheduled according to their available capacity [5].

One of the characteristics of a project is its uniqueness. Projects are set up for specific customers, demanding specific results. Due to this uniqueness only a small part of the scope of all projects within an organisation can be standardised. Obviously, the extent of standardisation varies by organisation and type of projects. By means of standards, templates, and copying from previous projects, one can reuse existing (standard) information of deliverables, parts of the project structures, and/or required resources. Information could be for example:

- About deliverables: relationships, lead times, sales values
- About project structure: hierarchical relations, activity networks, durations, earned value methods
- About required resources: cost/sales prices, quantities, product configurations

Nevertheless, a lot of time and effort is spent to define the parts that are not standard. For that part of the scope, one must maintain the remaining part of the estimate, budget and/or schedule. In order to reduce the steps in this process, reference models can be used. A reference model reflects all possibilities and is configured for the situation based on input variables.

2.2 Process Modelling

In virtual enterprises, processes need to be planned for each unique project. In this paper, we focus on the definition and adaptation of one-of-a-kind business processes, i.e. networks of activities. Before planning, co-ordination, and execution of the business processes of a virtual enterprise, they need to be defined, and during process execution they may have to be adapted due to new or changing requirements. This is particularly relevant for processes in the domain of complex system engineering [2, 4].

We define a *process* as a set of temporally or logically ordered activities intended to reach a goal involving resources. It can be regarded as a system where the elements are activities and resources and the relations are the sequential or logical dependencies between those elements. The set of relations describes the process structure.

A *process model* is a mental or explicit representation of original processes. In general, directed graphs are used for the explicit representation of processes. When we speak of process models in this paper, we mean semi-formal, computational representations in symbolic notation, i.e. general process elements like activities and their relations are represented by formal symbols (boxes and vectors) and additional information are attached non-formally, e.g. naming the symbols in natural language.

Process models capture know-how about ways of working in the past or intended in the future. However, a static process model contains no information about why work had or has to be done in a certain way. With our approach, we aim at creating generic, reusable representations of process configuration knowledge that can be applied across a variety of future process cases. The objective is to provide "guidance, suggestions, and reference material to facilitate human performance of the intended process" [1].

2.3 State-of-the-art

Current practice in virtual enterprises is to use a rather general process model and to manually fill in the details from all members from scratch for each project. This practice is time consuming, costly, and error prone. At the same time, global competition compels enterprises to drastically reduce their project duration in order to be competitive. For instance, to stay competitive in the power plant industry, an average project duration will have to be decreased from 24 months to 15 months in a near future.

Prevailing process models and supporting tools lack in the adaptation of actual processes. That is, there are sophisticated design, analysis, and management methods and tools, but methods for customising processes in the design and even execution phase are sparse. For example, one of the leading commercial process modelling tools, ARIS[i], allows to reuse functions that have been modelled and stored before. However, automated assistance for the context-specific deployment of such existing function descriptions is restricted to the matching of the function names. ARIS does not offer any support for the configuration of process models according to explicitly specified project requirements and constraints.

Some related research work on configuration and adaptation of process models and their restrictions are presented in [3].

3 APPROACH

3.1 Core Idea

Figure 1 shows the problem addressed in this paper. The business processes of a specific project are influenced by many factors, such as customer requirements, legal constraints, and resources. These requirements and constraints will be made explicit and represented in a model. A project manager has to take all these requirements and constraints into account when he/she initially performs project management activities, such as scope management, cost engineering, and scheduling. Likewise, all these factors have to be considered when process execution needs to be adapted to new circumstances. Nowadays, adaptation of business processes to specific project constraints and requirements is a manual, iterative, and difficult task. (Semi-)automatic configuration and adaptation will considerably ease this process. The configured project-specific models will be transferred to multiple project management applications that control the execution of the defined business processes. Each member of the virtual enterprise will receive its part of the configured models in order to execute its business processes as specified in the model.

3.2 Conceptual Framework

Our research will result in a methodology and a prototype for process adaptation based on configuration. The main pieces of knowledge for the process configuration prototype are reuseable process building blocks, requirements and constraints, and configuration rules.

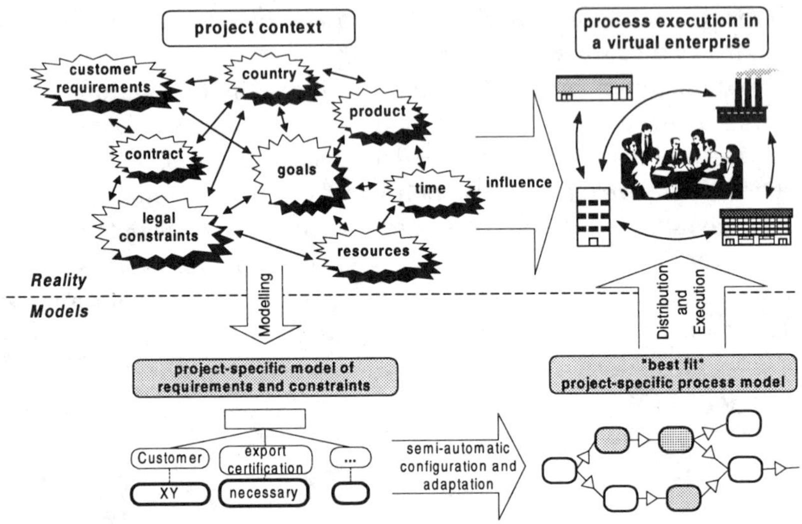

Figure 1 Core idea of process configuration and adaptation

For complex and innovative business processes, it is not sufficient to document a best-practice process once and follow this pattern forever in all future process cases. However, some smaller parts of a process are repetitive, i.e. for a certain task, the same activities are executed in the same sequence within different projects. For reasons of higher reusability and flexible connectivity, an extensive process model can be stripped down to temporally and logically isolated units called *process building blocks*. Such building blocks should have only few interfaces to other process models or building blocks, and they should "know" how they may be connected to other building blocks. In that case, a project-specific business process is created from combining such building blocks in unique combinations. In a virtual enterprise, each company keeps its set of process building blocks in a library.

Requirements and constraints have influence on the design of processes. They can be used to describe the project-specific context of a process and form the main input parameters to the configuration process. Options specify a requirement for a specific project, e.g. physical dimensions and expected due date of a deliverable. In order to use the set of requirements and their options in the configuration rules, the requirements can be derived from an existing information system whereas the options are entered at the moment of configuration. Taking into account dependencies between requirements, the configuration application also offers assistance for modelling the requirements in a specific project.

Experience in process modelling is broken down into single design decisions that are captured in terms of *configuration rules*. A configuration rule consists of a condition part which refers to constraints and an execution part which triggers configuration operators. Configuration rules are represented as dependencies among constraints and process building blocks. By applying these construction rules to a specific model of constraints and requirements, proposals for adaptations to the process model are generated. Together with the reuseable process building blocks, the set of configuration rules form generic reference process models. Generic means that the process models are automatically adaptable to a specific context via defined operators.

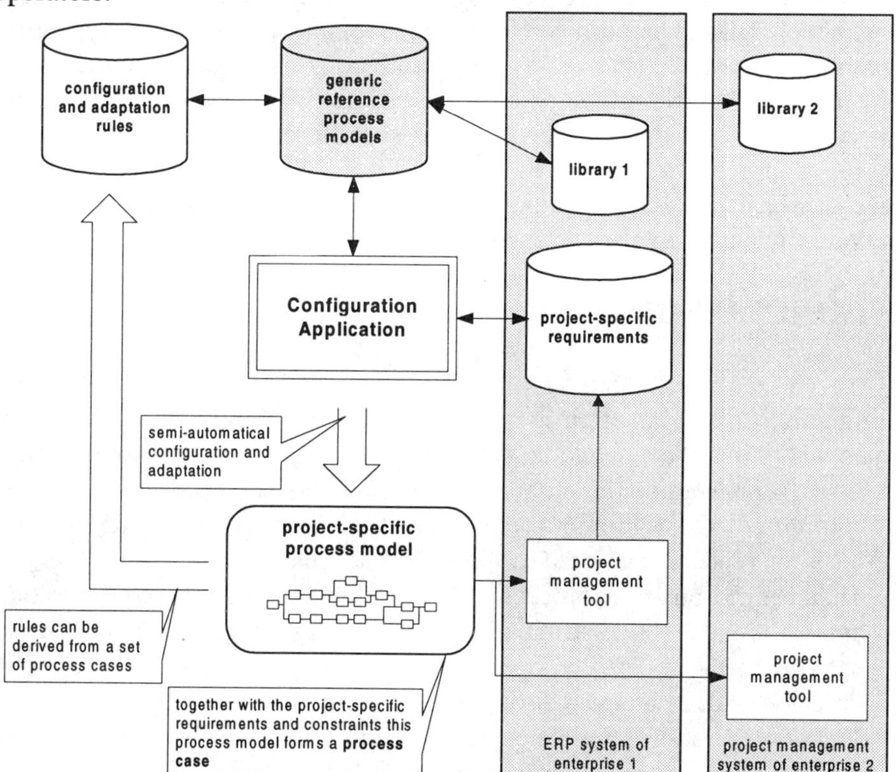

Figure 2 Basic architecture of a process configurator

Figure 2 shows an initial architecture for the process configuration prototype. The prototype will consist of a front-end (*Configuration Application*) in which reference process models will be defined, and in which models of project-specific business processes can be configured. The backbone will provide necessary data stored in libraries, and it will contain the requirements which are input to the configuration. Business processes can be planned, co-ordinated, and executed with the help of project management tools based upon the configured process models.

Modelling the process relationships between the members of a virtual enterprise is a key element of the configuration application. This will be achieved by de-coupling the configuration application from any other information system, and by incorporating data from libraries belonging to different members of a virtual enterprise. The configuration application defines with which information system it has to interface in order to use the library data in the reference process models.

The knowledge base is extendable, i.e. users can define new constraints, requirements, reference process models, process building blocks, and construction rules at any time on a generic level and add them to the existing constraints, models, and rules. By collecting and maintaining such knowledge from different engineers, process modelling experience (i.e. the reasons for process design decisions in a specific context) can be captured and reused for future process modelling cases. A set of process cases can serve as a basis for computational analysis, in order to make the knowledge about dependencies explicit in terms of configuration rules.

5 CONCLUSION

Our research will deliver a methodology and prototype for process configuration and adaptation in virtual enterprises, mainly concentrated on project industries. The technological approach brings four main innovations:

- In a (semi-)automatic way, one-of-a-kind business processes will be configured. Configuration technology will be used to define specific business processes tailored to specific requirements;
- The specific business processes that are generated in the configuration application can be distributed to multiple participants of the virtual enterprise;
- Processes can be adapted during the entire project lifecycle;
- The configuration application is open in order to receive process and product architecture data from multiple information systems.

For industrial enterprises, there are a number of benefits. For example, responsiveness during the bid stage is increased. Making a bid to acquire a project requires a lot of effort. Knowledge is very often fragmented over different people and hardly formalised. With process configuration, it should be possible to capture some of the knowledge, thereby reducing the amount of work and the duration of bid preparation phases.

Another benefit is the increased efficiency in setting up projects. Nowadays, templates automate a lot of the work in setting up a project. A process configuration tool can bring this one step further, since it has the capability to be more specific about a situation and determine more details than a template ever can. With reconfiguration capabilities, the changes in a project can be executed without a lot of effort.

Process configuration also increases an organisation's ability to learn. It is very hard to learn from projects due to their uniqueness. Trying to find out what parts of different projects can be compared is quite difficult. If projects are generated from the same reference model, it will be easier to point out the comparable areas.

9 REFERENCES

[1] Curtis, B., M.I. Kellner, and J. Over. (1992). Process Modelling. *Communications of the ACM*, Vol. 35, No. 9, pp. 75-90.

[2] Negele, H., E. Fricke, L. Schrepfer, and N. Härtlein. (1999). Modelling of Integrated Product Development Processes. In: *Proc. 9th Annual Int. Symposium of INCOSE Systems Engineering: Sharing the Future*. Brighton, UK.

[3] Rupprecht C., M. Fünffinger, H. Knublauch, and T. Rose. (2000): Capture and Dissemination of Experience about the Construction of Engineering Processes. In: *Proceedings of the 12th Conference on Advanced Information Systems Engineering (CaiSE*00)*. Stockholm, Sweden, *pp. 294-308*.

[4] Schott, H., A. Sieper, T. Rose, M. Fünffinger, C. Rupprecht, C. Schlick, and M. Mühlfelder. (2000). Process Knowledge Management in Concurrent Engineering. In: *Proceedings of the 2nd European Systems Engineering Conference (EuSEC 2000)*. Munich, Germany.

[5] Van Halm, E. (2000). *Project Configuration – Summary of the Market Requirements*. Internal Baan Development document.

Functional Requirements for Inter-enterprise Intranet Services

A.S. Kazi and M. Hannus
VTT Building Technology, Finland
Em: Sami.Kazi@vtt.fi

Keywords Inter-enterprise Communications, Intranet, Distributed Engineering, IMS-GLOBEMEN

Abstract A core requirement for inter-enterprise collaboration in a virtual enterprise (VE) is the availability of an inter-enterprise Intranet service that facilitates the seamless flow of data and information between the different actors involved in addition to fostering an environment for collaboration. This paper is based on explorations and findings of trying to find/develop an inter-enterprise Intranet service for the participants of a virtual enterprise. Some required functionalities/support that were identified include: concurrent engineering, suitability to project business conditions, open system architecture, electronic document management, product structure/configuration management, access control, redlining/mark-up, groupwork support, user profiling, version and revision management, workflow management, distributed databases, application integration, and application launching. The paper concludes by illustrating an envisioned system architecture for inter-enterprise intranet services.

1 INTRODUCTION

An evolving and currently under heavy research operational paradigm is through the formation of a virtual enterprise. This typically entails the efforts from different organisations participating together to resolve a unique problem or deliver a unique product. While it may be argued at length that this is the operational norm of many industries such as manufacturing, ship-building, construction, etc., a formal methodology for the process doesn't exist. Moreover, the concern lies more on the way that information is shared than others. With the participants of each organisation potentially using different proprietary systems, the sharing of information can become a bottleneck to the effectiveness of the virtual enterprise (VE). It may be argued that a common IT environment in terms of an application software platform should be enforced, this is not very practical when the reality that each participant may be involved in different virtual enterprises at a given time and adherence to a VE specific software application platform would seem uncalled for. The simple question then is, "How to share information between heterogeneous systems?"

While an exploration of the mechanism and alternatives for sharing information between heterogeneous systems is beyond the scope of this paper, it should suffice to mention that this is possible through compliance with information exchange standards (e.g. ISO-STEP).

The focus of this paper is on identifying possible solutions and implementation mechanisms for intranet services to enable inter-enterprise communications in a virtual enterprise setting. Core findings are related to early explorations within the IMS GLOBEMEN project, and the authors' experiences in several EU projects, especially CONCUR, PROCURE, and ToCEE.

2 INTER-ENTERPRISE COMMUNICATION

Initially, inter-enterprise communications in a virtual enterprise setting were based on document exchange between individual participants. This not only led to data/information redundancy, but a lack of control as to which information exists where and as to who is the owner of the information. Each participant either had to have the same application software installed as that of the sender's, or to alternatively use some document "viewing" tools. Furthermore, there was no central location where the information was stored.

Current intranet environments for VE settings are based on the concept of a shared information repository where project related data and information is stored. This resolves most of the issues raised in the previous form (see above) of inter-enterprise communications. One pitfall in this approach is that it is centred on the individuals participating within the VE without concern for their organisations, which in fact form the contractual basis for the VE. Furthermore, an individual/organisation may only wish to release a certain "amount" of information to the VE consortium whilst keeping the rest "internal". Last but not least, from a legal perspective, any document released to the "central repository" should imply endorsement from the organisation to which it belongs.

The road to the future is through the communication of organisation dependent intranets to the "central repository" and VE intranet. This would typically entail the transfer of "information packages" on a periodic basis from a participating organisation's internal system to the VE intranet. At the same time, this ensures compliance with legal contracts as in this form the information is coming from an organisation and not an individual. The two systems, organisation specific and VE would typically communicate through "interfaces" thereby eliminating/minimising the need for "compliance" with each other.

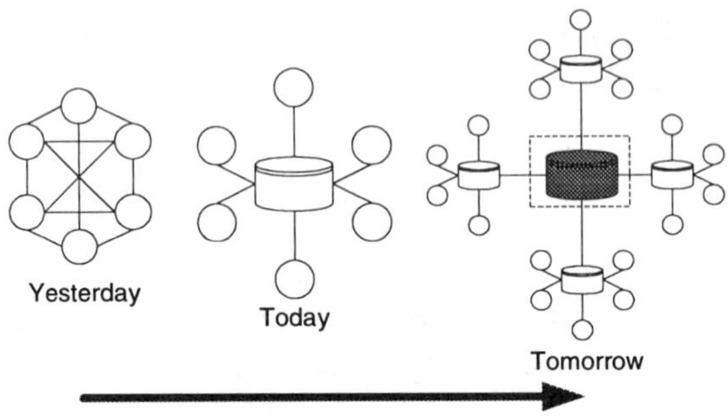

Figure 1 Inter-enterprise communication forms

We now proceed to presenting a summary of some functional requirements and their priorities within a VE setting. More information related to issues and needs resolution that led to the identification of the functionalities has been published elsewhere [1].

3 FUNCTIONAL REQUIREMENTS AND PRIORITIES

Table 1 summarises the perceived required functionalities of a VE intranet. Functionality priorities are identified from the perspective of inter-enterprise communication. Hence those that may be important at an organisational level alone are assigned a lower priority.

4 ENVISIONED SYSTEM ARCHITECTURE

One of the focus points of the IMS GLOBEMEN project [2] is related to the identification of requirements and specifications for a distributed engineering environment. This environment will in a way serve as a platform for inter-enterprise communications within a VE setting. In this section, we present the basic system requirements and illustrate a possible system architecture for the same.

Within dynamic VE conditions it is almost mandatory for inter-enterprise communication that a commonly accessible and usable communication infrastructure exists. Furthermore, each participant organisation would have its own legacy systems for internal use. For external communications with other members of the VE, the most feasible norm would be to communicate based on available standards. Organisation specific intranets and legacy tools could communicate with the VE intranet through interfaces enabling data/information mapping between the two intranets (Figure 2).

Table 1 Functional requirements and priorities for VE intranet services.

Function -ality	Explanation	Priority
Concur- rent Engin- eering	As part of the business process, information needs to be captured, (explicitly an implicitly), shared and controlled. Sources of information include: verbal instructions and information, applications, management tools, planning network analyses, reference specifications, computer model analysis, spreadsheets, drawings and layouts, three dimensional CAD models, sketches, calculations and supporting data, detailed designs, multi-media, etc. All this constitutes the complete project/product definition and needs to be captured and controlled.	High
Suit- ability to project	Internet-connectivity is very important. Accessibility via normal WWW browser (Netscape or IE). Ease of use. Quick set-up for new projects. Feasible licensing conditions for project conditions considering temporary use by several legal units. Back up of all project data. Usability of data after end/termination of VE (e.g. on a CD without reliance on a particular software)	High
Open system archit- ecture	Ability to connect different kinds of company specific applications and working environments. API for application integration. Interfaces must be based on standards: STEP / IFC, CORBA, PDM schema, Excel/Access etc.	High
Elect- ronic docu- ment manage ment	Procedures for document release, change, update and replacement. Publishing of distributed project information making it available to participants. Notify users about changes. Keep track on document dependencies. Version management. Document life-cycle management: support controlled change of documents from draft, working to release, change etc. Search mechanism based on attributes of meta-data and also on contents. Browser view support for many common data formats: CAD-files, IFC & STEP, MS Office documents and databases, HTML, VRML, SGML, XML, etc.	High
Access control	Manage access rights of users/roles in different projects/VEs. Maintain log (history) of all transactions. Assure authenticity of documents (objects): for legal reasons the originality of a released document (object) as submitted by the author must be guaranteed. Support for role specific views. Templates for fast configuring of the system. Support for IPR protection.	High

Product structure/ configuration management	Manage product structure (assemblies, subassemblies, components, relationships), related documents and data. Support information navigation according to product structure. Change and manage product structure with consequences to documentation, production planning and bills of materials/quantities (BOM/BOQ). Provide alternative and user specific breakdown structures (views). Manage alternative product configurations (designs). Search mechanism based on meta-data, attributes and relationships of objects.	High
Red-lining/ mark-up	Allow users to view different types (doc, xls, dwg, etc.) of documents (objects) and make comments to the subject just through marking / redlining.	Medium
Group-work support	Enable distributed team of persons to collaborate as if they were located in a common office: group-calendars, email, video / desktop conferencing, discussion groups, bulletin boards, shared document archives, common IT applications etc.	Medium
User profiling	Allow functionalities to be adjusted to the capabilities and needs of different users. Differentiate "user" and "role": access control is role-specific while profiling is user-specific.	Low
Version management	Manage new, parallel, alternative versions of documents (objects). Maintain change history. For practical reasons old versions may not need to accessible on-line. Back-ups should be stored.	Low
Work-flow management	Control transfer of documents (objects) between individuals in a shared working process for review, approval, change request etc. Support informal & formal processes. Highlight pending tasks to users.	Low
Distributed database	Managed interconnected data storages residing on different servers. Partial replication of selected data. The architecture should allow non-continuous connection between servers.	Low
App. integration	Provide tools for information sharing / data exchange between various IT applications. Typical means would be standards based interfaces, an object database and an API.	Low
App. launch	Allow end user to launch various applications on active documents / objects.	Low

Once the basic framework is established, it may be wise to explore the environment in more detail. Within the GLOBEMEN project, the work related to distributed engineering in a VE setting noted that an inter-enterprise intranet should be able to manage documents, product models, and processes/workflows. Therefore, in addition to the main functionalities identified in previous sections, there should be a document manager, product model manager, and process data manager. These basically provide different perceptual views to the data/information that is stored and shared trough the inter-enterprise intranet. The envisioned system architecture is shown in figure 3 followed by a brief explanation of its components.

Once the basic framework is established, it may be wise to explore the environment in more detail. Within the GLOBEMEN project, the work related to distributed engineering in a VE setting noted that an inter-enterprise intranet should be able to manage documents, product models, and processes/workflows [3]. Therefore, in addition to the main functionalities identified in previous sections, there should be a document manager, product model manager, and process data manager. These basically provide different perceptual views to the data/information that is stored and shared trough the inter-enterprise intranet. The envisioned system architecture is shown in figure 3 followed by a brief explanation of its components.

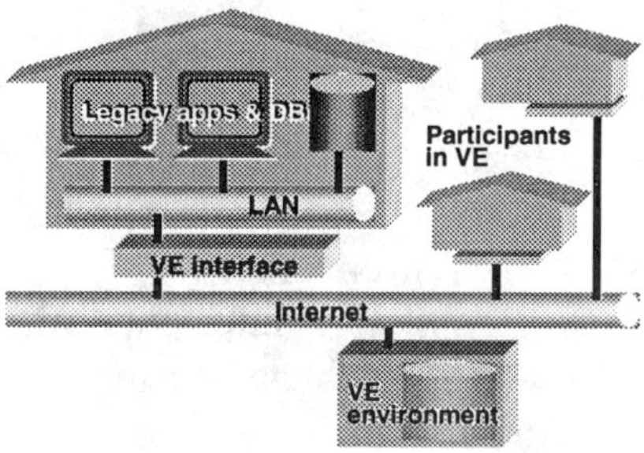

Figure 2 Inter-enterprise communications in a VE.

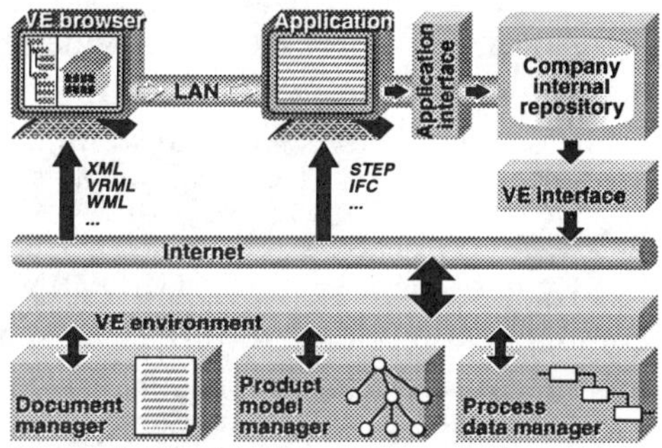

Figure 3 Envisioned system architecture for inter-enterprise communications.

As is evident from figure 3, the envisioned system architecture is quite generic in nature. The main components are:

- <u>VE environment:</u> A common data repository and sharing environment for organisation participating in the VE. It should provide support for distributed groupwork and sharing of information based on available (or agreed upon) standards. Different modules for the management and control of documents, products, and product models should be provided.
- <u>VE interface:</u> An organisation specific module, which provides a means for information exchange and communication between the organisation specific systems and the VE environment (intranet). The VE interface would be the communication bridge between the two systems and the control agent for the transfer of released information to the VE environment.
- <u>VE browser:</u> A tool that will enable users to search, view, and retrieve data from the VE environment. Note that transfers of information to the VE environment are in the form of information releases by a particular organisation and only allowed through the VE interface.

Other system components relate to the internal system and tools of each participant organisation:

- Applications: Tools that an organisation uses internally for performing and managing its core business.
- Application Interfaces: Services that enable applications to access the internal repositories of the organisation so users are able to store and retrieve data in line with the workflow of the organisation.
- Organisation internal repository: Organisation specific information storage and management system.

5 CONCLUSION

An envisioned system architecture was presented that focuses on the control and management of documents, product models, and processes/workflows. The system uses the Internet as a communications infrastructure and VE interfaces to exchange data based on standards between organisation specific systems and the VE environment.

Developments in GLOBEMEN to support inter-enterprise communications and information sharing for distributed engineering will include: integration of product model management with current document management technology, VE specific functionality to GroupWare tools, product model repositories with model merging and partial model extraction capabilities, VRML based user interface to access product model objects and related documents, and XML based product model access over the web.

More findings and developments will be reported and published as they become available during the course of the GLOBEMEN project (January 2000 – January 2003).

6 REFERENCES

[1] Hannus M., Aarni V. (1999). Requirements for Concurrent Engineering Environment in Concur Project, *Concurrent Engineering in Construction: Challenges for the New Millennium,* **CIB publication 236**, 25-27 August, Espoo, Finland, 99-110

[2] IMS GLOBEMEN web site: http://globemen.vtt.fi

[3] Hannus M., Kazi A.S. (2000) Requirements for Distributed Engineering, *Product and Process Modelling in Building and Construction (ECPPM 2000),* 25-27 September, Lisbon, Portugal, 41-48

The Architecture of an Internet-based Virtual Industrial Community

Mingwei Zhou
CSIRO Manufacturing Science and Technology, Australia
Email: Mingwei.Zhou@cmst.csiro.au

Key words Enterprise Integration, Enterprise Engineering, Virtual Industrial Community, Virtual Manufacturing Enterprise, Information Infrastructure

Abstract Virtual Enterprise (VE) has been increasingly seen as a promising solution for global manufacturing. This paper discusses the environment for supporting the lifecycle of virtual enterprises, from their creation, operation, to decommission. The concept of an internet based "Virtual Industrial Community" has been proposed to depict such kind of environment where potential member companies of virtual enterprises are facilitated with a virtual "community" or "industrial park" like infrastructure. The aims of VIC are to enable these companies to organise or participate in a virtual enterprise more conveniently and efficiently, and to stabilise the dynamics in the design and operation of a virtual enterprise.

INTRODUCTION

With the globalisation of economies, manufacturing enterprises are facing more and more rapid changes in marketplace, manufacturing practice, organisational structure, and information infrastructure. To seize and maintain its competitive advantage, a manufacturing enterprise must be able to quickly react to changes in their business. As a consequence, a new business paradigm of Virtual Enterprise (VE) has been evolved as a promising solution for global manufacturing. A Virtual Manufacturing Enterprise (VME) is usually referred to as a temporary consortium of independent companies which come together to quickly exploit fast-changing global manufacturing opportunities.

VME has been an active research topic in the last few years. A significant understanding on the architectures of VME has been achieved [1,2,6,11]. The tools and methods necessary for the design and operation of

VME have been proposed [3,4,7,10], and various software platforms for supporting the operation and management of VME have also been implemented [7, 12].

Current research has been mainly focused on the supporting of the operation of a VME, that is, the operation of a *virtual* enterprise in the *real* business environment. However, not much has been done on the supporting the operation of a *real* enterprise in a *virtual* business environment, which is the situation when a potential company is looking for opportunities to form or join a virtual enterprise.

We have found that the later (the operation of a *real* enterprise in a *virtual* business environment) in fact needs urgent attention, as it is the prerequisites for the former (the operation of a *virtual* enterprise in the *real* business environment). We identified that many obstacles for the effective formation and operation of a VME, such as the lack of confidence and trust, the lack of the identity and ownership, and the fear of losing uniqueness and know-how [5,13], are mainly due to the lack of a clear overall view of the virtual business environment.

The work on Virtual Industrial Community (VIC) is to investigate on the potential architectures of such a virtual business environment where the real companies can create and join a virtual enterprise, as well as the necessary information infrastructures for supporting the VIC.

VIRTUAL INDUSTRIAL COMMUNITY

The IMS Globeman'21 project [14] identified a common VE framework based on the GERAM (Generic Enterprise Reference Architecture and Methodology) [3]. The framework consists of three recursive lifecycles: network, VE, and product. These lifecycles indicate that the operation of the enterprise network leads to the creation, operation, and decommission of a VE, and the operation of the VE leads to the creation, operation, and implementation of the product. As show in Figure 1.

Here, the enterprise network is a very flexibly defined entity: it could be in the form of a simple directory service (such as a mailing list or a catalogue), or in the form of a fully incorporated temporary consortium. In most time, a network is only served as a candidate pool for a virtual enterprise. The network is usually non-exclusive, and a company may join multiple networks, therefore may end up with participation in competing virtual enterprises. There is very weak legal and financial binding, if there is any, within a network than that within a virtual enterprise. This often makes the operation of a network very dynamic, and sometimes, unstable.

The aim of a Virtual Industrial Community (VIC) is to establish a more stable and long lasting virtual business environment that can accommodate multiple networks and virtual enterprises, and provide the whole lifecycle support to the virtual enterprises.

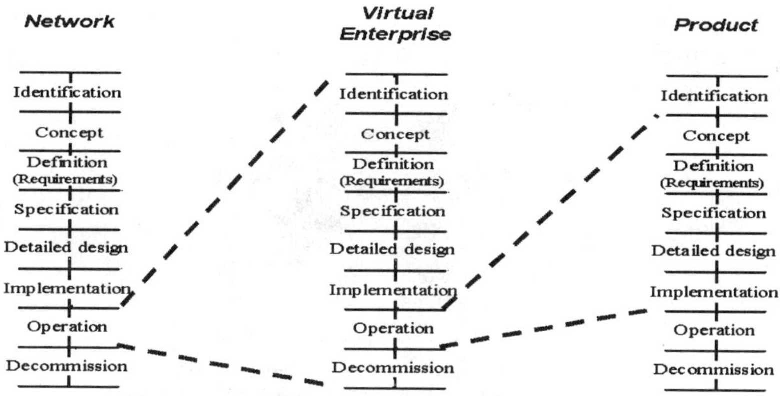

Figure 1: The Globeman'21 VE Framework [14]

A VIC, like a geographical community, is an open environment where any company may find a suitable place, and make connection with its neighbours by promoting and establishing its identity, and initiating a new network or joining any existing networks.

In addition to engineering and manufacturing companies, a VIC may have other companies providing various services such as IT, finance, accounting, legal, project management, and consulting. A VIC can be seen as an open network, or a network of networks, but in our opinion, a VIC should operate more like a normal community than a network. It provides services that a real industrial community such as the business centres and industrial parks.

ARCHITECTURE

A VIC has three major elements: the residents, the facilities, and the protocols. The residents of a VIC communicate to each other through their provided community facilities according to appropriate protocols (Figure 2).

The residents of a VIC consist of two types of companies: the core-competence providers and the service providers. Some residents may fall into both categories (e.g., a fastener maker plays the role as a core-competence provider when it is making a special purpose bolt, and as a service provider when it is making the standardised bolts). The core-competence providers are mainly the engineering and manufacturing companies who are usually the initiators and participants of a VME or a network. While the service providers may include a wide range of companies such as banks and ASPs (Application Service Providers) who are usually not a member of a VME or network, but they may play the role as the organiser, manager, or consultant of the VE or network in the VIC. As the names indicate, the service providers provide services to the core-competence providers.

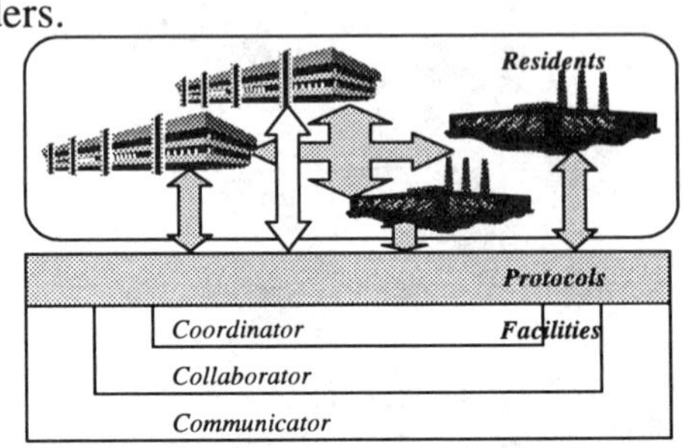

Figure 2: Architecture of a Virtual Industrial Community

The facilities of a VIC consist of three functional roles: the coordinator, the collaborator, and the communicator. These three roles were initially identified for supporting the lifecycle of a VME [12], and we find it is also applicable to VIC although their detailed functional specification may vary.

The VIC is managed through coordination instead of commanding. The coordinator is used to help establish trust, resolve conflict and make the VIC operate more efficiently. The coordinator provides the means for managing the company's identities, capabilities, capacities, and performances. It may also provide facilities for remotely monitoring and supervising the work-in-progress, as well as cross enterprise boundary scheduling and planning.

The VIC encourages and facilitates the collaboration among its residents. The collaborator will provide functions for exchanging and sharing of product and business information, enable the geographically distributed residents to work together more effectively, and provide the repository and warehouse for applications, models, data, and documents, serving as the memory of the VIC. The communicator provides a meeting place for the VIC. It also provides facilities such as identity verification, post office, chat room, bulletin board, and video conferencing.

The protocols are the spirit of the VIC. The protocols are those standardised or mutually agreed business processes that guide the coordination, collaboration, and communication among the VIC residents. The VIC protocols are implemented and supported by the VIC facilities (the coordinator, the collaborator, and the communicator).

CONCLUSION

The work presented in this paper contributes to the concept of a Virtual Industrial Community (VIC). The architecture of the VIC is discussed, and the proposed VIC is expected to provide a more stable environment for support the lifecycle of virtual enterprise. The findings reported here represent our first step towards the understanding of the social and organisational behaviour of a virtual manufacturing enterprise, which will enable us to develop a software platform for supporting the design and operation of a virtual manufacturing enterprise.

REFERENCES

[1] Afsarmanesh H et al, Towards an Architecture for Virtual Enterprises, Esprit Prodnet Project, 1997

[2] Burnes P, Nemes L, and Williams TJ, Architectures for Enterprise Integration, Chapman and Hall, 1996

[3] Burnes P and Nemes L, Modelling and Methodologies for Enterprise Integration, Chapman and Hall, 1996

[4] Camarinha-Matos, LM, Lima C, Configuration and Coordination issues in a Virtual Enterprise Environment, in *The Globalization of Manufacturing in the Digital Communications Era of the 21st Century: Innovation, Agility and the Virtual Enterprise,* 1998

[5] Westkamper, E, Manufacturing in Networks - Competitive Advantages for Virtual Enterprises, in *The Globalization of Manufacturing in the Digital Communications Era of the 21st Century: Innovation, Agility and the Virtual Enterprise,* 1998

[6] Hedberg B, Olve NG, Inside the Virtual Organisation, Strategic Management Society, October 1997

[7] Mills, JJ, Brand, M, and Elmasri, AeroWEB: An Information Infrastructure for the Supply Chain, in *Information Infrastructure Systems for Manufacturing II*, edited by Miles JJ. and Kimura F, Kluwer Academic Publishers, 1998,

[8] Neill H, Sackett P, The Extended Manufacturing Enterprise Paradigm, in *Management Decision*, 32(8):42-49

[9] NIIIP Consortium, U.S. National Industrial Information Infrastructure Protocols, Report No. NTR95-01, 1995

[10] Williams TJ, The Purdue Enterprise Reference Architecture, Instrument Society of America, North Carolina, 1991

[11] Zhou M., Enterprise Modelling Methodologies and Tools, IMS Globeman21 report, 1996

[12] Zhou M, et al, A framework for design a virtual manufacturing enterprise and its implementation as a workbench, in *Information Infrastructure System for Manufacturing*, edited by Miles J.J. and Kimura F., Kluwer Academic Publishers, 1998

[13] Zhou M, Neff P, Virtual Manufacturing Enterprises Design Using Enterprise Integration and Risk Management Methodologies, in *The Proceedings of the International Conferences on Manufacturing Engineering 2000*, Sydney, 2000

[14] IMS Globeman'21 Consortium, The Common Concept Report, 1999

From Single Enterprises to Complementary Networks

E.h. Hans-Peter Wiendahl, Arne Engelbrecht, Oliver Hamacher
Institute of Production Systems, University of Hanover, Germany
Email: hamacher@ifa.uni-hannover.de

Keywords Virtual enterprises, production networks, added value networks, complementary competencies, mass customization

Abstract Modern information infrastructure systems have strong effects on world wide operating enterprises, simplifying the access to global markets on the one hand and leading to a harder competition between enterprises offering similar products or services on the other. Additionally, the society of today asks for stronger customized products which cannot be produced economically with conventional production methods. In this context the early outmaking of upcoming needs of the society and their quick satisfaction with tailor-made products and services are future key factors to business success. Especially for small and medium enterprises the collaboration with partners possessing complementary competencies in so-called Complementary Networks offers a unique method for reaching this objective. The combination of complementary products and services adds to the value appreciated by the customer and opens new business fields and/or improves the market positions for the enterprises. This paper gives a brief description of the individual phases for the establishment and operation of these networks with practical examples showing the enormous potentials of Complementary Networks.

1 INTRODUCTION

The ongoing globalization process driven by the development of modern infrastructure systems has changed the local and global environment of producing enterprises dramatically. More and more rival companies are pushing into the existing markets, leading a strong price and performance competition. Improved productivity of up to double percentage figures by leading companies is characteristic of the challenges. The long-term result of the accompanying intensive automation in the industrial sector is that the requisite quantity of goods can be produced with ever fewer workers. The resulting products, in many cases with exchangeable functional features, mean that it is becoming more and more difficult to achieve cost advantages through more economical production [1].

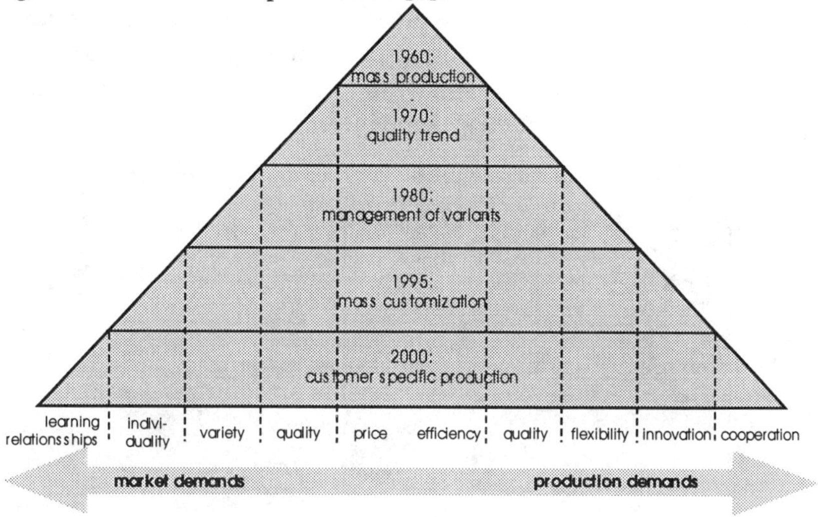

Figure 1: Increasing market demands require higher production demands [Piller]

Additionally the demands of the market rapidly change and increase (Fig. 1). In the industrial age customers were easy to satisfy with standardized and bulk products which were sold from stock. With the growing needs of the society, enterprises had to deal with a rising number of variants in the production and a higher complexity of the products.

2 NEW PRODUCTION STRATEGIES

These tightened circumstances force enterprises to find new strategies for realizing efficiency and individualization simultaneously. More and more enterprises today discover the Mass Customization as a unique way of combining the advantages of the cost reduction and the differentiation strategy, instead of trying to marginally improve their productivity. Mass Customization is defined as a production of large quantities with a strong customer focus to fulfill individual needs and demands, without a loss of production efficiency [2]. By tailoring fixed product elements to individual customer needs a maximum degree of product specific individualization can be achieved. In this context so-called "learning relationships" are of great importance, giving the enterprise a much better understanding of the customers needs as the relationship goes on.

On the other hand customers are always forced to choose from a fixed number of possible product variants which might only be a compromise between their true needs and the available product on the market. This is the approach for the customer specific production, where the customers describe their individual problem and then receive a unique manufactured solution for solving it. Such products, however, provide the opportunity not only for large enterprises but also for small and medium enterprises (SME) to differentiate themselves from their competitors.

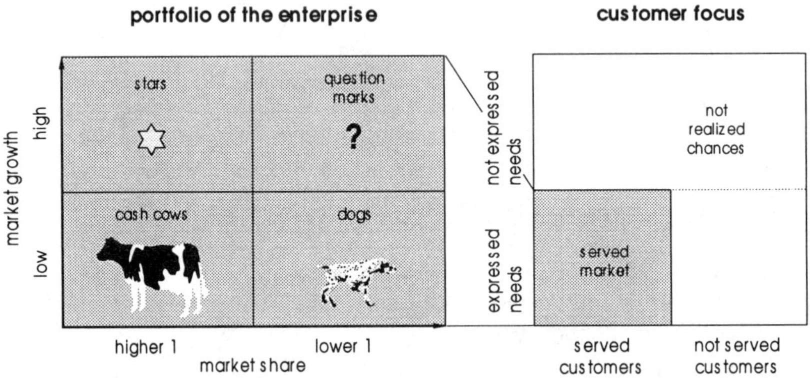

Figure 2: Possibilities of strategic differentiation for SME [BCG / Prahalad]

With their quest for customer orientation many enterprises merely use the potential of the market (Fig. 2). Since customers are generally remarkable for a notorious lack of far-sightness, only the articulated needs of the customers are served in the respective market segment [3].

There is often a failure to see the opportunities in the new business fields that can be purposefully opened up by focusing on a potential customer value outside the market segment. In one example, a large American automobile concern suffered an instructive experience when an assessment of the articulated needs of its regular customers did not lead to the desired success. A small family car was introduced in 1991 after a five-year development project. The design and specifications of the car were the result of the most intensive questioning of its customers that the firm had ever conducted. The problem emerged, however, that the newly produced car was extraordinarily similar to the three-year-old rival Japanese model. The customers had only formulated what was already known to them. Akio Morita, the visionary founder of Sony, put it as follows [3]: "Our plan is to lead the consumer to new products rather than to ask them what sort of products they want. The customers don't know what is possible - we do."

The not realized chances (blind spots) can be opened up through new core competencies which have the same customer value. SME cannot fulfil these requirements as easily as large enterprises because they seldom have sufficient R&D capacities at their disposal. One way of compensating for this deficit, however, is by teaming up with partners who already possess the requisite (complementary) core competencies [4].

Decisive success factors for SME therefore include adaptability, efficient structures and the management of relationships to other organizations leading to a common competence to open up new fields of business through the customized individualization of products and services. The only firms to survive will be those that recognize or generate new market needs in good time and quickly convert them into products and services tailored to individual customers with their specific partners [5].

3 DESIGN OF COMPLEMENTARY NETWORKS

3.1 Basic Structure of a Virtual Enterprise

Virtual Enterprises are praised to be the ideal organization structure for cooperation in the future [6]. They combine the adaptability of small units with the synergy effects of large organizations and build up on the idea to combine certain process units benefiting from synergy effects for the time of a specific project. This can lead to the extreme of pursuing only a single business idea, exploiting it and then splitting up again – an temporary enterprise which defines itself by its specific linking [7].

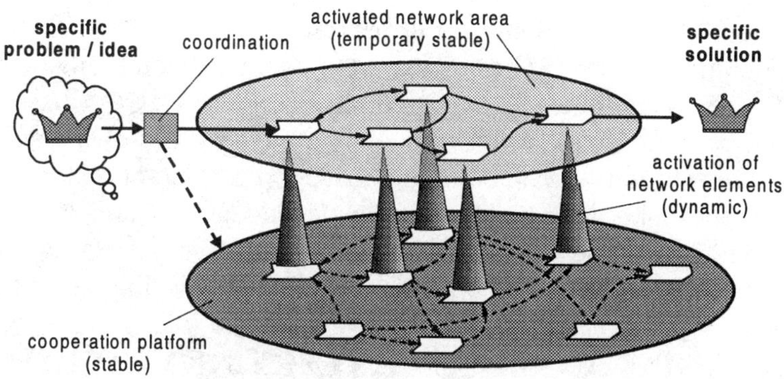

Figure 3: Basic structure of a Virtual Enterprise

Starting up a cooperation usually takes long preparation times including the search, the evaluation and the integration of potential business partners. The longer the cooperation lasts, the stronger is the relationship and the trust that builds up due to shared experiences and resources. For this reason, partners proved to be reliable and cooperative are integrated into a permanent cooperation platform from which they can be quickly activated to realize a specific idea or product (Fig. 3). Thus the cooperation platform shows a more stable character, the activated elements (single enterprises) only make up a temporary stable network [8].

3.2 Complementary Products and Services

One special form of the virtual enterprise is the so-called complementary network, whose potentials are particularly beneficial to SME. On the basis of the traditional product spectrum, cooperation partners are sought whose products or services and competencies serve the same generic customer value - the so-called added value concept [5].

In this context the terms added value, complement and core competency are defined as follows: added value comes into existence when the combination of two complementary elements creates a value which the customer rates more highly than the two single values. In other words it is the phenomenon of the whole being greater than the sum of its parts [9]. A complement to a product or service is any other product or service which makes the whole more attractive to the customers that they prize it more highly when they also own the complement [5]. A core competency is understood as an integrated totality of technology, know-how and processes coordinated through organizational learning processes and in possession of which an enterprise can claim its position in the market. It must make an above-average contribution to the value perceived by the customer, and create potential access to a number of markets [10].

Cooperating partners can use the joined complementary core competencies to combine existing product and/or service solutions or, with a rising degree of trust, to generate new solutions. Well known examples for added-value solutions are cinemas and fast-food restaurants (independence), mobile phones and accessories (single dependence), hard- and software (interdependence). The customers reward the newly generated value, for they see it as a competition-free system solution from a single supplier.

4 MANAGING COMPLEMENTARY NETWORKS

The life-cycle of complementary networks divide into three distinct phases: the development of the added value idea, the setting-up of the network and the operational phase.

The easiest way of finding added value is the reaction to the already existing wishes of customers. The customer value can be gathered concretely from expressed needs. Creative enterprises go further and take advantage of the possibility to infer future customer needs from current trends and new technologies, thus anticipating customer value. The customer value established in these various ways confirms an added value concept and prompts the further development of the network.

The description of the added value concept at this point provides no information at all concerning the resources that have to be determined and combined in order to realise the concept. A systematic procedure here facilitates the transition from the added value concept to the desired complementary network and creates new tasks for additional partners alongside the product and service providers. Firstly, the desired individualisation of the product must be reduced to an acceptable degree. The aim to satisfy all the requirements of all the customers can only be fulfilled up to a certain point, beyond which the cost and effort increase exponentially.

Figure 4 shows an acceptance profile which translates the added value concept into a string of success factors for the wooing of purchasers, and mirrors the strategic organization of the network. Consultation with customers ascertains the individual needs and preferences of the known circle of customers.

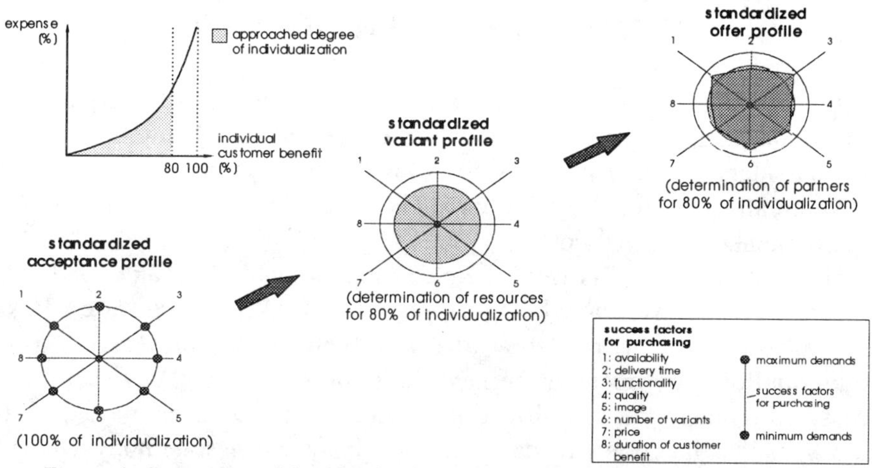

Figure 4: Conversion of the added value concept into a profile of success factors

The various opinions expressed about individual factors can now be used for a qualitative projection of the maximum requirements in the acceptance profile. The variant profile serves to determine what network resources are necessary. The degree to which the maximal requirements will be fulfilled is also decided upon. As a rule of thumb, an 80% fulfilment of requirements substantially saves effort and cost without severely affecting the number of potential customers. If the resource requirements established in the variant profile are assigned to specific network partners, the result could be a slight overshooting or falling short of the intended degree of fulfilment. It is therefore necessary to check whether the potential network partners' contribution will match the target requirement. Partners who meet this requirement or come very near are basically suitable for the network. With the aid of this information the network architect can now begin the negotiations in which those who can provide the competencies needed for the network are secured as partners. The continuously updated product and service profile supplies information as to how far the network partners' existing competencies match the catalogue of requirements. A poor level of correspondence between the variant profile and the product and service profile likewise indicates a problem for the future of the network.

In the operating phase, the most crucial consideration is the smooth co-ordination of products and services, thus to establish the proper relationships between the partners. These links concern not only logistics and the flow of materials and information but also the detailing of contractual relations. Order processing systems and marketing and sales activities are also planned during this phase.

5 REALIZATION OF A SPECIFIC SOLUTION

A successful example for a complementary network is the cooperation of a door manufacturer and electronic specialists. Bringing together complementary products and services allows them to realize the value added idea of creating an "intelligent door". This leads to a win-win situation for all participants in this network.

The customer benefits from the electronic door offering an added value compared to conventional door-systems. These doors are linked electronically and can thus show the way for a visitor to a certain room inside a building by opening the next door automatically. They can control access with integrated identification devices and can keep messages for persons inside the room, in case they are busy when somebody wants to enter.

During the phase of the product devolpment the participated enterprises profit from the exchange of their specific know-how and identifiy their own strenghts. By offering this unique product they open new market areas and come in contact with new groups of customers. This might give information about unserved customer needs and inspire the network partners for new product developments which an single enterprise is not able to create.

6 CONCLUSION

The change of the general conditions in the postindustrial age has drastic effects on producing enterprises. The distinctive force of this development are primary the globalization process, the ongoing rapid developments of infrastructure systems and the rising requirements of the society. Especially the individualization of the human needs requires an increasing number of variants and innovations from enterprises.

In this paper a framework has been presented to design and manage Complementary Networks, which can realize specific needs or ideas in a temporarily activated network. It distinguishes itself from other network types by the conclusion of complementary core competencies of SME in order to generate a specific solution for the customer giving him a added value and leading to a better market situation for the involved enterprises.

8 REFERENCES

[1] Kinkel, S; Lay, G. (1998). Der Leistungsstand der deutschen Investitionsgüterindustrie, Fraunhofer Institut Systemtechnik und Innovationsforschung, Nr.11.

[2] Piller, F.: Kundenindividuelle Massenproduktion - Die Wettbewerbsstrategie der Zukunft, Hanser Verlag, München 1998

[3] Hamel, G.; Prahalad, C.K.: Wettlauf um die Zukunft, Überreuter Verlag, Wien 1995

[4] Sydow, J.: Strategische Netzwerke - Evolution und Organisation, Gabler Verlag, Wiesbaden 1992

[5] Nalebuff, B.; Brandenburger, A.: Competition - kooperativ konkurrieren: Mit der Spieltheorie zum Unternehmenserfolg, Campus Verlag, Frankfurt 1996

[6] Delphi '98: Studie zur globalen Entwicklung von Wissenschaft und Technik, Fraunhofer - Institut für Systemtechnik und Innovationsforschung.

[7] Bellmann, K.; Hippe, A.: Management von Unternehmensnetzwerken, Gabler Verlag, Wiesbaden 1996

[8] Schuh, G. et al.: Virtuelle Fabrik. Neue Marktchancen durch dynamische Netzwerke, Hanser Verlag, München 1998

[9] Haken, H. (1995). Erfolgsgeheimnisse der Natur – Synergetik: Die Lehre vom Zusammenwirken; Rowohlt Taschenbuch Verlag.

[10] Wildemann, H. (1996). Produktions- und Zuliefernetzwerke, gmft-Verlag, München.

PART THREE

Modelling and Analysis of Virtual Enterprises

Use of GERAM as Basis for a Virtual Enterprise Framework Model

J. Vesterager, L.B. Larsen, J.D. Pedersen, M. Tølle
Technical University of Denmark
Em: jve@ipt.dtu.dk

P. Bernus
Griffith University, Australia

Key words GERAM, Virtual Enterprise, framework model, industrial cases

Abstract In the IMS-project Globeman21, the enterprise reference architecture GERAM was used as basis for creation of a virtual enterprise framework model. The model was used to map different industrial pilot projects, to classify virtual enterprise concepts, and as underlying structure for a virtual enterprise management methodology. The paper gives a survey of the use of GERAM and the results obtained.

INTRODUCTION

This paper builds on research work carried out in two IMS-projects: Globeman21 and Globemen. The results presented are primarily based upon the work of the so-called Common Concept group of Globeman21 as well as the EU work packages on "virtual enterprise methodology support" and "generic models". The work continues in the ongoing Globemen project with the aim of producing a Virtual Manufacturing Enterprise Guidelines Handbook.

GERAM AND THE VIRTUAL ENTERPRISE FRAMEWORK MODEL

The purposes of the framework model were: 1) to create a reliable VEF by the use of GERAM [1] – e.g. ensuring validity of basic concepts and a sound communication basis; 2) to extend the VEF with specific VE-concepts, e.g. by relating it to already existing literature, and in this way to prove a general use of the VEF as a synthesizing 'tool'; and 3) to support more Globeman21-specific uses including, 3a) establishing a common foundation for comparison (mapping) of Globeman21 industrial pilot projects for general collection of experience, 3b) development and communication of industrial reference examples for use by other companies, and 3c) use of the framework model as a supporting skeleton for a first VE Management Methodology (VEMM).

GERAM AND USED ELEMENT/COMPONENTS

Elements of GERAM used so far in a systematic way are: the Process Oriented Concepts of *Life Cycle* (LC) and *Life History*, and *recursiveness*. As regards the View Concepts of GERAM, the *Entity Purpose View* ('management and control', and 'customer service and product') has also been applied systematically. In addition, the other View Concepts have been used as follows: *Entity Model Contents Views* (function, information, resource, and organisation) by referencing when needed, assuming knowledge of these topics. Correspondingly for the Genericity Dimension. The *Entity Implementation View* has not been used explicitly but is nevertheless dealt with through systematic application of a traditional Industrial Engineering work preparation approach to the preparation of VEs. Lastly, the *Entity Physical Manifestation View* component "software" was dealt with due to the fact that all industrial demonstrators were software projects.

THE VE FRAMEWORK MODEL

In Figure 1 the Globeman21 VE concept is shown. Companies assign competencies to a network in order to be able to create customer focused VEs satisfying customer needs (creating deliverables/solutions for the customer). Depending on customer needs, more contractors or sub-suppliers can be included in the different VEs without being members of the network.

The network is based on a relatively long term cooperation, whereas the VEs are dissolved, transferring experience back to network members, when the customer need is satisfied. Correspondingly, the same network can form many VEs, and a VE can produce many deliverables. In short, a VE is defined as *"a customer solutions delivery system created by a temporary and reconfigurable ICT enabled aggregation of core competencies"*. The relationships between the partners in the network can vary from a loose relationship (by analogy with "Yellow Pages") to ownership as the other extreme. In Globeman21 the industrial partner cases were quite focused networks, implying that the networks in question were relatively formalised ones of the types: partnership, strategic alliances, and ownership. In the near future, other types of network collaborations will probably also be possible, resembling the "Yellow Pages"-situation. This is for example due to global virtual markets for competencies, information standards as STEP APs for information integration, standard/reference models for contracts and a corresponding model-defined set-up, e.g. of joint ERP and engineering systems. In one sense this means looser networks, in another sense more prepared ones, due to common standards and reference models.

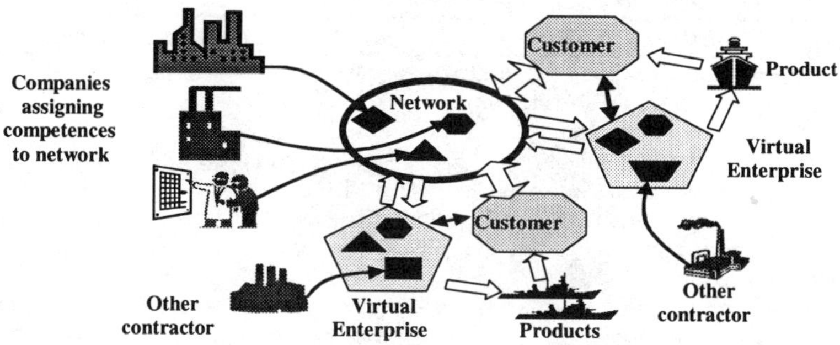

Figure 1 The Globeman21 virtual enterprise concept

In Figure 2 the GERAM based VE framework model is presented. The model consists of three recursive LCs: a network, a VE and a product LC (PLC). The double arrows between VE and PLC indicate that the deliverable corresponds to one or more phases of the PLC. The parts of the Entity Purpose View are indicated by shading of LC phases. For the sake of simplicity, the VEF does not include the LCs of the participating companies creating the network, for example by the formation of a common project.

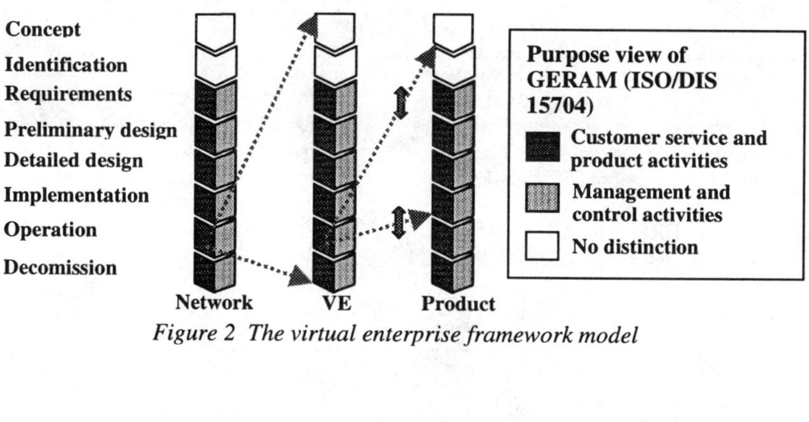

Figure 2 The virtual enterprise framework model

USE AND RESULTS

This chapter gives a summary of the results obtained. First, we present the way in which the framework can be used to classify concepts (description parameters) for VEs. Secondly, we show how the VEF was used to map Globeman21 industrial projects for comparisons and experience collection. Thirdly, we indicate how the VEMM was built by use of the VEF.

Table 1 Example of concepts (description parameters)

	Examples of situational factors	Examples of design parameters
Network	- The necessity of: increasing flexibility in different ways, specialising, decreasing risks, reducing future uncertainty, obtaining agility, increasing task frequency, decreasing lead time, etc. - Exploitation of ICT enablers (e.g. global distributed concurrent engineering and production) - Social and cultural issues, e.g. opportunity and barrier/threats - In general: effectiveness and efficiency considerations regarding competitive situation	- Member characteristics, ownership or otherwise, type of agreements (e.g. IPR, risk sharing), management structure (rules and roles in network), etc. - PLC coverage, types of VEs and corresponding business processes to cover (mission, vision, motivation) - Preparation issues: degrees of preparation to go for with respect to key business processes, contractors, etc. - Legal issues (e.g. leaving/entering network) - Handling of social and cultural issues
Virtual Enterprise	- Competitive VE lead time requirements - Global distribution of partners and customers - Types of competencies needed to meet customer demands - Obtainable frequency of different VE-types - Situation of potential contractors (e.g. ICT conditions, cultural issues) - Available ICT enablers (e.g. standards, vendor systems, communication systems)	- Types and degrees of preparedness; common models of VE creation processes including temporary contractors (e.g. how to qualify) and customers, and corresponding development of ICT applications - Rules for VE-management (e.g. roles, rules of leaving a VE) - Rules for exposure of partner competencies in operational delivery systems - Legal aspects (entering and leaving a VE)
Product	- Required PLC phases lead time - Requested degree of innovation to handle, stability of competency domains - Geographical distribution of partners - Available and usable ICT enablers (e.g. PDM for distributed concurrent engineering, ERP tools, standards) - Obtainable frequency of different PLC tasks - Necessary PLC types and phases to cover	- Types of PLCs and PLC-phases to cover - Rules for PLC management, common management and control models - Preparedness: common product, facility etc. models for instantiations (e.g. generic product model, model driven CE) - Relation: VE-type to carry out PLC-type - Rules for customer involvement, conflict resolution, exemption handling etc. - Use of standards (e.g. demands on temporary contractors)

Description parameters for virtual enterprises

Inspired by Mintzberg [3], description parameters were divided into situational factors (external conditions) and design parameters (options representing the solution space). Information sources were internal Globeman21 project questionnaires, workshops, general experiences from projects, and a literature search reported in [2]. Table 1 gives examples of general concepts characterising the three LCs making up the VEF.

The description parameters in Table 1 are 'high level' planning concepts. Note that the table says nothing about when to consider the concepts in a VE-network engineering project – this question relates to the methodology.

Several already existing theories and corresponding concepts on enterprise management and planning not included here can be reused in this context, in order to describe the "to-be" situation, e.g. the concepts in the SCOR model on SCM [5] such as "purchased materials", "engineer-to-order product", and "make-to-order product". Consequently, the concepts can be further extended and refined, especially when focusing on a specific industry as, for example, one-of-a-kind manufacturing.

Mapping of industrial cases

As mentioned above, the VEF was used to map industrial projects (pilots) in order to extract experiences and in a systematic way present industrial reference examples. Referring to LC-phases, a distinction was made between *where the pilots were engineered*, and *where they were used.* Figure 3 shows the results in an almost pictogram-like manner. The arrows point to the phases where the developed pilot tools are used. For more information, see [4]. Here we only wish to state that the VEF proved very useful, especially considering the big differences between the mapped pilots.

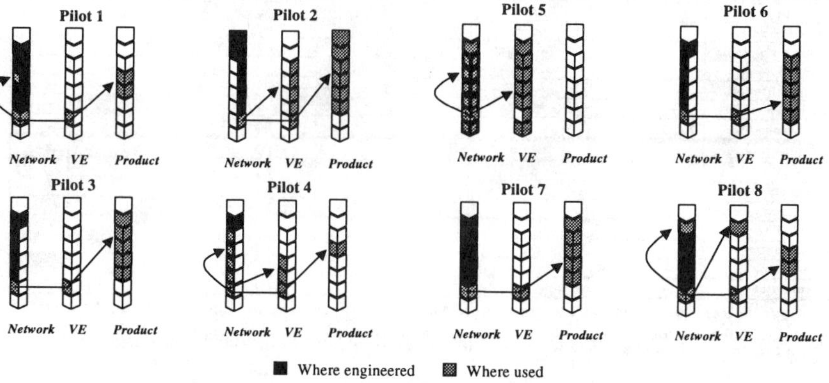

Figure 3 Mapping of Globeman21 industrial pilot projects

A VIRTUAL ENTERPRISE MANAGEMENT METHODOLOGY

In order to introduce the developed VEMM it is appropriate to demonstrate, how the VEF unfolds in a life history perspective. Figure 4 shows an example. Space only allows a short explanation here. For more information, see [6]. The 3 LCs of the VEF are shown on top of each other in order to introduce the time dimension. Numbers 1, 2, 3, and 3a concern the first phases of the network LC, which result in the setting-up of the management and control part of the network. Notice that the network entity so to speak boots itself because the VEF as mentioned does not include the entity producing the network. 4 relates to the execution of preparation projects creating, for example, reference models or ICT tools to use in the operation of the network – here shown as triangles. At 5 the network is in operation. 6 demonstrates a customer identifying a product need. 7 and 8 shows the network setting up a VE through preparation projects, one of which prepares a tool to be used in the PLC phase ‘preliminary design’. When operational, the VE creates a quotation, see 9 and 10. Subsequently, the VE is decommissioned – see 11. If the quotation is accepted, the network sets up a new VE producing the deliverable, see 12, 13 14. At 15 the product is handed over to the customer for operation and the VE is decommissioned. For the sake of completeness, 16 show that the network does not last forever.

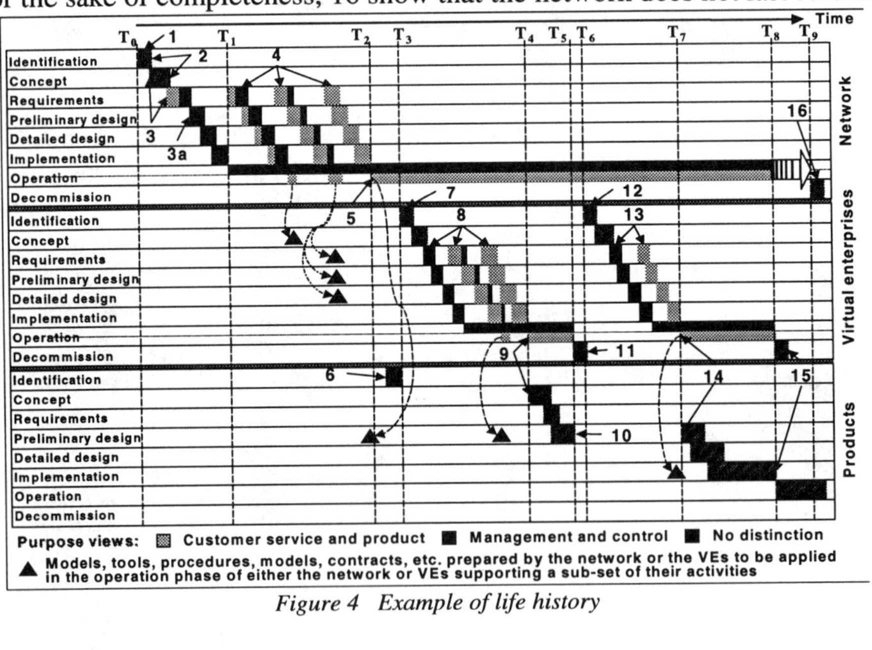

Figure 4 Example of life history

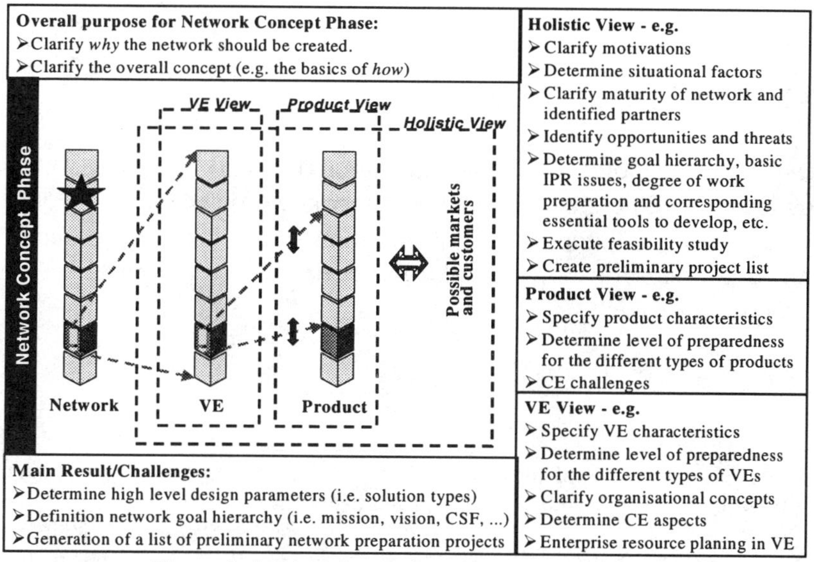

Figure 5 Methodology - the network concept phase

With this life history in mind, Figure 5 shows an extract of the VEMM. It gives examples of considerations to make in the concept phase of the network LC (see star). Well aware that planning and engineering considerations are iterative, the VEMM recommends first to take a holistic view followed by more focused analyses and decisions regarding the PLC and the VE.

CONCLUSION

Today much research work and many development projects address the problem of virtual enterprises. In our experience some of the key challenges are: establishment of joint reference models for a) global engineering, including its integrating infrastructure, and b) global management (e.g. resource and project planning). In our experience a unifying framework is crucial in order to ease communication across the many disciplines involved and across cultural borders, to enable a coherent and systematic collection and exchange of experiences, and to develop rewarding training programs.

The use of a virtual enterprise framework model mainly based on the enterprise reference architecture GERAM shows promise. The developed framework allowed – through a mapping of industrial projects – the demonstration of the relationships between the contributions of the large variety of seemingly unrelated projects within the consortium. Also, the framework served as a useful platform for the developed management methodology.

ACKNOWLEDGEMENT

The European Commission has sponsored this work. The authors wish to acknowledge the Commission for their support. Furthermore, we would like to express our gratitude and appreciation to all the Globeman21 and Globemen partners for their contribution during the development of results presented in this paper.

REFERENCES

[1] GERAM: *Generalised Enterprise-Reference Architecture and Methodologies*, ISO/DIS 15704, mainly developed by IFIP/IFAC "Task Force on Enterprise Integration" and IFIP WG 5.12 "Architectures for Enterprise Integration".

[2] Kaas-Pedersen C., Larsen L.B (eds.). (1998). *Modelling Overview* (public version), Report no D1.1, Globeman21, ESPRIT 26509, pp. 62. Department of Manufacturing Engineering, Technical University of Denmark.

[3] Mintzberg H. (1983). *Structures in Fives: Designing Effective Organisations*, Prentice Hall, 1983.

[4] Pedersen J.D., Vesterager J., Tølle M. Application of GERAM based Virtual Enterprise Framework – Results from IMS 95001/ESPRIT 26509, *Proc. of the 6th International Conference on Concurrent Enterprising*, Toulouse, France, 28-30 June 2000, p. 139-142.

[5] SCOR model. Supply-Chain Council Inc., 303 Freeport Road, Pittsburgh, PA 15215.

[6] Tølle M., Vesterager J., Pedersen J.D. A Methodology for Virtual Enterprise Management – Results from IMS 95001/ESPRIT 26509 Globeman21 project, *Proc. of the 6th International Conference on Concurrent Enterprising*, Toulouse, France, 28-30 June 2000, p. 119-127.

The Component-oriented Approach towards Complex Product Development

Fujun Wang, John J. Mills
Automation & Robotics Research Institute
The University of Texas at Arlington, Fort Worth, TX76118
{fwang, jmills}@arrirs04.uta.edu

Keywords Virtual Product Development, Collaborative Product Development, Software Component, System Architecture, Product Data Modelling

Abstract This paper proposes a new approach towards complex product development. The approach is based on the software component mechanism. A software component is a reusable software package which can run across heterogeneous platforms. This paper presents a Component-based Open System Architecture, in which the design of a product component is implemented into a software component. The product component data and operation methods are exposed and can be reused by others. The Collaborative Product Representation Model is proposed to represent the product component in this environment. It is a neutral, understandable, customizable and reusable model with self-management capability. The Product Model Processor helps to define this complex model data.

1. INTRODUCTION

This paper presents an approach towards the development of complex products. A complex product refers to one with more than one components, which is usually designed by one more persons or organizations. An airplane is a complex product. A gear box can also be looked as a complex product if its design needs more than one person. How to develop such a complex product in short time with reliable quality is always the pursued aim. The World-wide Virtual Product Development is an important methodology for the emerging global marketplace [11,12,20,21,22]. With this development mode, a complex product is developed collaboratively and virtually by a temporary alliance with world-wide scope. Product components developed by different alliance members in different locations are assembled into the final product at another location. Both the enterprise-wide collaboration and the world-wide development have the same development feature, i.e., the product components are developed by different persons or organizations in a distributed environment, and then they are assembled into the final product. This paper proposes an approach supporting this development feature. The key issue here is how to represent the product component and manage them to make them easily accessible and usable in a network environment. The requirements and previous research are outlines below.

- *Open System Architecture.* Since the product development alliance is possibly dynamic in the virtual product development mode, an open distributed system is required. Gauchel et. al. [8] proposed the concept of *Autonomy* and used it in product modelling. Cutkosky [5] proposed an agent based design system running on the Internet. Through a resource directory on the Internet, those agents can be found and then be accessed.
- *Customizability.* A customizable product component model can always provide much more freedom for reuse by others. Most of the research in this field is in three areas: the *Parametric Model*, the *Constraints-based Model* and the *Serialized Model.* The Parametric Model is a variables-based product data representation. Dai [6] presented a parametric feature-based model for integrated CAD/CAPP/CAM systems in his Ph.D thesis. The Constraints-based Model is always coupled with the Parametric Model. It uses the constraints to limit the scope of a variable and the relationships among one or more variables. Eastman [7] embedded integrity rules that apply between applications into his model. The Serialized Model refers to the representation of serial products or product families. Gero and Shi [10] built a product families model based on the phase transition in the biological world.
- *System Methods Reuse.* This might be the most important need and is different from other traditional models. In a traditional collaborative design session, what is exchanged between designers is the product model data and design knowledge. However, the system methods which operate on the product data can not be shared. In fact, system methods are an important design resource.
- *Neutral Model.* Since different product components are designed by different departments or persons, it is necessary to define those product components in some neutral model to promote their data exchange smoothly. One accepted approach to this problem is the Feature-based Modelling basing on some product data exchange standard as STEP [15,16].

- *Understandable Model.* The feature-based data is limited to geometry data and a few engineering properties, which is far away from the requirement of information exchanged between partners. Therefore, knowledge is often embodied in a product model called an *Ontology* [1]. Gero and Rosenman proposed the paradigm of '*purpose-function-behaviour-structure*' which describes the product engineering semantics [9,23,24,25]. Henderson [13] represented functionality and design intent in a meta-model. In addition, design rationale is captured in some systems to explain the design principle used [2,17].
- *Self-Management.* Since a product component can be serialized and parameterized, it is necessary to improve its robustness. This can be realized by enabling the product component to manage itself automatically. *Agent-based Management* is an AI approach to maintaining the product data automatically. Rosenman and Wang [24] proposed a component agent model for the representation and management of a product object.

2. COMPONENT-BASED OPEN SYSTEM ARCHITECTURE

2.1 Component-oriented Approach

As an approach to software development, *Object-Oriented* (OO) Analysis and Programming has been in a dominant position for one to two decades. Its approach is that of encapsulating attributes and methods in a class and permitting inheritance between classes (fig.1). Code reuse is available by the inheritance mechanism, but this reuse is limited inside an application, or on the class-level. Such a general application has two main parts, the Implementation and the Data Base (fig.1a). Communication between two applications is through the Data Base. That is, the interaction is limited to sharing data only. Therefore, the code reuse or system function sharing between two applications is impossible. However, the main task of distributed systems integration is the inter-operation between them: the code reuse.

Though the OO approach brought a giant revolution from traditional software development, the promise of large scale of code reuse did not become reality [26]. Recently, a new approach, *Component-Oriented* (CO) approach is becoming the focus in software industry. A component is a reusable software package. A *Component Based Application* (CBA) provides services containing the data operations and method operations in its implementation. They are the reusable attributes and methods which are shown in fig.1b by the small solid circles and boxes. An Interface File coupled with these services describes the exposed data and methods. If a CBA works as a server, the Interface File is the medium connecting the server and one or more clients. A traditional application can be wrapped into a component to expose its data operation [28] (fig.1c).

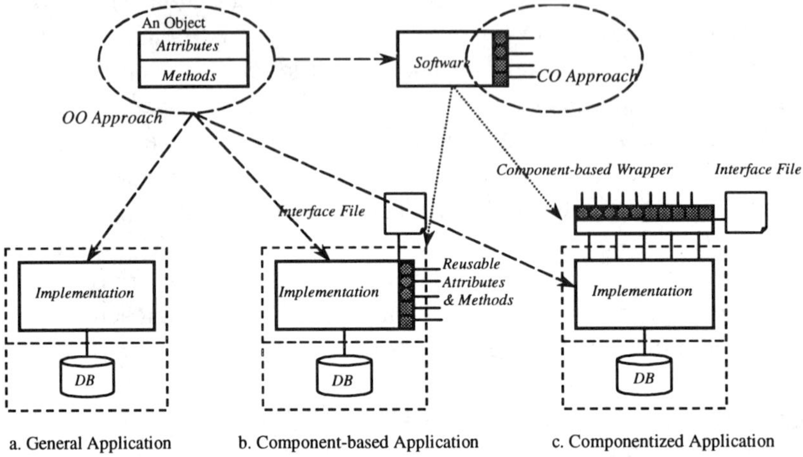

Figure 1. OO Approach and CO Approach

2.2 Component-based Open System Architecture

The reuse-ability of a component allows it to work as a *System Component*, or as part of a large system. CORBA, COM and EJB are sometime called distributed object 'middleware', because they mediate between components to allow them to work together, integrating them into a single, functional whole [14]. A distributed system made up of dozens of component-based applications can be called a *Component-based System.* Fig.2 indicates such a system. These System Components can be distributed in heterogeneous platforms. Each system component implements the design, analysis or manufacturing of a *Product Component* and supports to assemble high level product component or the final product. A system component might encapsulate some legacy application such as the CAD, CAPP, CAM, or some proprietary system. Because those system components are distributed, a central management facility—the *Web-based Design Resource Manager* manages the registering and accessing of those system components. This system is called an open system because of the distribution, independence and inter-operation of the component mechanism. The system component is not pre-defined and the number of the system components is not fixed. We call the system architecture the *Component-based Open System Architecture* (COSA).

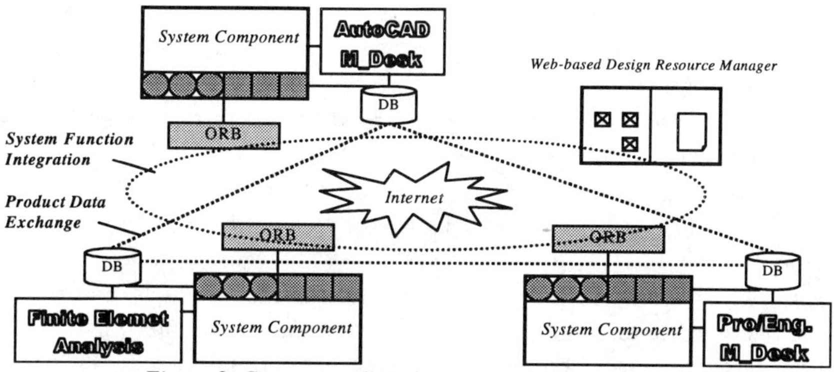

Figure 2. Component-based Open System Architecture

3. PRODUCT COMPONENT DESIGN

3.1 Product Component Model

The section covers the representation of a Product Component in the Component-based Open System Architecture (COSA). This paper presents a richer model to meet the five needs noted in section 1. The model here is called *Collaboration-oriented Product Representation Model* (CPRM). It is the combination of four data modules: *Functional Data*, *Embodiment Data*, *Management Data* and *Reusable Data*. Fig.3 indicates the model architecture. It should be noted that the Reusable Data is added and the Structural Data is replaced by the Embodiment Data. The Reusable Data represents the feature of a software component, i.e., the reusable attributes and methods. The Embodiment Data contains more semantics than Structural Data.

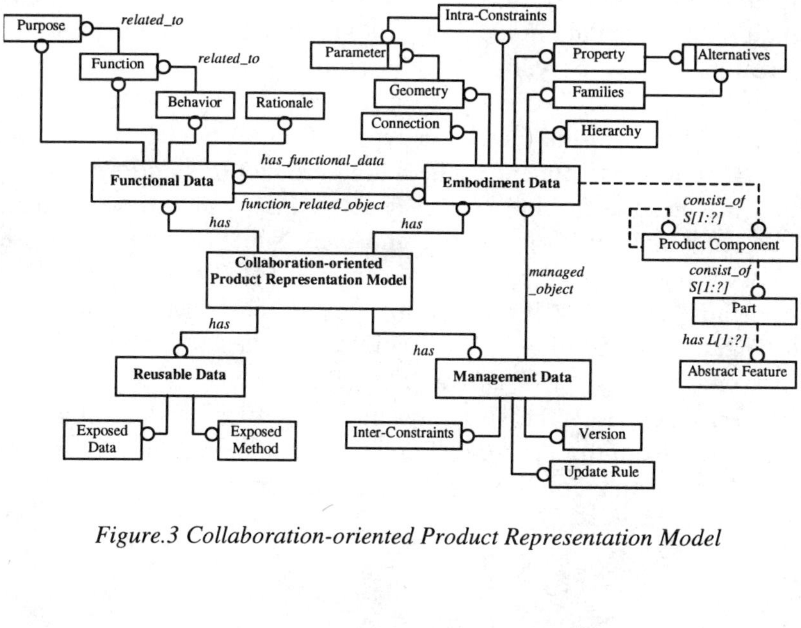

Figure.3 Collaboration-oriented Product Representation Model

Figure.3 Collaboration-oriented Product Representation Model

1). Functional Data

Functional Data describes the information related to what a product object is for, what it does and why it is what it is. It is decomposed into four types of UoF: *Purpose, Function, Behavior* and *Rationale*. The concept of UoF is from STEP [15,16], which is used to define the data unit with single function [23,24].

- *Purpose* is essentially the requirements for the products and describes the needs of clients and intentions of designers. It is the pursued aim of the design activity. Generally, it is more conceptual.
- *Function* describes what the product object does. Intended functions are those which are formulated as necessary to fulfill the purpose. Unintended functions may occur and it is important to note these too.
- *Behavior* illustrates the working principles of the object and logical actions or influence on other objects. The required behaviors are those which are necessary to produce the required or intended functions. The relationship between purpose, function, behavior and embodiment can be summarized as: PURPOSE enabled by FUNCTION achieved by BEHAVIOUR exhibited by EMBODIMENT [23,24]. Note that we have replace Rosenman and Gero's idea of "Structure" with "Embodiment", which is a term better suited to mechanical design.
- *Rationale* contains the reasons or justification of design decisions in terms of selecting values for structure variables to satisfy behavior constraints or values.

2). Embodiment Data

Embodiment Data describes the product structural or physical information, the core data module of CPRM. This data module is neutral and customizable. It has 6 types of UoF (*Unit of Functionality*): *Families, Hierarchy, Connection, Geometry, Property,* and Intra-*Constraints* (fig.3).

- *Families* represent the classes or types of an abstract feature, a product part, a product component or the whole product (called Product Object). For example, the turbine engine has subclasses of turbo-fan, turbo-jet, turbo-shaft and turbo-propeller, and each subclass engine has several types, even a special type has serial engines [4]. The CFM56 is a type of turbo-fan engine, but it has serial types as the CFM56-2, -3, -5A, -5B etc. [27]. The families relationship is a class hierarchy. It can be represented by "*Is_Type_Of*" or "*Has_Types*".
- *Hierarchy* describes the structural hierarchy of a product object, which is different from the *Families*. For example, a common turbo-fan engine is made up of fan, compressor, combustor, turbine and accessories. Each

of these component is also made up of sub-components. This relationship can be represented by "*Is_Component_Of*" or "*Has_Components*".

- *Connect* indicates the object relationship in a level of the product object hierarchy. For example, the fan is connected to the compressor. This relationship can be represented by "*Connected_To*".
- *Geometry* represents the geometry shape, dimensions and location of a product object. We use the application protocol STEP 203 [15] as our geometry standard. The geometry shape has many types and "*Alternatives*" can be listed for designers. In order to realize a parametric model, the dimensions can be represented by 'Variables', and the location should also be variable or relative to its related objects
- *Property* represents the engineering attributes of a product object such as the material and processing attributes. The application protocols STEP 203 and 214 [15,16] are used as our standards for this representation. To make the property customizable, it should be represented by "*Alternatives*", i.e., alternative properties.
- *Intra-Constraints* represents the constraint relationship inside a product object. The constraints here are limited to the geometry parameters. The parametric data with constraints realizes a real parametric object model. The constraints among different product objects are called *Inter-Constraints* which are classified into the Management Data.

3). Management Data

The information for maintaining and controlling the model data in the design process is classified as management data. It contains three types of UoF: *Inter-Constraints*, *Update Rule* and *Version.*

- *Inter-Constraints* define the geometry parameter constraints among multiple product objects in a product. They are in the form of formula, ranges, or rules. Inter-Constraints play an important role towards a parametric product model.
- *Update Rule* represents the action behavior on the change of model data. It is coupled with the data as the parameter constraints, product families, hierarchy etc., and it operates as a watch dog for those data. There are two types of update rules: *Condition/Action* and *Action/Condition/Correction* [18].
- *Version* serves to capture the product data evolutionary history, including not only the version number, but also the related designer, and updated time.

4). Reusable Data

The Reusable Data contains the UoFs of *Exposed Data* and *Exposed Methods* (fig.3).

- *Exposed Data* represents the Data Service which is realized by methods '*GetParameter()*' and '*SetParameter(...)*'. The '*GetParameter()*' gets the data value and the '*SetParameter(...)*' sets the data value. For example, if you want to get or set the length value of some object, the methods '*float L=GetLength()*' and '*void SetLength(L)*' can be used.
- *Exposed Method* represents the Method Service which exposes the system implementation methods which contain the data base accessing, performance calculation, constraints management and so on.

3.2. Product Model Processor

Section 3.1 proposes a complex product model with four data modules which are made up of fifteen UoFs and multiple relationships. Inputting these data becomes a major concern, especially when the geometry is perhaps entered using a commercial CAD System. To facilitate data entry and to help with converting of CAD system derived geometry, we are suggesting the concept of the Product Model Processor which always works with a CAD system. Generally, a designer designs the product object in a CAD system. The model is usually feature-based and the data is limited to the geometry data with some engineering properties. This data is called the Original Data. Then the original data is processed by the PMP. This process is in 5 steps. (I) Transferring the original data into the neutral data which is Abstract Feature-based, (II) Encapsulating more engineering semantics onto the neutral data towards to the Embodiment Data, (III) Customizing the model by parameterizing the geometry data, engaging constraints on them and serializing the model, (IV) Encapsulating Functional Data and Management Data onto the Model, (V) Activating the Enabling Service which exposes the model data and methods. This process makes it easy to define the complex CPRM data for a product object. To provide this process, the PMP has the following modules (shown in fig.4).

- **Model Broker**

The Model Broker transfers the model data between the special model data of any CAD system and the abstract feature based neutral data of CPRM. Since most of the data models of current CAD systems are feature-based, the model translation in the Model Broker is the feature.

- **Neutral Model Definer & Browser**

As the name shows, this system module has two functions: serving as a data browser and definer. The data coming from the Model Broker can be viewed in the browser. The definer is used to extend the data from the Model Broker or to define a product object into the neutral model. The definer contains a *Product Composition Definer* and dozens of *Abstract Feature Units*. The Product Composition Definer is used to define the composition of a PC including its sub-components, parts and their hierarchy or connection relationship. The Abstract Feature Unit allows users to define the neutral feature data which is independent of any shape representation required by a CAD application.

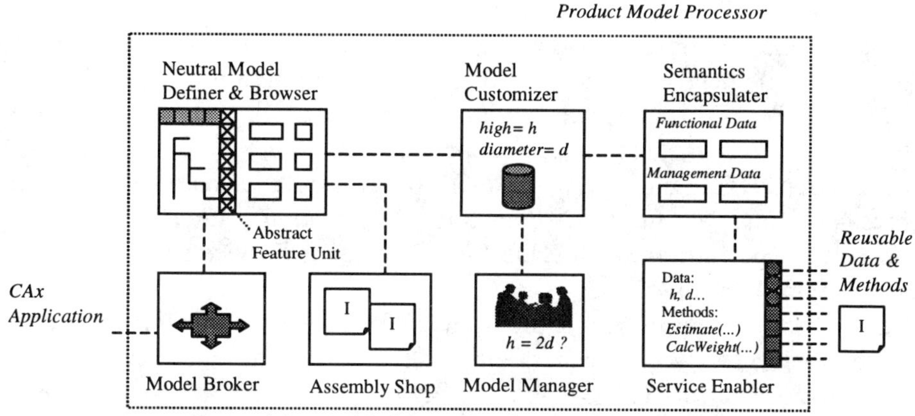

Figure.4 System Modules of Product Model Processor

- **Model Customizer**

The model data in the Neutral Model Definer & Browser is usually specific with specific values. This Model Customizer module is for transferring this special data into a class data which is customizable. This customizing process is in 3 steps. (I) *Parameterizing* the geometry data and extending the engineering properties. The special geometry data value is replaced by the related variable or parameter. The special engineering property as the material can be extended by alternatives, (II) Merging *Constraints* on those parameters. This work defines the limitation of a parameter and the constraint relationship among one more parameters, which enables the whole model become a parametric one, (III) *Serializing* the product families. It summarizes and classifies the product families. A special product in the product families is featured by its special function, structure and engineering performance, which is embodied by its components, parts even features. Therefore, the work here is not only to classify the product families and represent their featured data, but also to classify their components (parts, features) families and their distinguished data.

- **Semantics Encapsulator**

The Semantics Encapsulator helps a designer to define the Functional Data and some of Management Data, and combines to the Embodiment Data. The Functional Data contains the design purpose, object function & behavior, and decision rationale. The constraints have been defined in the Model

Customizer. The update rules can be attached to those constraints. The version of the model data can be updated automatically in the design process.

- **Service Enabler**

The Service Enabler serves to expose the model data & methods, i.e., produces the Reusable Data. The exposing process is a programming process usually implemented in the system development environment such as the Boland's JBuilder enterprise edition [3] and the Microsoft's Visual C++ enterprise edition [19] which supports the CORBA or COM based application development. To make it convenient for a user to expose the reusable data, a simple Application Programming Interface (API) is provided to perform the task. This API is called the Service Enabler.

- **Assembly Shop**

The Assembly Shop (AS) is an Application Programming Interface (API) environment for a designer to integrate the system function of related System Components (SC) which can be found from the Web-based Design Resource Manager. In this AS the reusable attributes and methods of related product components implemented in other system components can be reused if the designer wants to assemble those product components.

- **Model Manager**

The Model Manager maintains the model data consistency containing the product families, hierarchy, constraints, version and functional relationship. The product family management manages the creation of new product type, and the consistency between the product type and its special featured data. The hierarchy management manages the product composition. The constraints management manages the parameter constraints and update rules to keep the consistency of the model data. The version management captures and maintains the product design evolution history. The functional relationship management maintains the relationship between the purpose, function and product structure. The detail approaches to those management facilities will be described elsewhere.

4. CONCLUSION

A complex product is made up of multiple product components which are possibly designed by different persons or organizations at distributed locations. This paper proposes an approach towards the complex product development. This approach is based on the Component technology. A Component is a reusable software package. The design of each product component is encapsulated into such a (software) Component. The model data and operation methods of the product component can thus be reused by others because of the Component mechanism. The design of the product component is in a network environment implemented by a *Component-based Open System Architecture* (COSA) where a *Web-based Design Resource Manager* is responsible of managing those distributed components. In such an environment, the complex product's components can be designed in different places and then be assembled together. Word 'Assemble' refers not only to the assembly of product component data, but also that of the data operation methods.

In this COSA environment, the representation of the product component is another important issue. This paper presents a *Collaborative Product Representation Model* (CPRM) which is neutral, understandable, customizable and reusable model with self-management capability. It is the combination of *Functional Data*, *Embodiment Data*, *Management Data* and *Reusable Data*. The CPRM is a complex product data model with a few of units of functionality. In order to simply the definition of such a complex model data, this paper presents a *Product Model Processor* (PMP) helping a user to define the model data step by step or helps to wrap a special CAD model data into the CPRM data. The PMP is made up of six modules, a *Model Broker*, a *Neutral Model Definer & Browser*, a *Semantic Encapsulator*, a *Model Customizer*, a *Service Enabler*, an *Assembly Shop* and a *Model Manager*.

We believe that this approach is not only effective for product distributed development, but also valuable for the product data management in an enterprise. The reusability and customizability of the product model are valuable for the evolved product development. It will save the development cost and shorten the time to market.

REFERENCES

[1] Alberts, L. K., "YMIR: A Sharable Ontology For The Formal Representation of Engineering-Design Knowledge", see at http://wwwis.cs.utwente.nl.8080/kbs/publications/ontology-pubs.html, 1994.

[2] Ball, N., Matthews, P. and Wallace, K., "Managing Conceptual Design Objects", in J. S. Gero and F. Sudweeks (eds), *Artificial Intelligence in Design'98*, 1998, pp.67-86.

[3] Borland Corp., "JBuilder 3 Enterprise Developer Guide", 1999.

[4] Chen, G., Ma, M. and Xiao, L., "Structural Design of Gas Turbine Engine", Published by Beijing University of Aero. & Astro, 1988.

[5] Cutkosky, M.R., "Agent-based Concurrent Design", Advances in Concurrent Engineering, Aug.26-28, Toronto, Canada

[6] Dai, S., "Research on the STEP-based Integrated Product Model and the Implementation of an Integrated Design Platform", *Ph.D Thesis*, Beijing University of Aeronautics & Astronautics, 1997.

[7] Eastman, C. M., "Managing Integrity in Design Information Flows", *Computer-Aided Design*, Vol.28, No.6/7, 1996, pp.551-565.

[8] Gauchel, J., Vanwyk, S., Bhat, R. R. and Hovestadt, L., "Building Modelling Based on Concepts of Autonomy", in J. S. Gero (eds), *Artificial Intelligence in Design'92*, 1992, pp.181-197.
[9] Gero, J., "Design Prototypes: A Knowledge Representation Schema for Design", *AI Magazine*, Winter, 1990, pp26-36.
[10] Gero, J. and Shi, X.G. Design development based on an analogy with developmental biology, *CAADIA'* 1999. *Proceedings of the Fourth Conference on Computer Aided Architectural Design Research in Asia*, Shanghai, P.R.China. pp.253-264.
[11] Goldman, S., , Nagel, R.N., and Preiss, K., "Agile Competitors and Virtual Organizations", Pub. By Van Norstrand Reinhold, New York, 1995.
[12] Hart, D., Patterson, R, and Neal, R, "Next Generation Manufacturing: A Framework for Action", Executive Overview, The Agility Forum, Bethlehem, PA, January, 1998.
[13] Henderson, M. R., "Representing Functionality and Design Intent in Product Models", in 2nd ACM Solid Modelling'93, Montreal, Canada.
[14] Inprise Corp., "Distributed Object Computing in the Internet Age", see at http://www.inprise.com/visibroker/papers/distributed/wp.html, 1999.
[15] ISO/STEP, "10303-203: Product Data Representation and Exchange--Part 203: Application Protocol: Configuration Controlled Design", 1996 (a).
[16] ISO/STEP, "10303-214: Product Data Representation and Exchange--Part 214: Application Protocol: Core Data for Automotive Mechanical Design Processes", 1996 (b).
[17] Klein, M., "Integrated Coordination in Cooperative Design", *International Journal of Production Economics,* 1995.
[18] Mackellar, B. K. and Peckham, J., "Multiple Perspectives of Design Objects", in J. S. Gero and F. Sudweeks (eds), *Artificial Intelligence in Design'98*, 1998, pp.87-106.
[19] Microsoft Inc., "COM Home", at http://www.microsoft.com/com, 1998.
[20] Mills, J.J., "Information Infrastructures: Facilitating Global Design and Production", *9th International Conference on Production Technology (PTK 98),* Berlin, Germany, October 29-30, 1998a.
[21] Mills, J., "Introduction: Towards an Information Infrastructure for Manufacturing Industry", in John J. Mills and Fumihiko Kimura (eds), *Information Infrastructure Systems for Manufacturing II*, 1998b.
[22] Mills, J., Brand, M. and Elmasri, "AeroWEB: An Information Infrastructure for the Supply Chain", in John J. Mills and Fumihiko Kimura (eds), *Information Infrastructure Systems for Manufacturing II*, 1998c, pp323-336.
[23] Rosenman, M. A. and Gero, J. S., "Modelling Multiple Views of Design Objects in a Collaborative CAD Environment", *Computer-Aided Design*, 1996, Vol.28, No.3, pp.193-205.
[24] Rosenman, M. A. and Wang, F. J., "CADOM: a Component Agent Model based Design-Oriented Model for Collaborative Design", Submits for *Research in Engineering Design*, 1999a.
[25] Rosenman, M. A. and Wang, F. J., "A Component Agent Based Open CAD System for Collaborative Design", Submitted for *Automation in Construction*, 1999b.
[26] Sterling Software, "Component Based Development and Object Modelling", at http://www.cool.sterling.com/cbd/whitepaper/, 1998.
[27] Wang, F. J. and Chen, G., "The Design Evolution of CFM56 Families", in Preceding of Structure, Strength and Vibration of Aero-Engine, China, Oct., 1994.
[28] Wilkes, L., "Legacy Componentization and Wrapping", *Component Strategies*, 1(8), Feb'99, p50-57.

Supporting Partner Selection for Virtual Enterprises

Jens Dahl Pedersen
Technical University of Denmark, Denmark
Em: jdp@ipt.dtu.dk

Roelof J. van den Berg
Baan Development, The Netherlands
Em: rvdberg@baan.nl

Keywords Virtual enterprise, partner selection, partner qualification, competencies

Abstract This paper discusses partner selection for virtual enterprises. It provides a framework to define the notion of the virtual enterprise. It also defines the concepts resource, capabilities and competence and explains that two types of competence are essential to operate successfully in a virtual enterprise: functional competence and alliance competence. It discusses how they can be described and assessed.

1 INTRODUCTION

Current initiatives on business-to-business e-commerce tend to emphasise strictly technological requirements. While technological integration of distributed information systems is a key prerequisite for VEs, partner selection is another considerable obstacle. Evidence indicates that successful alliances rarely come to existence by chance, and that a main reason pitfalls often is poor partner selection resulting in mismatch between partners or unqualified partners for the particular job [5,7,13]. At the same time, the required agile formation of VEs leaves little room to sound out well-qualified and trustworthy partners. This is often a matter of years in project industries today. Based on a theoretical framework for VE concepts, this paper will define key concepts as well as highlight current challenges to successful partner selection for VEs.

2 THE VIRTUAL ENTERPRISE FRAMEWORK

A VE can shortly be described as a customer solution delivery system created by a temporary and re-configurable aggregation of core competencies [16]. A VE is thus an alliance, where individual enterprises join competencies in order to establish a customer oriented value system. This value system is carefully configured to meet a specific customer demand and when the demand has been fulfilled, the VE is dissolved.

The concept of VEs is captured and formalised by the framework for VE engineering and integration as is illustrated in figure 1 [9, 15, 16]. The framework illustrates that VEs originate in a network of co-operating organisations. The network represents a pre-established basis of competence carriers, for preparing and setting up VEs. In practice, it is not always easy to see who is within a network and who is not, because the closeness of the partnerships within the network varies. The network can create VEs in its operational phase and, correspondingly, a VE can create product(s) or service(s) in its operational phase.

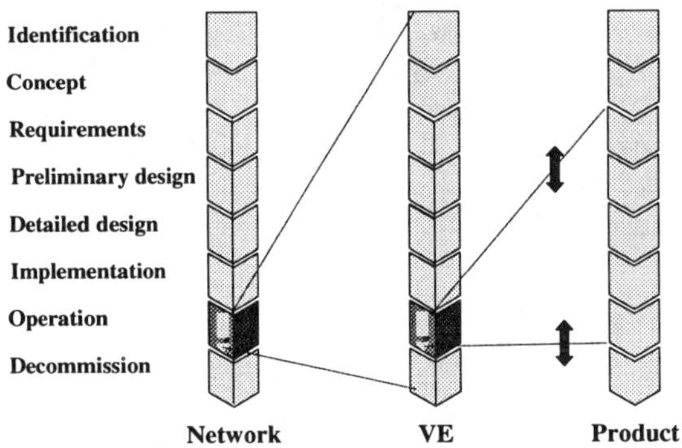

Figure 1: The Virtual Enterprise Framework

3 COMPETENCE DEFINED

The notions of resources, capabilities and competencies have gained considerable attention from scholars of strategic management research as they attempt to explain why companies differ in organisational performance. Particularly, what has become known as the *resource-based view* of the firm is dedicated to research on this question. The three concepts are however subjects of a rich variety of different definitions and often used highly synonymously.

For this paper a distinction between resources and capabilities is made based on the definitions set forth by Amit and Schoemaker. They define resources as "stocks of available factors that are owned or controlled by the company" [1]. Resources encompass both tangible and intangible assets [17] and can basically be categorised into three types: physical capital resources, human capital resources and organisational capital resources [2]. Physical capital resources include the physical technology, facilities, equipment, and materials. Human capital resources include education, experience, judgement, intelligence, and relationships. Organisational capital resources include formal reporting structure, formal and informal planning, controlling and co-ordinating systems as well as informal relationships [2].

In the view of Amit and Shoemaker capabilities refer to a company's capacity, or ability, to deploy resources. Capabilities are information- and knowledge-based processes deploying resources in an often co-ordinated way to effect a desired outcome as for instance products, services, organisational behaviour, trust and the like. Like resources, capabilities can be both tangible and intangible, but they are in essence limited to the information- and knowledge handling of human capital resources [1].

Competencies are complementarily a subset of resources and capabilities with the potential to lead to competitive advantages. The above characteristics of resources and capabilities also apply to competencies, but to constitute a competence, the resources and capabilities must meet the following criteria [2, 10]:

a) they must promote company efficiency and effectiveness and be perceived as valuable by the market and environment;
b) they should not be readily available at competitors;
c) they should be difficult to acquire for competitors;
d) they should not have strategically equivalent abundant substitutes

Note that a competence here is defined relative to competencies of other companies. Besides, a competence should not be confused with a specific product or service, rather a competence spans the ability to create several products or services.

4 COMPETENCE MODELLING

To create a successful VE it is necessary that its potential members have a thorough understanding of each other's competencies. For this purpose each one of them should be able to demonstrate the viability of its competencies to the others. In addition it should be able to show how its competencies add value to the planned VE by complementing those of others. Thus, it is not sufficient to have an intra-firm focus on competencies. Companies should be willing and able to look beyond their own boundaries as well.

Modelling of competencies should in this respect be perceived as a way for companies to qualify themselves to participate in VEs. Modelling competencies has many advantages, both intra- and inter-organisationally. Here we want to emphasise that the models facilitate much better communication about competencies with external parties, for the above mentioned reasons. This is especially true when some kind of standardised technique and format is used. It is not unlikely that eventually dedicated agencies will arise, which assist enterprises in modelling their competencies and also assess their suitability for a certain network or a particular virtual enterprise. See [3] for a description of the state of the art in audits for e-commerce and [4] for an outline of what is needed in this respect. To participate in a VE, companies should possess two types of competencies:

- *functional competence*

  This is a competence related to the product life cycle, therefore supporting the creation of the customer solution.

- *alliance competence*

  This is competence related to the VE life cycle, representing the ability of a partner to enter into and participate in VEs.

These two types of competencies are implicitly recognised by [6, 12] and will be discussed subsequently.

4.1 Functional Competence

Functional competencies are the competencies that partners bring to a VE to form an operational entity that can carry out the phases of the product life cycle, cf. figure 1. Definition and modelling of competencies are key issues for describing, communicating and justifying competencies offered for VEs. Very few companies have, today, an accurate and living model of their business processes let alone their wider capabilities and competencies. To facilitate an agile formation of VEs, allow a somewhat objective comparison of partners, and reduce misperceptions and dishonesty between partners, systematic and standardised methods for defining and modelling of competencies are needed [6]. Unfortunately, the competence literature rooted in the resource based view of the firm does not at the current state offer operational support for description of competencies. In an initial attempt, the dimensions mentioned below could be used relevant for description of functional competencies.

A) Solution addressed by the competence

This is an overall description of what market need the competence addresses.

B) Supporting resources and capabilities

This includes a more detailed model of the application sphere, constraints and uniqueness of the competence, which is useful for evaluation of the competence against more specific project requirements. It should indicate how the competence results in specific products or services, e.g. by describing basic assumptions, paradigms, theories, or methods.

C) Interfaces

It should be described what input, typically in terms of information and knowledge, the competence requires as well as what output it is expected to deliver. This should clarify dependencies and make expectations explicit.

D) Agility

This should describe how agile the resources and capabilities can operate (i.e. internal agility) as well as the degree of agility the competencies bring to the VE (i.e. external agility) [6].

E) Sensitivity

A description of sensitivity should describe how well rooted the competence is in the organisation and hereby how resistant it is to especially resignation of key employees.

F) Additional measures

Finally a competence model should be accompanied by a set of additional measures including measures for quality, capacity, and financial performance.

Each of the above dimensions needs to be complemented with specific metrics to make them sufficiently descriptive, measurable, and thus fully operational. Even though not all aspects of a competence can be modelled due to the inherent complexity and presence of tacit knowledge, it is possible to get a part of the way.

4.2 Alliance Competence

Possession of functional competencies is not sufficient to qualify for a VE. A VE is in essence a partnership between autonomous companies, which implies a high degree of interaction and integration to achieve common objectives. Therefore, companies must also possess ability to enter into and participate in VEs, or in other words, they must possess alliance competence. The notion of alliance competence is introduced by Spekman and MacAvoy, who characterises it as "partly a function of individual skills and capabilities and firm-level attributes that enhance, encourage, and support alliance-like thinking and behavior throughout the firm" [12]. Two main elements of an alliance competence are ability to manage and implement alliances and ability to display alliance spirit and behaviour.

An ability to manage and implement alliances includes thus both the formation, operation, and decommission of the VE while considering the product life cycle and starting from the partners in the network [15]. Naturally, the scope of resources and capabilities comprising an ability to manage and implement alliances should depend on the particular role(s) a company expects to have in VEs. For a design office or a production facility with a more secondary role in managing and implementing the VE this scope can be more limited than for the main contractor or an agent of competencies.

An ability to manage and implement alliances should preferably qualify a company to enter into and operate in VEs in an agile, plug and play fashion. It is possible to reuse and enhance the company's existing alliance knowledge and realise a certain degree of preparation for participation in VEs. Such preparation could be achieved by reference models for the VE formation process including partner selection criteria, standard contracts, models for risk and profit sharing, and ways of co-operating and communicating. When these elements are combined by powerful technology for process configuration in a virtual enterprise, the alliance competence increases significantly. This directly addresses the need for flexibility to partner with a variety of companies. Additional elements of an ability to manage and implement alliances are the ability to certify potential unqualified partners, acquire new qualifications, and to manage changes across the entire set of partners.

Ability to display alliance spirit and behaviour is largely a matter of organisational behaviour and relationship management. According to Spekman and MacAvoy alliance spirit builds on an atmosphere of flexibility, commitment to mutuality, sense of solidarity, and preference for harmony [11]. It should be stressed that displaying alliance spirit and behaviour relies on human skills and personal match between people.

Partners in VEs should display tolerant and respectful behaviour and in the common interest of the VE, thus avoiding opportunistic and power based behaviour. Partners should besides be willing to share information, respect IPR and trusted information, be open for teaching as well as learning, respect cultural differences, and understand motives of partners. Although alliance spirit builds very much on characteristics that are intrinsic to a potential partner, an atmosphere of trust in a VE is always the result of interaction between several partners. Business logic precludes an attitude of complete naivety in partnering and sharing, as a viable business strategy. Trust has to grow on joint successes [11].

5 CONCLUSION

Work on electronic business tends to emphasise the first half of the concept. It almost completely ignores that age-old challenges associated with commercial projects have to be addressed as well to truly realise a "new economy". One of these challenges is the selection of business partners. In the virtual enterprise this has to be done much faster than what is common today.

At this moment few companies are prepared to rapidly gear up for participation in a specific virtual enterprise. In case they have a clear comprehension of their competencies, it is used for internal purposes only. Techniques to systematically describe and communicate competencies need to be developed, as well as methods to objectively assess them.

Because these functions will easily distract most enterprises from their core business, it is not unlikely that they eventually will be covered by dedicated intermediaries. This applies to both functional and alliance competence. From this perspective it is likely that maturity models for VE partnering will emerge over time hand in hand with third party certification.

6 REFERENCES

[1] Amit R., Schoemaker P.J.H. (1993). Strategic Assets and Organizational Rent, *Strategic Management Journal*, **14**, 33-46

[2] Barney J. (1991). Firm Resources and Sustained Competitive Advantage, *Journal of Management*, **17**, 99-120

[3] Berg, R.J. van den, and Lieshout, van, J.M. (1999). Eliminating Hurdles to Trust in E-commerce, in Global Production Management, K. Mertens, O. Krause and B. Schallock (eds.), Kluwer, pp. 522-529.

[4] Berg, Van den, R. J. and Tølle, M. (2000). Assessing Ability to Execute in Virtual Enterprises,

[5] Brouthers K.D., Brouthers L.E., Wilkinson T.J. (1995). Strategic Alliances: Choose Your Partners, *Long Range Planning*, **28**, 18-25

[6] Goranson H. T. (1999). The Agile Virtual Enterprise: Cases, Metrics, Tools, Quorum Books

[7] Medcof J.W., (1997). Why Do Many Alliances End in Divorce, *Long Range Planning*, **30**, 718-732

[8] Prahalad C.K., Hamel G. (1990). The Core Competence of the Corporation, *Harvard Business Review*, **68**, 79-91

[9] Pedersen J.D., Vesterager J., Tølle M. (2000). Application of a GERAM based Virtual Enterprise Framework – Results from IMS 950001/Esprit 26509, Proceedings of the 6th International Conference on Concurrent Enterprising, Toulouse, 28-30 June 2000, 139-143

[10] Peteraf M.A. (1993). The Cornerstone of Competitive Advantage: A Resource-based View, *Strategic Management Journal*, **14**, 179-191

[11] Sabherwal R. (1999). The Role of Trust in Outsourced IS Development Projects, *Communications of the ACM*, **42**, 80-86

[12] Spekman R.E., Isabella L.A., MacAvoy T.C. (2000). Alliance Competence – Maximizing the Value of Your Partnerships, John Wiley & Sons

[13] Spekman R.E., Isabella L.A., MacAvoy T.C. (1996). Creating Strategic Alliances which Endure, *Long Range Planning, 29, 346-357*

[14] Talluri S., Baker R.C. (1996). A Quantitative Framework for Designing Efficient Business Process Alliances, IEMC, 656-661

[15] Tølle M., Vesterager J., Pedersen J. (2000). A Methodology for Virtual Enterprise Management – Results from IMS 95001/Esprit 26509 Globeman 21 project, Proceedings of the 6th International Conference on Concurrent Enterprising, Toulouse, 28-30 June 2000, 119-129

[16] Vesterager J., Larsen L. B., Gobbi C. (1999). Architecture and methodology for creating virtual enterprises – results from Globeman 21, www.vtt.fi/aut/projects/gm21/demo/index2.thml

[17] Wernerfelt B. (1984). A Resource-based View of the Firm, *Strategic Management Journal*, **5**, 171-180

Specifying Interactions in Integrated Manufacturing Systems

David Flater
National Institute of Standards and Technology, U.S.A.
Email: dflater@nist.gov

Keywords Component, Integration, Modelling, Specification

Abstract Systems of manufacturing software are often constructed by integrating pre-existing software components. Accurate specification of the component interactions in these systems is needed to ensure testability and maintainability. Moreover, standards for manufacturing systems must specify the interactions to achieve interoperability and substitutability of components. In this report we discuss approaches for specifying component interactions and examine a number of potentially useful specification techniques.

INTRODUCTION

Systems of manufacturing software are often constructed by integrating pre-existing software components. The interactions between the components are central to the operation of the system, yet without a specification that unambiguously explains the expected behavior, we have only intuition to tell us whether the observed behavior is correct. STEP, known informally as the Standard for the Exchange of Product model data, has demonstrated the value of rigorously specifying data exchange interactions (for the full story, read STEP: The Grand Experience [12]). But data exchange only scratches the surface of the interactions that are possible. Even if we have complete and consistent specifications for the functions provided by two components in a system, these do not necessarily define how the components would work together to achieve a specific goal. If interoperability and substitutability of components is a goal of the specification, then the interactions must be specified completely.

In the following sections we discuss different models of interaction and summarize the features of potentially relevant tools and techniques.

INTERACTION MODELS

Nested control-flow models

Interaction with nested flow of control [16] is a special case of synchronous communication. In these models, communication behaves like a procedure call. Borrowing terminology from the Common Object Request Broker Architecture (CORBA) [4], we would say that the component making a request remains in a blocked state until the response is complete.

Various communication infrastructures following in the footsteps of Remote Procedure Call (RPC) [23] are predisposed to the production of systems having a nested flow of control. Nested control-flow models, hereinafter referred to as nested models, enable us to analyze such systems without the complexity that would be introduced by more flexible models. The amount of nondeterminism in a nested model is limited by the fact that a component will not receive extraneous events while it is awaiting the response to a previous request.

Flat control-flow models

In systems having a flat flow of control, components are designed around a dispatch loop that never blocks. Because new requests can be processed while previous operations are still pending, the ordering of events in the system becomes more random than it would be with a nested flow of control. Whether you are using "asynchronous messaging" or "event handling" as your model, the architectural ramifications are the same. This is independent of attributes such as payload, indirection mechanisms (subscribe/notify), or support of multicasting that distinguish the various approaches to communication in a "flat" system.

The need for manufacturing systems to be responsive and deadlock-free encourages the use of a flat model. However, flat models are more difficult than nested models from a testing perspective because the actual state of any executing component no longer has a direct relationship to any distributed process that may be in progress. It is also more difficult to specify the intended behavior of event-driven systems because what we see as independent processes may run concurrently in the same component and interact in unintended ways.

The challenge for semantic modelling of flat systems is to identify the behaviors that are intended to be created by the interactions of components, rather than the behaviors of components as seen from their own limited points of view.

Mixed models

Practical considerations sometimes get in the way of taking a purely flat-model approach to system design. Unintended interactions between separate processes, such as competition for a shared resource, force the designer to place restrictions on the sequences of events that the system may process. Resource locking and transactions are two features that are commonly used for this purpose. The introduction of locking and transactions into a flat system can produce a system that sometimes exhibits a nested flow of control.

In the other direction, the introduction of multi-threading into a system having nested control-flow can also produce mixed behavior. A designer might use multi-threading if non-blocking transactions are the exception rather than the rule; this would be simpler than defining all of the explicit locking behavior that would need to go with a flat approach.

SPECIFICATION LANGUAGES AND TECHNIQUES

Architecture Description Languages (ADLs)

ADLs appeared in the 1990s as a promising new formalism in the software domain. However, as the cited reference describes, "There is... little consensus in the research community on what an ADL is, what aspects of an architecture should be modeled by an ADL, and what should be interchanged in an interchange language." [13] The point of commonality in all ADLs is support for modelling the architectural features of a software system at a high level of abstraction.

The cited reference compares ten ADLs and finds that the feature sets vary significantly. Some are designed to enable particular forms of automated analysis while others are primarily intended for abstract modelling. ADLs that enable much automated analysis necessarily have more of the flavor of a programming language or algebra than those designed only for human consumption. The kinds of automated analysis that are available with various ADLs range from deadlock detection to schedulability analysis.

Some ADLs support the modelling of system-level interactions, though it is not their primary focus. Rapide includes built-in support for modelling both synchronous and asynchronous interactions; other ADLs make use of process algebras for modelling components. Process algebras are discussed later in this document.

Component Interaction Specifications (CIS)

While rigorously specifying the behavior of a distributed system in general is very difficult, specifying this behavior for a specific scenario is more tractable, as is demonstrated by the Component Interaction Specification (CIS) [5] based method supported by the Manufacturer's CORBA Interface Testing Toolkit (MCITT) [6], and by UML Sequence Diagrams [17] and Message Sequence Charts [21]. Unlike the other notations, CIS was designed to be translatable directly into test scaffolding for CORBA systems, but this produced disadvantages that will be discussed below.

CIS is a derivative of the integration testing method that was being used by NIST's industrial partners in the Advanced Process Control Framework Initiative (APCFI) [19]. This method, in turn, made use of ideas that are also used in UML Collaboration Diagrams [18].

A CIS interaction scenario consists of a tree of requests having specified inputs, outputs, and/or return values. The tree is rooted at a test client that initiates the entire chain of events. In order to capture the tree structure of the interactions in a flat ASCII script, an outline numbering convention similar to that of UML Collaboration Diagrams is used.

CIS assumes a nested flow of control for interactions. For CORBA-based systems having a nested flow of control, CIS is a simple and powerful tool. It enables automatic generation of run-time assertions to verify that the inputs and returns for each interaction are as specified in an actual running system. Unfortunately, although the CIS syntax is expressive enough to describe an entire tree of interactions through a distributed system, it assumes a single source of activity. To remove this limitation from the CIS syntax would be easy, but removing it from MCITT's test code generation would require a switch to a more intrusive testing approach.

Finite state machines

Finite state machines (FSMs) are simple abstractions of component behavior comprised of a set of states and transitions between them, with defined criteria for triggering the transitions.

FSMs are the cornerstone of many different specification languages and techniques. For example, Specification and Description Language (SDL) [20], a standard of the International Telecommunication Union (ITU), is now used by a popular software product for realtime modelling, simulation, and analysis. Some impressive software products for FSM analysis are also available free from universities and research laboratories.

Like process algebras (below), finite state machines are a high-level abstraction. They provide a straightforward way to model and analyze networking protocols, embedded control, and other algorithms that lend themselves to finite state analysis without unnecessary implementation detail. The tool support for finite state analysis is very good, enabling properties of systems to be proven or refuted automatically.

With FSM approaches, the focus is on specifying the behavior of individual components, rather than on specifying interactions directly. However, given a complete set of FSMs, all possible interactions and emergent system behaviors can be deduced. They are therefore an efficient tool for modelling systems with flat control-flow, where the interactions of components having simple state spaces give rise to countless possible behaviors depending on the interleaving of events.

FSMs are less useful for applications having a complicated flow of control because the task of identifying finite state spaces for the components of the system becomes complex and error-prone. The most frequently cited reason for failure of finite state analysis is an explosion in the size of the state space resulting from an attempt to model a complex system.

Process algebras

A process algebra is a formal language for specifying and reasoning about the behavior of interacting processes. Many process algebras exist; here we will discuss only a few prominent examples.

Communicating Sequential Processes (CSP) [8] was published in 1978 in Communications of the ACM by C.A.R. Hoare. Since then it has been used as the basis for several parallel programming and specification languages. Various software exists for simulating and/or checking CSP specifications for properties such as deadlock freeness.

Milner's Calculus of Communicating Systems (CCS) [14] was emerging around the same time and also became the inspiration for much later work. Both CSP and CCS, in their original forms, assume that processes synchronize on communication, so the domains of systems that they can model are basically the same. However, they have subtle semantic differences 0.

Language of Temporal Ordering Specification (LOTOS) [10] is an International Organization for Standardization (ISO) standard which was used in the Open Systems Interconnection (OSI) project [11]. It inherits some ideas from both CSP and CCS and is generally preferred for formal specification and verification of networking protocols.

A process algebra can help to separate the essence of an interaction-driven system from the details of any given interaction. Using process algebras, it is sometimes possible to construct formal proofs of properties of distributed systems. In cases where the pattern of interactions between software components becomes complex, process algebras can be employed to analyze or even define the specification at a high level of abstraction.

Different process algebras vary in their capacity for modelling time and synchronization, and the level of software support for using different algebras varies as well. Given an algebra with the correct feature set, it is possible to model nested, flat, or mixed control flows.

In general, process algebras permit more direct modelling of communication and synchronization than is possible with FSM-based techniques.

Unified Modelling Language (UML)

UML is really a collection of modelling languages and techniques that have been harmonized with one another and bundled under a common name. It began as a unification of object models, but quickly expanded to enable modelling of many different aspects of a software system. With the support of the Object Management Group and leading software firms, UML has become established as the canonical language for software models.

UML encompasses Static Structure Diagrams, Use Case Diagrams, Sequence Diagrams, Collaboration Diagrams, Statechart Diagrams, Activity Diagrams, and Implementation Diagrams, as well as the recently canonized Object Constraint Language (OCL). OCL first appeared under that name in 1997 as part of a joint IBM and ObjecTime Ltd. response to the first Analysis and Design RFP [9][15]. OCL appears to have evolved at IBM from the Integrated Business Engineering Language (IBEL) and/or the Syntropy method [3]. Other parts of UML are easily recognizable as evolved and assimilated versions of popular, pre-existing computer science abstractions, including Entity-Relationship Diagrams [2], Harel Statecharts [7], Petri Nets [22], and classical flowcharts.

Such an assemblage of modelling techniques clearly aspires to provide every feature that could reasonably be needed. Nevertheless, UML is constantly being extended. Because the UML core has achieved a "critical mass" of industrial usage, it is generally easier to communicate using an extension to UML than using a completely different language.

For specifying interactions, the most applicable part of UML is the Sequence Diagram, which provides sufficient syntax and semantics to model flat, nested, or mixed control flows. There is even a software product that can verify whether the interactions in a simulated system conform to a Sequence Diagram. The syntax is easily extended to include special cases of control flow, such as balking and time-outs, that are not handled by most modelling techniques. These extensions are not understood by automated tools, but they are valuable for specifications that could not be expressed using a more formal, more restrictive notation.

SUMMARY

Having reviewed the state of the art in techniques for specifying interaction-driven systems, we find that there are many ways to do it, but always with a tradeoff between rigor and flexibility or ease-of-use. For a system having difficult kinds of interactions, one can obtain a precise characterization of a simplified view of the system that may not be accurate, or a less formal characterization of the system in all of its complexity that may prove to be insufficiently precise or incomplete.

The Unified Modelling Language has achieved a "critical mass" of usage. Although some tools use it in a rigorous way, UML favors flexibility and ease-of-use. In some contexts, we are obliged to sacrifice ease-of-use to achieve a level of rigor that is appropriate to the task at hand. This paper has discussed the different methods and tools that are available to meet different requirements, and the reader is encouraged to choose wisely for his or her own context.

REFERENCES

[1] Stephen D. Brookes, "On the relationship of CCS and CSP," in *Lecture Notes in Computer Science #154: Automata, Languages, and Programming, 10th Colloquium*, Springer-Verlag, 1983, pp. 83-96.

[2] Peter P. Chen, "The Entity-Relationship Model – Toward a Unified View of Data," *ACM Transactions on Database Systems*, vol. 1, no. 1, 1976.

[3] Steve Cook and John Daniels, *Designing Object Systems: Object-Oriented Modelling with Syntropy*, Prentice Hall, 1994.

[4] CORBA/IIOP 2.3.1 Specification, http://www.omg.org/cgi-bin/doc?formal/99-10-07, 1999.

[5] David Flater, Component Interaction Specifications, http://www.mel.nist.gov/msidstaff/flater/mcitt/cis.html, 1998.

[6] David Flater, MCITT home page, http://www.mel.nist.gov/msidstaff/flater/mcitt/, 1999.

[7] David Harel, "Statecharts: a visual formalism for complex systems," *Science of Computer Programming*, vol. 8, 1987, pp. 231-274.

[8] C.A.R. Hoare, "Communicating Sequential Processes," Communications of the ACM, Volume 21, Number 8, August 1978, pp. 666-677.

[9] IBM/ObjecTime Ltd. joint submission on AD RFP1, http://www.omg.org/cgi-bin/doc?ad/97-01-18, 1997.

[10] ISO 8807:1989, *Information processing systems — Open System Interconnection — LOTOS — A formal description technique based on the temporal ordering of observational behaviour.* Available from ISO, http://www.iso.ch/.

[11] ISO/IEC 7498, *Information technology — Open Systems Interconnection — Basic Reference Model.* Available from ISO, http://www.iso.ch/.

[12] Sharon J. Kemmerer, editor, *STEP: The Grand Experience*, NIST Special Publication #939, U.S. Government Printing Office, Washington, D.C., 1999. Available at http://www.mel.nist.gov/msidlibrary/summary/9920.html.

[13] Nenad Medvidovic and Richard N. Taylor, "A Classification and Comparison Framework for Software Architecture Description Languages," *IEEE Transactions on Software Engineering*, Vol. 26, No. 1, January 2000, pp. 70-93.

[14] Robin Milner, *Lecture Notes in Computer Science #92: A Calculus of Communicating Systems*, Springer-Verlag, 1980.

[15] Object Analysis and Design PTF — RFP1, http://www.omg.org/cgi-bin/doc?ad/96-05-01, 1996.

[16] The concepts "nested flow of control" and "flat flow of control" as applied to interactions are canonized in OMG Unified Modelling Language Specification version 1.3, http://www.omg.org/cgi-bin/doc?ad/99-06-08, 1999, Section 3.62, "Message and Stimulus."

[17] *Ibid.*, Notation Guide, Part 7, Section 3.59: Sequence Diagram.

[18] *Ibid.*, Notation Guide, Part 8, Section 3.65: Collaboration Diagram.

[19] Project Brief: Advanced Process Control Framework Initiative, http://jazz.nist.gov/atpcf/prjbriefs/prjbrief.cfm?ProjectNumber=95-12-0027, National Institute of Standards and Technology, 1996.

[20] Recommendation Z.100 (03/93) – CCITT specification and description language (SDL). Available from ITU, http://www.itu.int/.

[21] Recommendation Z.120 (10/96) – Message Sequence Charts (MSC). Available from ITU, http://www.itu.int/.

[22] W. Reisig, *Petri Nets: An Introduction*, Springer-Verlag, 1985.

[23] RFC 1831, "RPC: Remote Procedure Call Protocol Specification Version 2," http://www.faqs.org/rfcs/rfc1831.html, 1995.

Modelling Semiosis of Design

V.V. Kryssanov
Dept. Computer and Systems Engineering, Kobe University,Rokkodai, Nada-ku, Kobe 657-8501, Japan
E-m: kryssanov@al.cs.kobe-u.ac.jp

J.B.M. Goossenaerts
Faculty of Technology Management, Eindhoven University of Technology, The Netherlands
E-m: J.B.M.Goossenaerts@tm.tue.nl

Keywords Design, Semiosis, Information Infrastructure

Abstract This paper addresses the modelling of the cognitive and evolutionary aspects of engineering design in support of two broad goals: the better understanding of product life cycle processes, and the development of relevant information infrastructure services. The modelling elements are derived from the Framework of Industrial Semiosis. The semiosis of evolutionary design is explained, and general principles of the evolution of the product concept are formulated.

1 INTRODUCTION

As products and production processes become more specific to individual requirements, there is a need to expand and diversify information infrastructure services for design and configuration of products. The development of these services requires a thorough understanding of the process in which the product evolves as driven by the customer's needs and desires, the environment of use, the designer's imagination, and the enterprise strategic goals and capabilities.

Our expectations for the information infrastructure services for design processes have, over a long time, been guided by evolutions in supply chains. These show a clear trend towards strongly differentiated logistics processes that are realized after reconstructing the sales and fulfillment cycle, whereby the traditional production process is broken down and reconstructed in a manner that maximizes the overall efficiency of the chain [18]. This trend will be enhanced by advanced electronic commerce systems, virtual organizations, and e-markets, enabling enterprises, small and large, to offer their capabilities in multifarious commercial models [1].

The large variety of products, customer desires, and use-environments calls for a similar differentiation in design semiosis processes [6]. In these processes, the product is determined (conceived-represented-realized) by devising a product structure and the possible relations between the product and its operating environment, which together aggregate the possible product configurations [13]. Most naturally, the product and its environment are defined in terms of a product language subject to the semiosis processes. The dependence of the semiosis of design both on the entire product life cycle and on the dynamic (economical, social, technical, etc.) environments is critical for organizing efficient computer-aided support for the development of modern high tech products.

The goal of the present work is to advance the modelling of design semiosis. The resulting models serve as an input for the creation of infrastructure services for evolutionary design. They also offer an improved understanding of the design process in a context of the product life cycle.

2 PARADIGMS OF DESIGN

A paradigm may be seen as a disciplinary matrix that (*a*) has three main properties: common symbolic abstractions, model beliefs, and a system of values; and (*b*) that generalizes some examples, which illustrate the properties. Taking this definition as a basis, we will briefly review three most influential design paradigms learnt from the literature.

In the *Empirical Paradigm*, a design theory is to be founded on empirical knowledge and factual data. Such a theory usually has a methodological character and focuses on systematizing, structuring, and ordering of both design problems and problem solving methods. The Empirical Paradigm is based on a quite naïve assumption that all the design problems are well structured, and that there exists ideal knowledge for every given problem. The works [19] and [10] are typical representatives of this approach: their strong point is that if a task can be classified and related to the accepted system of beliefs and values, the formulated principles can guide the designer throughout the problem solving process. Unfortunately, the class of design tasks that would be solved in this way, although important, is very limited [2].

The *Logical Paradigm* attempts to compensate for the limitations of the empirical approach. Its fundamental assumption is about the possibility of decompositing a complex design problem into a number of sub-problems with relatively independent goals. In this paradigm, the design process starts from analysis of requirements, passes through one or more stages of synthesis directed by the decomposition, and is completed by evaluation. If the obtained outcome is found unsatisfactory, the process may be repeated. [11] and [16] present design theories that propose some principles for systematization of the synthesis-analysis interaction while designing. The fallacy, however, is that design problems are often incomplete as the goals are only partially understood, and solutions are emergent rather than decomposed. Besides, the requirements may be inconsistent, and during the synthesizing of the decomposed solutions, second order requirements may arise due to the interaction of the solutions. Many design theories, which come from Artificial Intelligence (e.g. [2]), reveal similar drawbacks: in practice, not only the initial problem state, but also the goal state is frequently ill defined, and it is generally infeasible to *a priory* assign a system of operators (rules, synthesis-analysis procedures, etc.) for the transitions from the initial state to the goal state. Overall, the Logical Paradigm suggests a rather straightforward view of the design process and can only deal with design problems of a low complexity.

The *Evolutionary Design Paradigm* explicitly recognizes the social, evolutionary, and error-prone nature of product development. It postulates the tentative and conditional character of design solutions and strives first to develop the main scientific body of design philosophy. The life-cycle design approaches realize some of the important aspects of this Paradigm by making the adequacy of the design dependent on the requirements, which are not constant but change over time [17]. Evolutionary Design is more general and assimilates the earlier theories. However, because of the complexity and multi-disciplinary nature, Evolutionary Design is still not properly framed, while the different 'evolutionary' approaches remain isolated at the level of computer applications and cannot deal with the dynamic aspects of the design process [9]. In the following section, an approach is proposed that could address the principal weaknesses of the Evolutionary Paradigm.

3 MODELLING EVOLUTIONARY DESIGN

3.1 Semiotics

Thinking in engineering design requires and is based on the representation and processing of information in the mind, prior to any physical realization of the design [12]. Semiotics – a science about signs, signification, and sign systems – provides a powerful and elegant tool for the investigation of thought and evolutionary processes involving products and their operational environments. Semiotics deals with an '...action, or influence, which is, or involves, an operation of three subjects, such as a sign, its object and its interpretant, this tri-relative influence not being in any way resolvable into an action between pairs' [3].

Semiotics studies how signs mediate meaning through the processes of semiosis, in which something functions as a sign. Semiosis is classically defined on and affects the *signifier* (the sign vehicle or token; usually a sign), the *signified* (the designatum, e.g. an object, whether physical or abstract), and the *interpretant* (i.e. the sense made of the sign or that which follows semantically from the process of interpretation). The semiosis of design is modelled by sign systems standing in for conceptual spaces and physical phenomena. These sign systems are evolved, blended, and analyzed, resulting in the creation of new concepts and the design itself [12].

3.2 The Product in its Environment

In the semiotic approach, human cognition at any stage of the product life cycle is characterized as a structuring of experience and perception to provide structured information – a language – that is to deal with the product. A design is seen as a text 'written' in such a product language with a *syntax*, which constrains the product's topological organization, *semantics*, which defines the functional structure and product-environment interaction, and *pragmatics*, which manifests physiological, psychological, and sociological effects associated with the product. This language is a *sign system* that influences and, conditional on the usage context, determines the meaning conveyed with signs. The sense made of a sign is understood as also a sign in the interpreter's mind.

A product is normally perceived through its distinctions that are revealed as *technologic*, *contextual*, and *ergonomic* relations between product parts and between the product and its environment, which includes consumers. The relations are represented in a language. In every situation, the language (that is a system of signs) may be different but has a common ground – the reality – that fundamentally constrains the relations. The language does not prescribe a particular model, but instead provides us with a universe of models for physical and mental phenomena.

The *Artifactual Wheel-Work* (AWW) classifies product life-cycle related models and techniques [7]. The framework postulates that, as one observes different levels of progression and repetition in the human dealing with artifacts, a natural distinction should be made between 3 kinds of existence in the life cycle of products and their groupings as follows: the individuals or *ens* (e.g. a single product), giving rise to the *ontogenic wheel*; *types* (e.g. the type of the product) or generations (in the case of self-reproducing species), giving rise to the *typogenic wheel*; and *tribes* or families (*phylos*), e.g. the totality of previous and actual types and occurrences of the product, or of all products, giving rise to the *phylogenic wheel*.

The distinctions between *ens*, *types*, and *phylos*, explain the existence of different models: an *ontogenic model* presenting life history data and future data about an occurrence of an artifact type; a *typogenic model* combining architectural and developmental aspects of an artifact type; and a *phylogenic model* integrating typogenic models of different generations, along with the steps that have been taken to move from one generation to another (e.g. reuse, production problem solving, etc). All three models matter for both the product and the environment. Design semiosis processes can naturally be classified as contributing to *typogenesis*, *ontogenesis*, or *phylogenesis* for the artifact, and resting on the genesis of 'resources' in the environment.

Design expectations are defined as stable relation patterns existing in the product language and realized in the product (see [13] for details). Violation of design expectations is the 'driving force' for design semiosis. As such, violations of functionality-related expectations control *typogenesis*, while violations of environment-related expectations activate *phylogenesis* in the product life cycle. Complementary to these, the *law of Weber-Fechner*, which tells us that in order for a series of quantities to appear 'regular' in our perception, it should be based on a geometric scale, characterizes the dynamics of the *ontogenesis* processes, where relation patterns are originally detected to be further abandoned or realized in the product [15].

Complementary to the AWW, and generalizing the concepts of resource and enterprise, the *Enterprise wheel-work* (EWW) describes the space-time-matter situated industrial system of resources and their structures in enterprises [6]. These resources, during their ontogenic life cycles, play rôles in tasks moving the ontogenic, typogenic, or phylogenic wheels of artifacts.

With the artifact life cycle in focus, the resource concept can be broadened to also cover objects interacting with the product in its use-environment. Thus, the EWW serves as a basis also for the modelling of elements in the product's use-environment. This environment can be considered a stable but evolving combination of 'resources' with which the product will interact. The environment offers the context within which the 'customer' exploits the product, and which – through its own evolution – forms a source of new opportunities and threats for the product's evolution.

3.3 Semiosis processes

Consumers' needs are first detected as stable patterns of conceptual relations. An initial product concept is nothing but a composition of related signs. The signs can be arbitrary, while the evolution of the concept throughout the product life cycle can be described in terms of transitions between different relation patterns by means of a *distinction dynamics* [14].

The first fundamental law of design semiosis states that in the product life cycle, the distinction dynamics always seeks to decrease the number of possible relations between the product and its environment [13]. A more precise characterization of design semiosis can be done, however, by introducing four types of elementary transformations (translations) of the product language, reflecting *conservation*, *destruction*, *creation*, and *creation and destruction* cognitive processes [8]. Semiosis processes of the first type result in preserving certain elements of the product language and are completely controllable, predictable, and reversible. Instances of such processes are selection, optimization, any kind logical or causal inference, methodologies, and deterministic algorithms. The second type of process is, in contrast to the first, non-distinction conserving: there is an information loss affecting the language elements. Such a process is always not reversible but still predictable, e.g. abstraction, habituation, adaptation, and the like. Next, semiosis processes behind creation can well be described as adoption from 'the outside' (acquisition) of new elements to the product language. These processes may only statistically be predictable, but they can, in principle, be reconstructed in the present-to-past direction (e.g. analogical or metaphorical reasoning, discovery, or any type of emergence). Finally, the last class of semiosis processes involves both the creation and destruction ('forgetting') of product language elements. Such processes are not systematic in any way (e.g. trial-and-error experiences), and they are not reconstructable. While all these different semiosis processes can formally be defined in terms of Algebraic Semiotics at the quite computational level [4] [12], we limit our discussion to qualitative characterization of the product life cycle processes.

The *Framework of Industrial Semiosis* [6] distinguishes four activity layers – *observation*, *operations*, *improvement*, and *invention and innovation* – and corresponding contexts for industrial semiosis. This framework permits us to classify the AWW semiosis processes. At the observation layer, life cycle *ontogenesis* has to be, in the main, a conservation process. Indeed, each particular artifact is associated with certain distinctions; distinction change usually leads to consideration of a different (even if slightly) artifact. Conservation is also important in *typo-* and *phylogenesis*, although to a less extent. At the operational layer, only *ontogenesis* has a plain and unequivocal characterization: this type of semiosis processes have to be non-distinction conserving. By denying the latter, almost any artifact would become unmanageable at the cognitive level (due to the natural limitations on the interpreter's side), as its handling would every time require possessing the artifact' perfect and complete knowledge. The improvement layer, though it allows for some destruction, preserves artifact distinctions throughout the semiosis processes. The *onto-*, *typo-*, and *phylogenesis* differ there by the degree of the permitted destruction (from less to higher, respectively). Obviously, as we come to the invention and innovation layer, we can unambiguously consider only the case of *ontogenesis*, where the creation semiosis processes play the major rôle.

4 CONCLUSION AND FUTURE WORK

The semiotic theory of evolutionary design has been developed into a more detailed form. The theory pays significant attention to differentiation and optimization of all the life cycle processes. The models proposed do not focus on a particular product or even a family of products – instead, they are oriented on the creation of a ubiquitous, broad and universal information infrastructure for design and life cycle engineering.

Concentrating on the theoretical aspects, this work contributed to the development of a general, interdisciplinary frame of design philosophy. To facilitate understanding of the main ideas, formal and computational details of the semiotic approach were not presented in the paper; the interested reader is however referred to [12] and [14]. Besides, some implementation issues as well as descriptions of case- and pilot- studies of the semiotic theory of evolutionary design can be found in [13] and [5].

In future work, the authors aim to build a complete classification of design and life cycle semiosis processes and investigate their properties. The further elaboration of the information infrastructure for the design semiosis services is also expected and should lead to the development of next generation information support tools and industrial technologies.

5 ACKNOWLEDGEMENT

This research has been supported in part by the Japan Society for the Promotion of Science (project No 96P00702). Dr. Goossenaerts would like to acknowledge the support of the ESPRIT IMS Working Group, EP 21995. We thank Prof. John J. Mills for valuable comments on the draft.

6 REFERENCES

[1] Aerts, A.T.M., Szirbick, N.B, Goossenaerts, J.B.M. (2000). Flexible Infrastructure for Virtual Enterprises, In: *DIISM'2000* (this volume)

[2] Arciszewski, T., Michalski, R.S., Wnek, J. (1995). Constructive induction: The key to design creativity, In: *Preprints "Computational Models of Creative Design," Third International Round-Table Conference*, pp. 397-426. Heron Island, Australia
[3] *Collected papers of Charles Sanders Peirce, (in 6 volumes).* (1960). C. Hartshorne and P. Weiss (eds.), The Belknap Press of Harvard University Press, Cambridge, MA
[4] Goguen, J.A. (1999). An introduction to algebraic semiotics, with applications to user interface design, In: C. Nehaniv (ed.), *Computation for Metaphor, Analogy and Agents, Springer Lecture Notes in Artificial Intelligence*, vol. 1562, 242-291
[5] Goossenaerts, J.B.M. (1999). *A Systematisation Framework for Legal Drafting and Norm Implementation in the Information Society.* Preprint. Eindhoven University of Technology, the Netherlands
[6] Goossenaerts, J.B.M. (2000). Industrial Semiosis: Founding the deployment of the Ubiquitous Information Infrastructure. *Computers in Industry*, vol. 43, 189-201
[7] Goossenaerts, J.B.M., Thoben, K.-D. (1999). The Artefactual Wheel-Work and the Information Infrastructure, In: H. Yoshikawa, R. Yamamoto, F. Kimura, T. Suga, and Y. Umeda (eds.), *Proceedings of the First International Symposium on Environmentally Conscious Design and Inverse Manufacturing "EcoDesign'99,"* pp. 674-679, Tokyo
[8] Heylighen, F. (1990). Non-Rational Cognitive Processes as Changes of Distinctions, *Communication & Cognition*, vol. 23, No.2-3, 165-181
[9] van den Hoed, R. (1997). An exploration of approaches towards Sustainable Innovation and their implications on the product development process, In: *Proceedings of The Greening of Industry Conference "Developing Sustainability: New Dialogue, New Approaches."* University of California, Santa Barbara, CA
[10] Hubka, V., Eder, W.E. (1996). *Design Science: Introduction to the needs, scope and organization of Engineering Design knowledge*, Springer-Verlag, London
[11] Klein, R. (1998). A Knowledge Level Theory of Design and Engineering, In: G. Jacucci, G.J. Olling, K. Preiss, M.J. Wozny (eds.), *Globalization of Manufacturing in the Digital Communication Era of the 21st Century: Innovation, Agility, and the Virtual Enterprise*, pp. 271-286, Kluwer Academic Publishers
[12] Kryssanov V.V., Tamaki, H., Kitamura, S. (1999). Understanding Design Fundamentals: How Synthesis and Analysis Drive Creativity, Resulting in Emergence, In: K. Ueda (ed.), *Proceedings of the International Workshop on Emergent Synthesis*, pp. 37-48, Kobe
[13] Kryssanov, V.V., Goossenaerts, J., Goncharenko, I., Tamaki, H. (2000). A Semiotic theory of evolutionary design: concepts and an illustration. In: *Proceedings of the 7th CIRP International Seminar on Life Cycle Engineering*, pp. 68-75, Tokyo
[14] Kryssanov, V.V., Tamaki, H., Ueda, K. (1999). Agents for Assessing Requirements in Dynamic Environments to Support Evolutionary Design, In: *Proceedings of the 6th CIRP International Seminar on Life Cycle Engineering*, pp. 151-160, Kingston, Canada
[15] van Onck, A. (1994). *Design. Il Senso Delle Forme Dei Prodotti.* Lupetti Editori di Comunicazione, Milan (in Italian)
[16] Suh, N.P. (1990). *The Principles of Design.* Oxford University Press, NY
[17] Tomiyama, T. (1999). The Post Mass Production Paradigm, In: H. Yoshikawa, R. Yamamoto, F. Kimura, T. Suga, and Y. Umeda (eds.), *Proceedings of the First International Symposium on Environmentally Conscious Design and Inverse Manufacturing "EcoDesign'99,"* pp. 162-167, Tokyo
[18] Wouters, M.J.F., Sharman G.J., Wortmann H.C. (1999). Reconstructing the Sales and Fulfillment Cycle to Create Supply Chain Differentiation, *International Journal of Logistics Management*, vol. 10(2), 83-97
[19] Yoshikawa, H. (1981). General Design Theory and a CAD System, In: T. Sata and E.A. Warman (eds.), *Man-Machine Communication in CAD/CAM*, pp. 35-58, North-Holland

Modelling for Designing, Managing and Improving Virtual Enterprises in One-of-a-kind Business

Lauri Koskela, Abdul Samad Kazi, Matti Hannus
VTT Building Technology
PO Box 1801, FIN-02044 VTT, Finland
Em:lauri.koskela@vtt.fi

Keywords Production Theory, Virtual Enterprise, One-of-a-kind Production

Abstract The paper starts with an exploration of the current conceptual underpinnings, paradigms and theories of production from the point of view of modelling. It is argued that modelling should bear on designing, controlling and improving production systems. Modelling should orient towards the existing three conceptual views on production: transformation, flow and value generation. Furthermore, modelling should cover both the physical process and the control information process. Finally, modelling should respond to the specific needs of virtual production in one-of-a-kind business, for example, the design of the production system accentuates. Two sets of modelling tools are analyzed against this framework: comprehensive modelling approaches and specific, more partial tools for virtual production. These results suggest that, at the level of research, the search for a unified conceptualization of production should be emphasized; only based on such a foundation can truly comprehensive and integrated modelling tools be constructed. At the more practical level, modelling efforts - in lack of tools based on a unified conceptualization - have to be partial; however, these efforts should be structured so that various modelling approaches could more easily be interfaced with each other, and the limitations of the particular tools in use should be clearly recognized.

1 INTRODUCTION

The goal of this presentation is to present a comprehensive view on the requirements to be set for modelling virtual enterprises in one-of-a-kind business, and thus to contribute to the definition of a generic architecture for these kinds of enterprises. This is essentially a background paper by nature and does not contain concrete proposals for a generic architecture. However, existing modelling methods are analyzed against the requirements set.

The plan of the paper is as follows. First, we discuss modelling needs attributable to the purpose of modelling, namely design, control and improvement of the production system. Secondly, we treat the requirements with respect to different theories of production. Thirdly, requirements derived from the subject of modelling are derived. Next, we compare existing modelling methods to the set of requirements produced. Finally, we summarize our findings and discuss their implications.

2 DIFFERENT LEVELS OF ANALYSIS OF PRODUCTION

There are three generic actions, which we would like to be guided by modelling production:

- Design of the production system
- Control of the production system to get the intended production realized
- Improvement of the production system.

Production modelling efforts can be divided into two distinct classes: physical material focus vs. control data focus [10]. With physical material focus, the aim is to represent the product flow through the production facility (network). The purpose of modelling is to improve operations. In contrast, with control data focus, the aim is to represent the relationships and flows of control data and the data processing logic. The purpose of modelling is to improve information management so as to improve manufacturing processes.

Thus, in summary, we require that modelling is capable of representing the design, control and improvement of the production system. Also, both the physical flow and the control information flow should be covered.

3 THEORIES OF PRODUCTION

In the following, the novel TFV (Transformation, Flow, Value) framework of production, as presented in [12], is summarized.

The primary characteristic of a theory of production is that it should be prescriptive: it should reveal how action contributes to the goals set for production. Production has three kinds of goals. Firstly, the goal of getting the intended products produced in general (this may seem so self-evident that it is often not explicitly mentioned). Secondly, there are goals related to the characteristics of the production itself, such as cost minimization and level of utilization (internal goals). Thirdly, there are goals related to the needs of the customer, like quality and flexibility (external goals).

Furthermore, the theory of production should cover all the essential areas of production, especially production proper and product design.

Throughout the 20th century, the *transformation* view of production has been dominant. In the transformation view, production is conceptualized as a transformation of inputs to outputs. There are a number of principles, by means of which production is managed. These principles suggest, for example, decomposing the total transformation hierarchically into smaller transformations, tasks, and minimizing the cost of each task independently. The conventional template of production has been based on it, as well as the doctrine of operations management. The transformation view has its intellectual origins in economics, where it has remained unchallenged up to this day. The popular value chain theory, proposed by Porter [16], is one approach embodying the transformation view. A production theory based directly on the original view on production in economics has been proposed by a group of scholars led by Wortmann [22].

However, this foundation of production is an idealization, and in complex production settings the associated idealization error becomes unacceptably large. There are two main deficiencies: it is not recognized that there are also other phenomena in production than transformations; it is not recognized that it is not the transformation itself that makes the output valuable, but that the output conforms with the customer's requirements. The transformation view is instrumental in discovering which tasks are needed in a production undertaking and in getting them realized. However, the transformation view is not especially helpful in figuring out how to ensure that the customer requirements are met in the best manner or how not to use resources unnecessarily. Therefore, production, managed in the conventional method, tends to become inefficient and ineffective.

There has existed, already in the framework of early industrial engineering, another concept of production, namely the view of production as *flow*. The flow view of production, firstly proposed by the Gilbreths [5] in scientific terms, has provided the basis for JIT (Just In Time) and lean production. This view was firstly translated into practice by Ford [3]; however, the template provided by Ford was in this regard misunderstood, and the flow view of production was further developed only from 1940'ies onwards in Japan, first as part of war production and then at Toyota. As a result, the flow view is embodied in lean production. In the flow view, the basic thrust is to eliminate waste from flow processes. Thus, such principles as lead time reduction, variability reduction and simplification are promoted. In a breakthrough book, Hopp and Spearman [7] show that by means of the queueing theory, various insights, which have been used as heuristics in the framework of JIT, can be mathematically proven.

Still a third view on production has existed from the 1930'ies. In the *value generation* view, the basic thrust is to reach the best possible value from the perspective of the customer. The value generation view was initiated by Shewhart [21] and further refined in the framework of the quality movement but also in other circles. Principles related to rigorous requirements analysis and systematized flowdown of requirements, for example, are put forward. Cook [2] has recently presented a synthesis of a production theory based on the flow view.

Thus, there are three major concepts of production, and each of them has induced practical methods, tools and production templates. Nevertheless, except a few isolated endeavors, these concepts – as candidate theories of production - have raised little interest in the discipline of operations management. As stated above, there has not been any explicit theory of production. The consequential problem is that the important functions of a theory, as outlined, have neither from the viewpoint of research or practice been realized.

What, then, should be held as the theory of production? There are three partial theories, and because they have not yet been unified, we have to use them simultaneously. As argued at more length in [12], such a TFV theory of production provides a framework for analyzing production (Table 1).

Table 1. The framework for analyzing production, based on the TFV theory of production.

	Transformation view	*Flow view*	*Value generation view*
Design of production	What is the structure of activity decomposition?	What is the structure of the flow?	What is the structure of the value generation process?
Control of production	How are production resources allocated to activities?	How is the productive flow controlled?	How is the stagewise transformation of requirements into products controlled?
Improvement of production	How are activities improved?	How is the flow performance improved?	How is the value generation performance improved?

4 ONE-OF-A-KIND PRODUCTION BY VIRTUAL ENTERPRISES

One-of-a-kind production is characterized by two issues [22, 18]. Firstly, product design is an integral part of production (that is, product design or development beyond mere selection of options or configuration design). Secondly, there is uncertainty, which is critical especially in regard to customer order acceptance. This uncertainty is, of course, caused by the lack of experience on the one-of-a-kind features of the product.

A virtual organization has been defined as a new customer-oriented and opportunity-based organizational model that uses technology to dynamically link people, assets, and ideas [6]. In a virtual organization, core competencies from several real organizations are integrated.

Thus, this kind of production sets several special requirements to modelling:

- product development and design is an integral part of production, and has to be modeled as a part of the production process
- design of the production system is crucial
- control of the production system is made difficult by the associated uncertainty
- improvement of the production system has to take place on-line; there are few possibilities for continuous, long-term improvement
- control information modelling accentuates in those cases where chunks of physical production are treated as black boxes.

5 ADEQUACY OF CURRENT MODELLING METHODS

5.1 Introduction

Are current modelling methods adequate in view of the requirements defined in the preceding section? This question is approached by analyzing representative examples, first, of comprehensive modelling tools, and, second, of partial modelling tools suitable for virtual, one-of-a-kind production.

5.2 Comprehensive modelling tools

A number of comprehensive process modelling tools have been developed in the 1990's. In the following, we discuss three of them: the enterprise engineering model, the electronic process handbook, and the unified process specification language. They all are purported, among other things, for design of production systems.

Enterprise engineering models, like GERAM, provide a generalized framework for describing the components needed in all types of enterprise engineering/integration processes, such as the formation of a virtual enterprise [4]. The development of these models has started from efforts to integrate industrial automation. Unfortunately, this background still seems to be a restricting factor. This is evident in the motivation for a standard in this area [8]: "A standard for enterprise models should enhance interoperability by establishing the elements that must be present in an enterprise model". Thus, although extended to enterprise engineering, the focus of this discipline is largely on integration, e.g. control information, rather than the physical process itself.

The initiative towards an *electronic process handbook* [14] is based on the functional view (this equals to transformation-centered view) on processes, but adds a number of important features. In addition to the hierarchical abstraction (decomposition-composition), the dimension of generalization and specialization is added to the description. Also, the types of dependencies between activities are analyzed. It is claimed that there are three types of basic dependencies: flow, sharing and fit. Flow dependency is basically the supplier-consumer relationship. Sharing dependencies occur when multiple activities use the same resource. Fit dependencies arise when multiple activities collectively produce a single resource. Coordination mechanisms for different types of dependencies are correspondingly analyzed and classified. The handbook is intended for redesigning existing organizational processes, inventing new processes and sharing ideas about organizational practices.

The National Institute of Standards and Technology (U.S.) is promoting a project developing a *unified process specification language* [20]. The motivation is to create a language by means of which various applications, ranging from project management to manufacturing process planning, could exchange information on processes, understood as collections of activities. Requirements for such a language have been grouped into four categories: core, outer core, extensions and application specific. Although the main focus is on manufacturing processes, product development processes are also covered.

All these three frameworks are new, and few, if any, practical applications exist. Analysis reveals that they all are biased towards the transformation model of production. Furthermore, the enterprise engineering model is mainly focused on control information, whereas the unified process language and the electronic process handbook are focused on the physical production process.

5.3 Partial modelling tools

5.3.1 Tools based on the transformation model

Generally, project management methods, like the Work Breakdown Structure (WBS) and critical chain network, provide examples of tools based on the transformation model. Such tools have been integrated into *Simo-2* [13]. It consists of a modelling tool, a multi-user project planning and simulation package, and a separate analysis tool. Modelling divides into three phases: product, business and project modelling. First, all partners model the product structure. Then, each partner creates a business plan, that is, a non-public operational plan of how they will execute their part of the project. Then, project modelling can start: main and subcontracts are defined and the schedule created. Several users can co-operate in this phase for iteration of the contract structure and schedule. With all the models finished, the project can next be simulated to determine if all targets are being met.

However, this model is not without limitations, due to its conceptual basis. It is not possible to model the flow of information and material in detail by this model; also the interdependencies between different contracts are not treated. Likewise, design iteration cannot be modeled.

5.3.2 Tools based on the flow model

There are several generic tools that are based on the flow model, like IDEF3, Design Structure Matrix and the traditional flow analysis methods of industrial engineering. However, it has been a challenge to provide a user-friendly interface to such models. In this regard, Rose [19] describes an interesting tool called PROVE for engineering process representation and assessment. A visualization tool is used for animating process graph structures, and this animation offers the main interface vis-a-vis users. The process is primarily modeled in regard to activities and documents. This tool is used for process design from two view points. First, process assessment for judging the accuracy, performance and quality of the process and its description. Second, process synchronization for reconciling the contributions of related processes.

Again, there are limitations due both to the conceptual basis selected and to the fact that only parts of that conceptual basis have been realized. For example, in this PROVE model, it is not possible to model design iterations, neither it is possible to analyze the implications of value generation issues of the flow of the design process.

5.3.3 Tools based on the value generation model

Value-oriented modelling is well represented by requirements traceability, which refers to the ability to describe and follow the life of a requirement, in both a forward and backward direction, ideally through the whole systems life cycle [9]. Various commercial and in-house traceability models are currently in use, especially in high-tech domains [17]. Maybe the oldest generic traceability tool is Quality Function Deployment (QFD) [1].

Such traceability models support the design, control and improvement of the product realization process from the value point of view. Again, natural limitations of value-oriented models can be recognized: they are not very helpful for managing activities or flows.

6 CONCLUDING DISCUSSION

It has been argued that modelling should bear on designing, controlling and improving production systems. Modelling should orient towards the existing three conceptual views on production: transformation, flow and value generation. Furthermore, modelling should cover both the physical process and the control information process. Finally, modelling should respond to the specific needs of virtual production in one-of-a-kind business, where, for example, the design of the production system accentuates.

Two sets of modelling tools were analyzed against this framework: comprehensive modelling approaches and specific, more partial tools for virtual, one-of-a-kind production. The analysis reveals that the comprehensive approaches tend to be biased towards particular aspects of production, especially the transformation model. Specific modelling tools, corresponding to all three views of production, can be found, which proves that these three views have been recognized as necessary and important from the practical point of view. These results suggest that, at the level of research, the search for a unified conceptualization of production should be emphasized; only based on such a foundation can truly comprehensive and integrated modelling tools be constructed (examples of such work provide [10, 11, 15]). At the practical level, modelling efforts - in lack of tools based on a unified conceptualization - have to be partial; however, these efforts should be structured so that various modelling approaches could more easily be interfaced with each other, and the limitations of the particular tools in use should be clearly recognized.

7 REFERENCES

[1] Akao, Yoji (ed.). 1990. Quality Function Deployment. Productivity Press. Cambridge, Ma. 369 p.

[2] Cook, H.E. 1997. Product Management - Value, quality, cost, price, profit and organization. Chapman & Hall, London. 411 p.

[3] Ford, Henry. 1926. Today and Tomorrow. Doubleday, Page & Co., Garden City. (Available as reprint edition: Productivity Press, Cambridge MA. 1988.) 286 p.

[4] GERAM: Generalised Enterprise Reference Architecture and Methodology. IFIP-IFAC Task Force. Version 1.6.3 (March 1999).

[5] Gilbreth, Frank B. and Gilbreth, L.M. 1922. Process Charts and Their Place in Management. Mechanical Engineering, January, pp. 38 - 41, 70.
[6] Goldman, Steven L., Nagel, Roger N. & Preiss, Kenneth. 1995. Agile Competitors and Virtual Organizations. Van Nostrand Reinhold, New York. 414 p.
[7] Hopp, Wallace & Spearman, Mark. 1996. Factory Physics: Foundations of Manufacturing Management. Irwin/McGraw-Hill, Boston. 668 p.
[8] ISO TC184/SC5/WG1. 1999. Industrial Automation systems - Concepts and rules for enterprise models. 1999-April-14 version. Captured from http://www.mel.nist.gov/sc5wg1/std-dft.htm, March 3, 2000.
[9] Jarke, Matthias. 1998. Requirements Tracing. Communications of the ACM, Vol. 41, No. 12, pp. 32-36.
[10] Kang, H.W., Kim, J.W. & Park, S.J. 1998. Integrated Modelling Framework for Manufacturing Systems: A Unified Representation of the Physical Process and Information System. The International Journal of Flexible Manufacturing Systems, 10, 231-265.
[11] Karhu, Vesa. Forthcoming. Proposed new method for construction process modelling. International Journal of Computer Integrated Design and Construction.
[12] Koskela, Lauri. 2000. An exploration towards a production theory and its application to construction. Espoo, VTT Building Technology. 296 p. VTT Publications; 408 ISBN 951-38-5565-1; 951-38-5566-X WWW: http://www.inf.vtt.fi/pdf/publications/2000/P408.pdf
[13] Laurikkala, Heli, Tanskanen, Kari, Nevalainen, Paavo, Vainio-Mattila, Markus. 1999. Consortium as a Virtual Enterprise in Project Planning. The 5th International Conference on Concurrent Engineering, The Hague, The Netherlands, 15-17 March 1999.
[14] Malone, Thomas W., Crowston, Kevin, Lee, Jintae, Pentland, Brian, Dellarocas, Chrysanthos, Wyner, George, Quimby, John, Osborn, Charles S., Bernstein, Abraham, Herman, George, Klein, Mark, O'Donnell, Elissa. 1999. Tools for Inventing Organizations: Toward a Handbook of Organizational Processes. Management Science, Vol. 45, No. 3, pp. 425-443.
[15] Park, H. & Cutkosky, M.R. 1999. Framework for Modelling Dependencies in Collaborative Engineering Processes. Research in Engineering Design 11:84-102.
[16] Porter, M. 1985. Competitive Advantage. The Free Press. 557 p.
[17] Ramesh, Balasubramaniam. 1998. Factors Influencing Requirements Traceability Practice. Communications of the ACM, Vol. 41, No. 12, pp. 37-44.
[18] Riis, Jens, Mortensen, Harry & Johansen, John. 1992. A new concept for managing one-of-a-kind production. In: 'One-of-a-kind production': New approaches. Hirsch, B.E. & Thoben, K.-D. (Ed.). Elsevier Science, Amsterdam. Pp. 195 – 208.
[19] Rose, Thomas. 1998. Visual Assessment of Engineering Processes in Virtual Enterprises. Communications of the ACM, Vol. 41, No. 12, pp. 45-52.
[20] Schlenoff, Craig, Knutilla, Amy, Ray, Steven. 1996. Unified Process Specification Language: Requirements for Modelling Processes. National Institute of Standards and Technology.
[21] Shewhart, W.A. 1931. Economic Control of Quality of Manufactured Product. Van Nostrand, New York. 501 p.
[22] Wortmann, J.C. 1992. Factory of the Future: Towards an integrated theory for one-of-a-kind production. In: 'One-of-a-kind production': New approaches. Hirsch, B.E. & Thoben, K.-D. (Ed.). Elsevier Science, Amsterdam. Pp. 37 - 74.

An Adaptive Process Management System (APMS)

Christopher Menzel and Perakath Benjamin
Knowledge Based Systems, Inc., U.S.A.
Em: cmenzel@kbsi.com

Keywords Knowledge sharing, integration architectures, enterprise integration, process management

Abstract The increasing complexity of manufacturing systems coupled with the emergence of truly global manufacturing enterprises requires the use of novel approaches to facilitate enterprise integration. Specific challenges include (i) the lack of theoretical foundations for enterprise integration and (ii) the absence of mechanisms for knowledge sharing and semantic interoperability. This paper presents a process-centered approach to address these problems. Specifically, the paper describes the theoretical foundations and concept of operation of an integrated architecture for process management, the Adaptive Process Management System (APMS). APMS facilitates information-integrated management of the Product Realization Process (PRP) using a life-cycle perspective.

1 INTRODUCTION

The rapid increase in the speed and sophistication of computers in recent years have led to a corresponding increase in the number computer-based enterprise modelling applications. Among the most important of these are applications that support the creation of *dynamic* models: models that represent information about how things within a portion of an enterprise change, or ought to change, over time – in a word, information about enterprise *processes*. Such models are especially critical to a business' ability to respond rapidly and creatively to change.

Cross-functional processes are the focus of many different business functions. Consequently, there are a wide variety of process-oriented applications for the creation and maintenance of enterprise process models that support those functions: simulation, project management, process and production planning, scheduling, workflow management, and so on. Thus, to alter its enterprise processes, a company must alter the models it maintains across these applications. But there is a serious problem: like the builders of the Tower of Babel after the confounding of tongues, these applications cannot talk to each other. If, however, the applications that harbor an enterprise's models cannot communicate, then there is no reliable way for the implications of changes in one model to be propagated appropriately to all related models; one must rely upon human agents managing the processes to communicate and to do so quickly, knowledgeably, and accurately – attributes not often associated with human communication within a modern enterprise.

The idea of an interlingua is not new – such a language for exchange of information between (typically static) knowledge bases was in fact developed as a part of the DARPA-sponsored Knowledge Sharing Effort (KSE) [1]. More recently, however, two projects have turned their attention to the development of an interlingua that is tailored specifically to address the problem of integration among dynamic modelling applications: the Process Interchange Format (PIF) project [2] and the NIST Process Specification Language (PSL) Project [http://www.nist.gov/psl].

2 FRAMEWORK FOR PROCESS KNOWLEDGE SHARING

2.1 The Process Specification Language (PSL)

The PSL consists of several components, or *theories*, based on first-order logic. The fundamental theory is known as *PSL core*. The idea behind PSL core is to provide a theory that characterizes a basic ontology of dynamic information consisting of four basic types of entity: activities, activity occurrences, objects, and timepoints. The essential characteristics of these entities are captured in a series of *axioms* expressed in a precise logical language. Examples of such principles are that timepoints are linearly ordered (by the *temporal precedence* – a.k.a. *before* – relation), and that the ending timepoint of an activity occurrence cannot precede its beginning point. Extensions of PSL capture the properties of activity resources, subactivities, and complex activities.

2.2 Process Specifications for Knowledge Sharing

When one describes the general structure of a process—whether graphically, or by means of a structured text file or a well-defined representation language—one is giving a *process description*. A primary function of PSL is to provide a single framework that is capable of expressing the content of any process description in any reasonable application representation language. Any two applications that have translators into and out of PSL will then be able, as far as possible, to share information despite the fact that neither "speaks" the other's "native tongue." The Adaptive Process Management System (APMS) is an integrated process-oriented architecture that implements the PSL framework. Section 3 outlines an example application of the APMS architecture.

The current implementation of PSL approaches the specification of processes by identifying processes with a certain class of "complex" activities—activities whose instances can vary greatly in form and which are specified by means of complex logical descriptions. A typical example of such an activity would be one that is specified by an if-then-else conditional: "If condition φ holds, then activity A occurs, otherwise B occurs." An occurrence of the complex activity so described could consist in either an occurrence of A or an occurrence of B depending on whether or not the precondition holds. A process specification in a given application language would be translated into set of logically complex PSL statements that jointly define a complex activity. Complex activities and their description, however, are indeed rather complex, and the extension that introduces them into PSL involves quite a lot of heavy theoretical machinery. For many purposes, when one needs to capture the content of highly detailed and fine-grained process descriptions, this additional power is necessary. However, as often as not, a simpler framework that builds only upon PSL core will suffice. More exactly, on this approach, there are no complex activities; i.e., there are no processes, *per se*. Rather, there are only PSL-based process specifications; i.e., certain syntactic constructions that can be given a semantics in terms of the semantics of PSL core. To define a PSL process specification, we must first define the notion of an *activity role specification*. To define this idea precisely, we need a complete characterization of the notion of a PSL language, which would take us rather too far afield.

A process specification, then, is simply a set P of activity specifications that satisfies two conditions: (i) no two activity role declarations in P can have the same numerical identifier (i.e., the same value in their :id fields), and (ii) every identifier in the :successors field of an activity role declaration in P must be the numerical identifier (i.e., the value of the :id field) of some activity role declaration in P. Condition (i) is the formal mechanism that enables us to represent activity roles (as this allows us to specify the same activity in the name fields of distinct activity roles. Condition (ii) ensures that there are no "dangling" successor relations in a process—i.e., cases where an activity role is declared to have a successor in the structure of the specified process by specifying an activity ID in the "successors" field, but where there is, in fact, no corresponding activity role with that ID in the specification.

3 APMS APPLICATION SCENARIO

We will now outline an application of the APMS integration architecture for knowledge sharing. We use a simple example to illustrate how APMS facilitates knowledge sharing through the use of the PSL. The knowledge sharing occurs in this example through *reuse*: the ability to import the information from a model M1 in one application A1 into a model M2 from another application A2 so that the information garnered in the first needn't be reconstructed from scratch in the second. The application scenario is as follows (Figure 2). Mr. Dawn, ACME Inc.'s production manager, would like to validate and implement a new Widget manufacturing process. He performs the following activities: (i) develop a process plan for making the Widget, (ii) analyze the plan to validate process performance, (iii) formulate an implementation schedule and dispatch resources, and (iv) implement the process and monitor performance. PSL is used in this scenario to enable the transfer of information between (a) a process planning model, (b) a simulation model and (c) a scheduling model.

3.1 The Process Plan

The process plan is developed using the IDEF3 process modelling language (http://www.idef.com).

Figure 1: Portion of the Widget-Making Process Plan

This graphical aspect of the model is self-explanatory (Figure). The planned Widget-making process involves turning a part and then drilling it, followed by an inspection. If the part passes the inspection, it is sent immediately to the plating shop. If not, it is sent for rework before being sent to the plating shop.

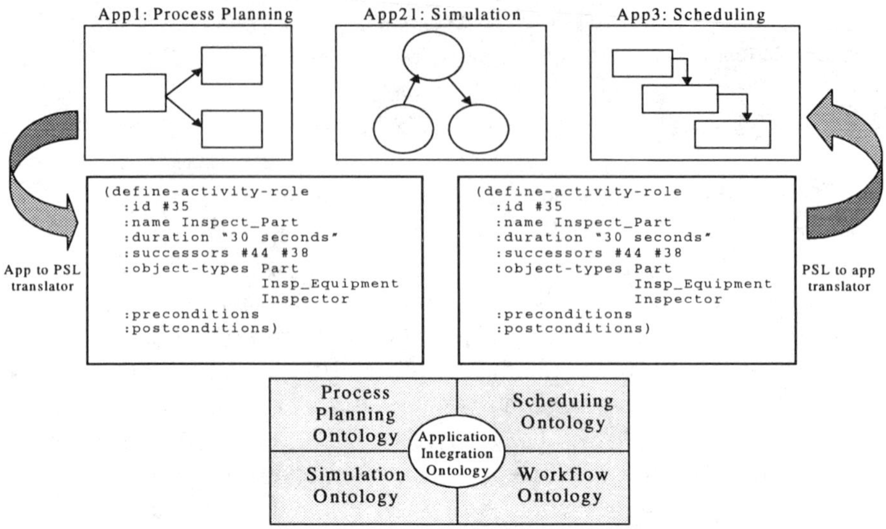

Figure 2: APMS Facilitates Integration through Knowledge Sharing

In addition to the graphical information, an IDEF3 model also enables one to store information about the resources needed for each activity in the process as well as its duration. The type of information that can be stored is shown in Table 1.

Table 1: Activity Specification

Activity Name	Resources	Processing Time
Turn Part	Lathe, Turning Operator, Turning Tool	20 seconds
Drill Part	Drilling Machine, Drilling Operator, Drilling Tool	15 seconds
Inspect	Inspection Equipment, Inspector	30 seconds
Send to Plating Shop	Transporter	5 seconds
Rework Part	Rework Tool, Rework Operator	25 seconds

3.2 The Schedule

The process plan is now used as the basis for generating an implementation schedule. The implementation schedule includes (i) a calendar of tasks, (ii) start and finish dates including an identified critical path, and (iii) a resource utilization chart. The schedule is used as the basis for dispatching work to the widget manufacturing shop floor resources (people, machines, and tools). The schedule is also integrated with the shop floor work status monitoring systems. The schedule is updated regularly based on (a) changes in execution status and (b) changes in the manufacturing plan requirements (Figure 3).

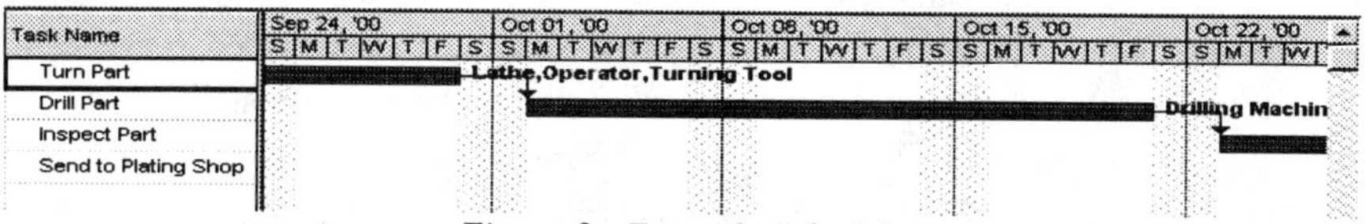

Figure 3: Example Schedule

Additional information is required by the schedule model over and above the information captured in the IDEF3 process planning model: the *scheduled plan instantiation paths*. In this example, the IDEF3 "X" Junction involved a decision (based on the quality of the part) that would result in the selection of one out of two possible instantiation paths. Recall that the simulation model required the specification of the decision rule for the decision. Scheduling requires the representation of a detailed schedule (tasks and resources represented on a calendar) for each instantiation path. Moreover, some widely used scheduling tools (such as MSProject) do not allow for the representation of multiple, mutually exclusive paths in a schedule.

4 HOW PSL FACILITATES KNOWLEDGE SHARING

In our scenario, a variety of application-specific tools from difference vendors are used in the development and implementation of a process. This is the typical case. It is also typical that there is no way to share the information in one application with another. Thus, if the above scenario were played out along the usual lines, Mr Dawn would have to recreate information in each successive model that is present in the preceding models. Such recreation is wasteful of time and resources, and is also prone to error and inconsistency. It would be preferable if information, once recorded in a given model, can be *reused* in successive models, thus saving the time and resources that are lost in information recreation and avoiding the high potential for error and inconsistency.

For reasons discussed in Section 1, the development of pairwise translators between applications is not feasible. PSL provides the representational interlingua between applications that brings the benefits of knowledge sharing without the liabilities of pairwise translators. How is this accomplished? The interlingua architecture is predicated upon the existence of application representation languages. However, many process-oriented modelling applications, such as IDEF3 and MS Project, are graphically based. How can PSL then be used?

The answer to this question is that most all modelling applications have the ability to translate the contents of a model into some sort of text-based file format. Such formats can be thought of as rudimentary representation languages. Consider the IDEF3 representation of a process plan for making widgets, generated by means of the PROSIM process modelling tool. To represent the information in the model in a non-graphical form, PROSIM can translate its graphical models into a well-defined textual representation language. Thus the above model—together with information about participating object types and instantiation constraints not pictured in Figure —translates into a structured text.

In essence, this process description is stated in a perfectly respectable representation language, albeit one that is tailored toward the eccentricities of IDEF3. There is, in fact, no explicit mathematical semantics for this language (or for IDEF3 generally), but one could clearly be given. It is also reasonably clear how this language can be translated directly into a PSL process specification. For instance, from the information for `Inspect_Part`:

```
Process 34
  Name: Inspect Part
  ID: #35
  Duration: 30 seconds
    Object Occurrence List
      Object Occurrence 54 2 (Part)
      Object Occurrence 106 1 (Inspection Equipment)
      Object Occurrence 107 1 (Inspector)
    End Object Occurrence List
End  Process 34
```

It is possible to extract the contents of the :name, :id, Duration:, and :object-types fields of a PSL activity declaration. And, while links do not correspond to anything in a PSL process specification but instead from information from the IDEF3 links, it is possible to extract information for the :successors field. We would thus have the following activity role declaration:

```
(define-activity-role
  :id #35
  :name Inspect_Part
  :duration "30 seconds"
  :successors #44 #38
  :object-types Part Inspection_Equipment Inspector
  :preconditions
  :postconditions)
```

The logic of the XOR junction would be expressed as preconditions on the Send to Plating Shop activity with ID #38 and on the Rework Part activity. For example, in the case of the former we would have

```
(define-activity-role
  :id #38
  :name Send_To_Plating_Shop
  :duration "5 seconds"
  :successors
  :object-types Part Transporter
  :preconditions (PassedInspection (the ?x (Part ?x)))
  :postconditions)
```

whereas the `:preconditions` field for `Rework Part` would be the negation `(not (PassedInspection (the ?x (Part ?x))))`.

As it happens, PROSIM uses a logical language to express constraints that is very similar to PSL, but in general this will not be the case. Hence, for applications that include a logical language, part of the translation process will have to include a "compiler" for rendering statements in the application language in PSL.

Microsoft has developed a text-based form for the information in a Project schedule known as MPX. Again, as with PROSIM, this can be viewed as a tailored representation language. Thus, reuse of the PROSIM process plan can be achieved via translation from PSL process specifications into MPX format. This, in fact, can be easily accomplished. Adding certain pieces of scheduling boilerplate (currency formats, work hours, etc), the PSL representation of the PROSIM schedule can be translated into two MPX files corresponding to the two possible instantiations that can arise from the XOR of the process model.

5 APMS BENEFITS

5.1 APMS Facilitates Semantic Integration

We conclude from the previous section that the APMS architecture solves one of most critical problems for semantic integration: *conflicts in terminology*. These conflicts can be of two sorts. First, the same term can be used to denote different concepts (the *ambiguity* problem). Second, different terms can be used to denote the same concept (the *synonymy* problem). The ambiguity problem is eliminated because different concepts are represented differently in PSL. Since application representation languages with ambiguous terms do not directly "speak" to each other, ambiguity will be filtered out via intermediate PSL translations. Similarly, the synonymy problem will be avoided because the different terms will be mapped to the same representation in PSL, and hence translators will ensure that synonymous terms in different applications are properly mapped to one another.

5.2 APMS Provides a Robust Integration Ontology

There is a further important element to the PSL integration architecture that deserves mention. PSL is part of a growing, global effort to develop large ontologies of fundamental clusters of concepts for next generation knowledge management applications. Part of this effort consists in the development of a number of general process-oriented ontologies targeted by different classes of process-oriented applications. Thus, in addition to PSL's general ontology of processes, there are also more specialized ontologies for process planning, simulation, scheduling, and workflow that are being developed. However, these specialized ontologies alone are still not enough for robust knowledge sharing. In terms of a logical connection between a process plan and a corresponding schedule, is not something present in an ontology of either process planning or scheduling. Rather, it is what's known as an *integration axiom*—it relates concepts across an ontological divide. A collection of such axioms for a given set of ontologies is known as an *integration ontology*. Integration ontologies are critical for ensuring that the maximum possible information in a model of one sort is available for sharing and reuse in a model of another sort. Consequently, the use of integration ontologies forms a critical part of the overall PSL-based architecture for knowledge sharing among process-oriented applications.

6 CONCLUSION

This paper described a process-oriented theoretical framework and architecture (APMS) for facilitating enterprise integration through knowledge sharing. The role of the PSL in facilitating semantic interoperability between process-oriented applications was described. An example application scenario was used to illustrate the role of PSL for knowledge sharing. APMS provides important enabling technology for the management of information-integrated manufacturing enterprises.

7 REFERENCES

[1] Neches, R., Fikes, R., Finin, T., Gruber, T., Patil, R., Senator T., and Swartout, W., (1991), "Enabling Technology for Knowledge Sharing," *AI Magazine* 12(3) 36-56.

[2] Lee, J., Grunninger, M., Jin, Y., Malone, T., Tate, A., Yost, G. (1996). "The Process Interchange Format (PIF) and Framework Version 1.1," MIT Center for Coordination Science, Working Paper #194.

Modelling Requirements for Self-integrating Manufacturing Systems

Peter Denno
National Institute of Standards and Technology, U.S.A.
Email: peter.denno@nist.gov

Keywords Self-integrating systems, Information Infrastructure, Self-describing systems

Abstract This paper discusses modelling requirements to support self-integrating manufacturing systems. Integration requires reconciliation of mismatch in various forms of context. Four forms of mismatch involving software components and aspect of the enterprise are identified. Although the paper stops short of describing a complete solution, aspects of the solution, as they relate to the four forms of mismatch, are discussed.

INTRODUCTION

Self-integrating manufacturing systems (SIMS) is a new area of research, so it is likely that there are various visions for what it should be. Some goals that might be attributed to SIMS are: implementation of virtual enterprises, supply chain integration, and plug-and-play factories [8]. At an extreme end of the spectrum, dynamic relationships between functional entities may be formed through resource brokering. From the standpoint of this paper, however, relationships are relatively static and the goal is simply to reduce the cost of business process re-engineering through automation of much of the re-engineering activity. This goal is, of course, not unique and even some aspects of the solution are not new [10], [2], [8]. For example, an essential aspect the SMART [2] architecture is self-descriptive software components, that is, components that provide a profile of their functional and interface characteristics. These descriptions may be used by higher-order systems to generate a middleware solution to inter-component communication.

This paper assumes that the integration process of self-integrating systems will rely on a comprehensive enterprise model. An instantiation of the model provides a context under which a self-describing component may be integrated into the enterprise's manufacturing operations. The processes and components involved with integration, as envisioned, are illustrated in *Figure 1*. As this figure suggests, the principle processes involved are (1) *coherence matching*, that is, identifying the relationship between the activity performed by the candidate software and the goals of the enterprise and (2) *technical reconciliation*, that is, overcoming differences in protocol, semantics and information structure. Both of these processes rely on the enterprise model, which describes the goals, processes, organizational structure, resources, interfaces and information entities that are subject to modification in a business process re-engineering project.

Enterprise models and modelling disciplines such as identified by CIMOSA [1] and ARIS [11] identify much of necessary content and methodology. However, in representing the relationship among goals, agents and interfaces, self-integrating systems have an additional requirement: it is necessary to make apparent the mismatch of technology, semantics and function that may occur between the existing and future work process. Whether integration of process is feasible, and whether the technical and semantic differences can be bridged, depends on the characterization of mismatch that can be made between the existing and future enterprise model instances. The remainder of the paper provides a characterization of the forms of mismatch after considering the aspects of communication and context that gives rise to the mismatch.

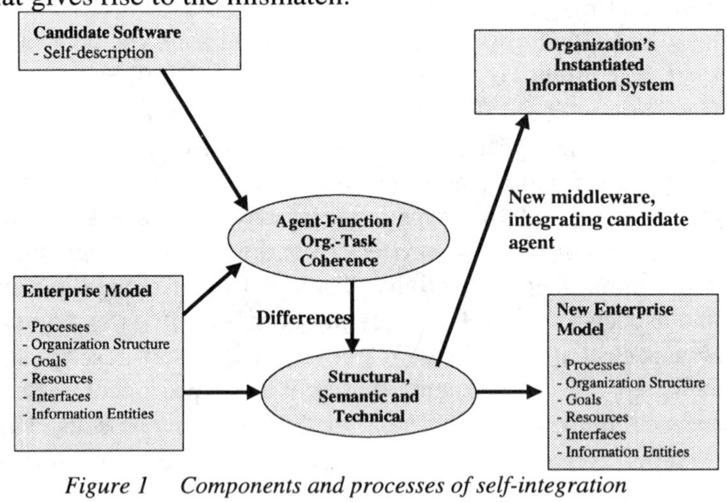

Figure 1 Components and processes of self-integration

INTEGRATION, COMMUNICATION AND CONTEXT

In this paper, *integration* refers to enabling communication among humans and software entities for the purpose of achieving a goal. That is, integration is the enabling of useful communication. *Self-integration* is integration where human intervention is not required.

The principle challenge of self-integrating systems is that of getting components to communicate with each other. Broadly speaking, an agent communicates with another for the purpose of influencing the behavior of the recipient. There is no other purpose. With respect to communication among humans in particular, the speaker, in choosing how to word his utterance, must have in mind the recipient's background knowledge and goals as well as the environment (time and place) in which the utterance is to occur. *Context* is a word for those things (background knowledge, recipient's goals and environment) the speaker must consider in designing an utterance.

The problem of communication is the problem of attending to context well enough to design an utterance that achieves, through the recipient, a desired behavior.

This definition of integration, as well as the use of "communication" applies as well to human agents as it does to software components. This is not an accidental coincidence but rather a design requirement for self-integrating systems. Because the mix of automation versus human endeavor varies from organization to organization (and component integration often changes the mix) it is valuable to have a language to model these notions that is neutral to whether a human or an automaton is committed to the task.

The problem and challenge of self-integrating manufacturing systems centers on the notion of context and the fact that the various, disparate components that might play a role in the manufacturing enterprise were designed in relative isolation; each built with a certain, idiosyncratic context in mind. Whereas the notions of communication and integration are neutral to whether human or automatons are involved, the notion of context is decidedly not. The context in which software agents communicate is narrow and artificial, not the 'lived experience' that we must grapple with as communicating humans. Context is, for the creator of artificial agents, a design problem; in communicating software agents, the crucial design question is that of how agents establish a context for communication. Assuming that the designer of the agent did not expect to exploit some unusual and extraordinary ability the agent might possess to behave usefully in ambiguous circumstances, the narrowness and artificiality of context among communicating agents is a fortunate accident. This narrowness facilitates integration.

CONTEXT AND FORMS OF MISMATCH

Persons managing a manufacturing enterprise have the responsibility of specifying the basic architecture of the enterprise. Broadly speaking, this entails identifying goals, defining processes, committing resources, and selecting systems for automating tasks. Through time, commitments of this sort are revised to better address the changing business environment. With respect to production equipment and software, management is faced with the choice of (1) building its own solution to address the unique requirements that management has imposed by other commitments, or (2) applying an existing solution such as a commercial off-the-shelf (COTS) software package. The latter is generally the less expensive choice, at least in terms of initial cost. It is, increasingly, the more common choice.

When a choice is made to accomplish a task with commercial off-the-shelf software, the enterprise is immediately faced with the difficulty that the selected software was quite likely designed in relative isolation from this enterprise's commitments; that is, it was designed with a different context in mind. The nature of the mismatch of context and how it may be expressed is the subject of the remainder of this paper.

FORMS OF MISMATCH

Generally speaking, *mismatch* (to resemble or harmonize unsuitably or inaccurately) applies to many forms of mismanagement: hiring the wrong persons, providing him with conflicting or ambiguous goals and so on. However, here the concern is specifically with the mismatch between components of the enterprise's architecture (component to component) and mismatch between component function and agency goal (function to goal). Four forms of contextual mismatch will be discussed: semantic, structural, functional and technological mismatch.

Semantic mismatch

A principle challenge of system integration is that of identifying sense differences that occur between information elements at points of communication. We envision that the self-description of an agent to be integrated would include assertions regarding how its information elements relate to terms defined by the enterprise model. In defining these terms the enterprise model is arbitrarily authoritative (it must make *some* ontological commitment) as there is no ultimate ground on which industrial terms might be founded.

An important design concern with regards to the definition of industrial terms in the enterprise model (and even speaking more generally about language [9]) is that meaning is a less useful notion than equivalence of meaning. That is, it is more useful to know whether my notion of "part" is equivalent to your notion of "detail", your notion of "assembly" etc., than it is to know what "part" is, in some fundamental sense.

The self-describing component may define relationship among the information elements of its interface and the terms defined by the enterprise model using sense relationships [3] and other characterizations of semantic proximity [12]. These are discussed below.

- *Synonymy* – having the same or nearly same meaning, substitutable
- *Hyponymy* – taxonomy, 'is kind of' relationship, relating narrower terms to broader terms
- *Antonymy* – having opposite meaning. Crystal [3] identifies three forms of antonymy that may be useful here: (1) *gradable*, permitting the expression of degree, such as good/bad/very good; (2) *non-gradable*, not permitting degree of contrast, such as single/married; and (3) *converse*, two-way contrasts that are inter-dependent, such as buy/sell
- *Incompatibility* – terms that are "mutually exclusive members of the same superordinate category" [3] such as red/green

Wiederhold [12] cites distinctions that follow from the encoding of information:

- *Value semantics* – the choice of threshold values where a quantity takes on another meaning
- *Temporal grain* – the quantum of time on which the value is based (*e.g.* versus monthly salary)
- *Abstraction grain* – a quantum (by some less fundamental metric than time) on which the value is based (*e.g.* production by lot versus monthly production)

The sort of distinction listed here are commonly found in an ontology (a set of terms and definitions in a formal logical language which connect the terms). We envision that the enterprise model would embody an ontology for the purpose of relating the information elements of the self-describing component to the terms of the enterprise model. The ontology can serve to define the nature of the narrowing between hyponymous terms (*e.g.* through membership predicates or other forms of constraint) for example.

Structural mismatch

Structural mismatch refers to differences in encoding or organization between information entities whose semantics are similar. Resolving structural mismatch presupposes that semantic equivalence has been recognized. In fact, structural mismatch often cannot be separated from semantic mismatch. Value semantics, temporal grain and abstraction grain can be viewed as structural problems arising out of the encoding of information, as they are semantic discriminators. The important distinction between semantic and structural mismatch is the manner in which it is addressed in the integration process: structural mismatch may be resolved through the generated middleware integrating the component or through an information mapping engine such as Express-X [4].

Functional mismatch

Functional mismatch refers to the degree to which the behaviour of an agent fails to achieve an expected effect (an enterprise goal, presumably). Functional mismatch is a concern first during the coherence matching process where an assessment of the feasibility of using the agent can be made, and second during technical reconciliation, where useful behaviours and results can be isolated from useless ones. ('Useful' and 'useless' are assessments made relative to the business context).

Concepts of function often rely on more general concepts of business practice and business resources. As described above, we envision that these later concepts and terminology would be addressed in the enterprise model. Therefore the modelling of concepts of function are tightly coupled with modelling of general enterprise terminology used in reconciling semantic mismatch. We envision that much the same relationship would exist between the component to be integrated and the enterprise model as it does in semantic reconciliation, the enterprise model defines terms of function by which the self-describing component describes itself.

Technical mismatch

Technical mismatch refers to differences in the software technology under which components provide interfaces (*e.g.* CORBA, COM, message queues). Components of the enterprise's information infrastructure might communicate through a framework (an architecture where components share a communication technology) or communication channels may be heterogeneous. General consensus in middleware technology is unlikely in the near future. For these reasons, self-integration will often involve bridging technical dissimilarity. The enterprise model, therefore should possess the ability to recognize the nature of the mismatch. Middleware technology can be categorized roughly as based either on message passing or remote procedure calls [5]. Subtle differences between technologies exists and must be made apparent. The development of an ontology of software technology is perhaps the least mature of requirements enabling self-integrating systems.

CONCLUSION

Self-integration, like business process re-engineering through traditional means, requires reconciliation of differences in semantics, function, structure and technology between the component to be integrated and the enterprise's existing information infrastructure. The complete context of the integration, as is understood by business- and technically-minded analysts undertaking a business process re-engineering project, involves aspect which are not traditionally subject to enterprise modelling. These include models of middleware technology, enterprise goals and component (or agency) function. Development of these aspects of the enterprise model defines a complete context for communication between components of the enterprise, enabling self-integration.

Commercial equipment and materials are identified in order to adequately specify certain procedures. In no case does such identification imply recommendation or endorsement by the National Institute of Standards and Technology, nor does it imply that the materials or equipment identified are necessarily the best available for the purpose.

REFERENCES

[1] AMICE, (editors) (1991). *CIMOSA: Open System Architecture for CIM* Springer-Verlag, 2nd edition.

[2] Barry, J., et al. (1998) *NIIIP-SMART: An Investigation of Distributed Object Approaches to Suport MES Development and Deployment in a Virtual Enterprise*. In The Proceedings of The Second International Enterprise Distributed Computing Workshop (EDOC98), 2-5 Nov.

[3] Crystal, D. (1997) *The Cambridge Encyclopedia of Language* Cambridge University Press.

[4] Express-X (2000) Product data representation and exchange — Description methods: Part 14: The Express-X Language Reference Manual. International Organization for Standards, Committee Draft.

[5] Grasso, M., P. (2000). Getting the Message. *Application Development Trends*, 101communications, LLC, August.

[6] Hunt, V., D. (1996). Process Mapping : How to Reengineer Your Business Processes, John Wiley & Sons.

[7] Liebowitz, J. and Khosrowpour M. (1997). Cases on Information Technology Management in Modern Organizations, Idea Group Publishing.
[8] NEMI, *NEMI Homepage*, http://www.nemi.org/
[9] Quine, Q. V. (1980) From a Logical Point of View : Nine Logico-Philosophical Essays, Harvard University Press.
[10] RosettaNet, *RosettaNet Homepage,* http://www.rosettanet.org/
[11] Scheer, A.. W. (1998) *Business Process Engineering* Reference Models for Industrial Enterprises, Springer-Verlag, Study edition.
[12] Wiederhold, G. (1992) Mediators in the Architecture of Future Information Systems, *IEEE Computer*, March, 38-49.

PART FOUR

Supply Chain Management

Ergonomic Concerns in Enterprise Resource Planning (ERP) Systems and Its Implementations

Ram R. Bishu
Department of Industrial and Management Systems Engineering, University of Nebraska-Lincoln
Rbishu@engvms.unl.edu

Brian M. Kleiner
Department of Industrial Engineering, Virginia Tech
Bkleiner@vt.edu

Colin G. Drury
Department of Industrial Engineering, University at Buffalo
Drury@buffalo.edu

Keywords ERP systems, Implementation, Ergonomics

Abstract Enterprise Resource Planning (ERP) systems promise to improve the overall effectiveness of organizations through integration of all the functionalities within the organizations. Further, within the context of "managing the supply chain", ERP systems promise to include even more coverage, in essence automating the entire Chain. This is achieved through all encompassing software. Over the last four years, success stories of ERP implementations have been few and far in between. Besides high initial start-up costs and high implementation costs, the implementation process have been problematic due to lack of due consideration to the 'human' component. Consequently, many companies have minimized their losses by abandoning their projects mid-course. This paper takes the perspective that ERP is in desperate need of ergonomic research, design and implementation to minimize the financial and human costs currently being experienced.

INTRODUCTION

Enterprise Resource Planning (ERP) systems are commercial software packages that promise seamless integration of all the information flowing through a company. It attempts to integrate all departments and functions across a company onto a single computer system that can serve all those different departments' particular needs (Davenport, 1998). ERP systems have evolved from earlier production planning and control approaches, such as Materials Requirements Planning (MRP) and Manufacturing Resource Planning (MRP II) systems. These helped plan and schedule production by expanding a product order into a bill of materials, and scheduling ordering and manufacture of components and sub-assemblies. The newer ERP systems integrate a number of additional functions, such as product design, sales and distribution, order entry, billing, and even human resource functions. In other words ERP system integrates all the functions in an organization, at all business units, across all global regions. Major Contemporary ERP vendors are SAP, BAAN, PEOPLESOFT, and ORACLE. Each of these ERP systems has evolved from their respective specialized software programs. For example SAP evolved out of an integrated finance and cost module, while PEOPLESOFT evolved out of a specialized human resources module. The best known among these is SAP [Taylor, 199X].

ERP systems typically have three layers of architecture (Figure 1). At the top layer is the user interface containing all the application modules. The system software is the middle layer, while the database is located at the bottom. These systems are transparent to the underlying database. The application modules for SAP include human resources (HR), quality management (QM), materials management (MM), production planning (PP), sales and distribution (SD), and plant maintenance (PM).

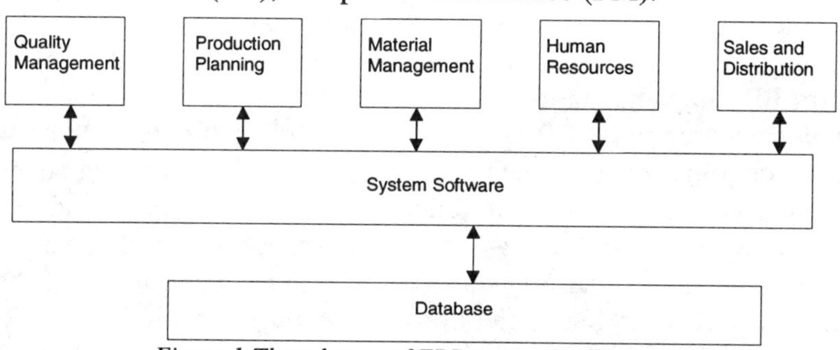

Figure 1:Three layers of ERP systems architecture

ERP

Three major reasons can be provided for the need of such systems. First, is the global nature of business itself in the 21st Century. Businesses have gone global to make use of differences in economy and cost structure among nations. For example, the trend in manufacturing is to move from first world countries to second and third world nations. Global production and global distribution implies a clear understanding of the various ethnographies, cultural differences, regulatory differences and other such differences. There is certainly a need for a global information system. Such a system would manage all these differences in a timely and reliable fashion and provide solutions in real time. Second, is the proliferation of the internet, intranet and extranet across the globe. A few years ago, fax and telephones were the primary mode of communications among facilities located in different parts of the globe. With the internet, the communications environment has changed. Intranets and extranets have provided the needed security in communications. Hence, an organization headquartered in, Detroit, USA, can have transactions immediately at its plant in, San Paulo, Brazil. Intranets and extranets provide both the media and the correct amount of security for plant-to-plant transactions within a business. Third, software is needed which would integrate all functions within a unit, all units within plant and all plants within an organization. ERP systems are company-wide software systems that integrate all the functions within an organization. ERP combines all of them into a single, integrated software program that runs off a single database so that the various departments can more easily share information and communicate with each other. However, the integration can have tremendous effect if the software is installed correctly. The ERP suppliers, such as SAP, BAAN, Oracle and others have shown that such integration can be the basis for large cost savings in many different organizations from manufacturing, through service and maintenance companies, to government departments. Indeed, cost savings driven by customer demand are typically cited as the main reason for ERP implementation.

Implementations of ERP systems are struggling throughout the world. They take too long, cost too much and fail to deliver the promised benefits of competitive advantage and cost reduction (Buckhout, Frey and Nemee, 1999). In a study to look at the effectiveness of implementation in companies with more than 500 million sales, it has been reported that the average cost overrun is 175%; the average schedule overrun was 230%, and the average slide in functional improvements was an astonishing 59% deficit (Buckhout, Frey and Nemee, 1999). By its nature, software that promises to integrate all the functions within a unit and all the units within an organization will be complex. In addition, ERP software developers claim that the software, with all its suite of functional modules, is sufficiently generic to be adaptable to the needs of a number of diverse organizations. The only thing more complex than the software itself is the process of its implementation. The steps in the implementation are development of an implementation project plan, determination of existing processes ("as is status"), gap analysis and business process reengineering, develop a conceptual design, design interfaces and enhancements, detailed design and systems set up, preparations for going live, productive operation and post implementation support. The scales of business process reengineering and customization tasks involved in implementation process are reported as one of the major reasons for ERP dissatisfaction. It has also been reported that the ratio of implementation costs to software costs is in the range of 5:1 to 7:1 (Scheer *et al*, 2000).

Development of an implementation project plan would include laying out a time line, resource plan and other project details for the implementation process. Determination of the existing process can range in complexity depending on the organization size and type.

While the basic functionality such as material management, MRP, Quality control and others are common across all organizations, their depth and breadth will vary considerably between organizations. Given that the basic system software is generic to be used across all industries make the implementation process much more complex. The main process in implementation is configuring the software to suit the client organization. There are two issues of concern here. Organizations need to make the correct strategic choices to configure the system, and need to address the ergonomics of implementation process. It is not difficult to visualize implementation difficulties if one realizes that the same generic software is being configured for two large organizations that have little, except size, in common as in Monsanto, the chemical giant and John Deere a farm equipment giant.

The costs of ERP implementation are high, as would be expected for such a pervasive system. For large companies, costs often exceed $100 million, with installation costs typically over twice software licensing costs (Hughes, 1999). Installation costs are so high because in most cases the ERP software demands that particular business practices be changed before the software can be installed. In addition, the migration to ERP can take several years.

If costs are high, so too are potential benefits. The software suppliers and customers claim annual savings of 10% to 50% of installation costs, and payback periods as low as two year. Major savings come from elimination of legacy software systems, reduction in software maintenance costs, and adoption of the business practices necessary to run the software.

IMPLEMENTATION OF A TYPICAL ERP SYSTEM

Phase 1: Definition Phase

In this phase the organization's objectives for using ERP system are defined, as are its current processes. An initial comparison is made between existing processes and ERP system's functionalities. The project team is assembled and its training needs are detailed. Typically the end product at this juncture is a workable project plan. Integration is the keyword in ERP systems. They operate on a hierarchical level with the highest level defining the organization as a whole and the next level defining the different functional divisions or plants. Each of the plants may be in different geographic regions in a global economy. The complete system environment with its respective settings is decided in this phase.

The scoping phase is an important part of the implementation process. A complete list of existing processes are developed and compared against existing functionality of the system. The comparisons often result in one of the three options: (1) either the system functionality matches completely, in which case prototype solution development is the next stage; (2) there is a gap which may result in system modification (very rare); or (3) there may be a gap that may result in business process reengineering. In short this is the 'problem definition and solution development phase' of the implementation process. Templates for design solutions, interfaces, and enhancements to be developed are defined. The conceptual design solution thus created is tested for completeness and functionality.

Phase 2: Detailed Design and System Set-Up

This is one of the critical phases of implementation. In the previous phase the design solution was conceived. In this phase the actual design is performed initially on some trial (or training) system. The global system parameters are defined. Functional designs are configured. Configuration is setting all the definable parameters in the software to meet clients' needs. Often the implementation process spins out of control here, because the company has not made the required strategic choices here, or the software needs major modification to meet the clients' needs. Master data sets are decided and entered. Data transfer from legacy systems are detailed here. Interfaces are programmed, as are enhancements, and modifications. Design decisions on gaps between available functionality of ERP systems and required functionality of clients are made in this phase. Reporting strategy decisions are also made. Report needs are established, formats designed and programmed. Archiving concerns and change management concerns are also addressed at this stage. Finally the prototype design solution, one that has been constructed on a training (often called SANDBOX) environment is reviewed with users, validated and tested.

Phase 3: Preparations for Going Live

Typically ERP systems implementation processes proceed from one environment to others, before finally going live. For example, the first construction of the design solution will be performed in a 'free-for-all-play-with-it-as-you-wish' environment. Subsequently, the development will be done in a more restricted sense with some real time data in a cleaner environment (often called PRISTINE environment). Validation, testing and quality checks are then performed in this environment. Once it is ensured that the system works, construction of the solution begins in what is called the DEVELOPMENT environment. This is the equivalent of a robust working model that is ready to be used in the shop floor, in a manufacturing sense. This system has all the master data transported from legacy systems. The next step is creation of a "Go-Live Plan".

Production systems, as they are called, are the final versions of the ERP systems that are to be used eventually by users. The configuration of the production system, procurement of the production system and creation of user documents are performed at this stage. It is relevant to note that the system, in all its totality, is transported electronically from the development to production environment. The development version is an actual working model. Future changes will be made first to the development system, verified for accuracy and working and then transported to the production system. Training users is an important step in this phase. It includes creation, preparation and delivery of the training program. The last function to be performed in this stage is design and creation of user documentation.

Phase 4: Productive Operation

This is typically a post implementation operation. It involves providing support to users and provision of help desk etc. In many ERP implementations there is significant "rework" due to necessary plant-specific customization or because the vendors do not adequately train users, to effectively use the software.

THE PROBLEMS IN IMPLEMENTATION

As stated earlier in this paper, implementations of ERP systems are struggling throughout the world. They take too long, cost too much and fail to deliver the promised benefits of competitive advantage and cost reduction. Why is this so? Some of the problems lie with the enormous technical challenges of rolling out enterprise systems, which are large complex pieces of software. Business problems, i.e., lack of corporate ability to take tough business decisions needed for implementation are the other reasons quoted (Davenport, 1998). Finally implementation processes are complex and need careful management attention, planning and resources. Companies need to establish guidelines for involvement-or 'rules of engagement' for different levels in the organization so those team members get to participate in an effective way.

RESEARCH AGENDA FOR ERGONOMICS

There are a number of ergonomic research issues with the design and implementation of ERP systems. However, the importance of the ergonomist's role has not been realized either by the ERP community or addressed by the ergonomic community. There are three main related pitfalls in system design; (1) technology–driven design; (2) leftover approach to function and task allocation and (3) failure to attend to sociotechnical system characteristics (Hendrick and Kleiner, 2000). ERP evolution has exhibited all three pitfalls. Clearly, the software advances coupled with aggressive marketing tactics have dominated the system design process. Essentially, an implicit design goal has been to maximize the level of automation. If a function could be automated, it was done. This has lead to a leftover role for the human decision-maker. In many cases, for example, inadequacies in the software resulted in a human endeavor to program patches or additional programs. What is needed is a deeper understanding of the socio-technical characteristics of the supply chain and an answer to the question, "what *should* be automated". This has also been raised by Taylor (199X), who suggests using a socio-technical systems approach to ERP implementation.

In a macroergonomic sense there are mismatches among roles of members on the implementation team as well. From a user viewpoint, user training warrants attention and finally, from a human-computer interaction standpoint the system itself is a maze of screens navigable through a series of hierarchical menus. While some attention has been given to the actual screens, cognitive engineering considerations in the design of the information architecture underlying these screens appear to be missing. Overall the system and the implementation process itself appear to be technology-centered rather than human- centered.

Implementation issues

Conceptually, the implementation process follows an interaction model comprising three teams: (1) a team of functional experts (users), (2) a team of implementation experts (primary implementing vendor) and (3) a team of

ERP experts (Figure 2). Further the implementers themselves are from two groups-staff from a primary implementing vendor organization and consultants.

The ERP professionals are experts in software. Usually they belong to the 'product support group' or the 'after sales service group' of the organizations that make and sell the ERP software. They interact with the users who are the proverbial 'bill payers' in the whole scheme. Their interaction with the users is at the highest level during the implementation stage. They have very good knowledge of the software, but lack functional knowledge or implementation knowledge. For example they are adept at answering 'how an aspect of the code works', or ' what does it take to modify a code'. They cannot answer the questions on 'what does it take to make the code work for this client', or 'to what extent the code (or the process needs to be modified to make it work'.

The implementers are the experts who are supposed to know the complete functionality of the software. They are supposed to understand the required functionality of the clients' organization from the users and establish the compatibility between the system functionality and client functionality. This is established in one of the three ways-either there is a perfect match between the two functionalities, or the system is modified (this route is least preferred by the implementers), or the business process in reengineered (much to the chagrin of the users). This team is supposed to be the 'know-it-all-experts' in the implementation process and are the most expensive. They interact with all the other players in the team. The only useful methodology that combines these is sociotechnical systems. The users are the 'bill payers' and functional experts. They are supposed to define the required functionality to the implementers.

From an ergonomic viewpoint the main concern here is in the process of interaction itself. The team members have different expectations, different skill levels, different understanding of what the system capabilities are vis-à-vis required functionality and speak different languages (in a professional sense). This is not uncommon among other teams such as 'Quality improvement team' or 'Trouble shooting team'. However, these teams have a much narrower focus in problem area, problem definition, solution development and in solution implementation. ERP system implementations, however, are much larger in scope, cost and in all other aspects. Typically ergonomists specialize working on the interactions among the human-machine-task-environment system framework (Wilson, 1999). Extension of that expertise into the realm of information systems should be natural. Ergonomists should investigate interaction processes in these large-scale implementations and develop the required paradigm. As reported by Wilson

(1999), good understanding of the interactions will go towards a better human centered implementation methods, instead of technology centered implementation methods.

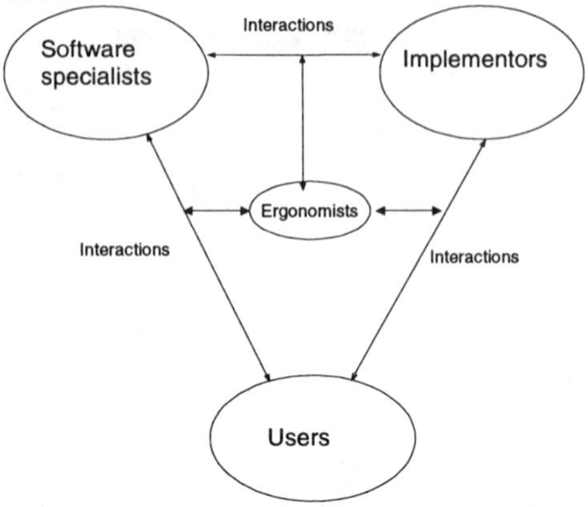

Figure 2: Interactions among the team members

User training concerns

Training is the next major area of concern in the implementation process. Typically in the implemented stage, a number of processes end up reengineered, to suit the system functionality. The user training are, unfortunately, not designed from a user need viewpoint, but are designed from a system viewpoint. This results in general frustration among all users. Further many vendors use third-party training organizations to support their software installations. To make things worse, typically trainers are often novice users themselves of the ERP package. Most of the training -is off-the-shelf as well which means trainers have not taken the time to understand the particular contextual environment in which implementation will occur.

CONCLUSION

In summary, ERP systems have not lived up to their initial promise and hype. The main reasons are costly implementation, and a technology driven paradigm. A company was quoted as saying, "and we really believe this is an organizational and a reengineering revolution and not a software exercise" (Hughes, 1999).

The best practices defined by the software are typically lean solutions based on a business process reengineering approach (Hammer and Champy, 1993). As a final quote, a spokesperson for another aerospace manufacturer plans to "lean those rascals down" before an ERP computer system is installed (Hughes, 1999). Such an approach means that the major costs are in the organizational side of implementation, rather than in the costs of computer hardware and software. In the current climate of mergers between large organizations, legacy processes as well as legacy information technology need to be changed anyway, so that the ERP solution is seen as particularly timely.

However lean solutions have been questioned in the ergonomics literature because of their implications for Tayloristic job design, and their tight coupling of previously buffered processes. In addition, there has been a long and generally successful effort at implementing information technology from a human factors and sociotechnical perspective (Eason, 1988). Neither of these considerations has been found in the current ERP literature.

REFERENCES

Buckhout, S., Frey, E., and Nemee, J. (1999). Making ERP succeed: Turning fear into promise, *IEEE Engineering Management Review,* Fall 1999, 116-123.

Davenport, T. H. (1998). Putting the enterprise into the enterprise system, *In Harvard Business Review* 76 (4), 121-131.

Eason, K (1988). Information technology and organizational change. London: Taylor and Francis.

Hammer, M. and Champy, J. (1993). Reengineering the corporation. New York: Harper Business

Hendrick, H, and Kleiner, B (2000). Macroergonomics: An Introduction to Work System Design.

Hughes, D (1999) Aerospace invest billing in ERP software, processes. *Aviation Week and Space Technology*, Jan 25,1999, 68-76.

Scheer, A.W., and Habermann, F. (2000). Making ERP a success, *Communications of ACM,* April 2000, 43(4), 57.

Taylor, J.C. (199X), Participative design: Linking BPR and SAP with an STS approach, Journal of Organizational Change Management, 11(3), pp

Wilson, J. (1999), Interactions as the focus for human centered systems, Proceedings of the Conference on TQM and Human Factors, Linkoping, pp 35-43.

Low-cost System for Supply Chain Management

K. Fürst, T. Schmidt
Institute of Flexible Automation – University of Technology Vienna, Austria
Em: kf@infa.tuwien.ac.at

Keywords Internet, EDI, XML, Supply Chain Management

Abstract In this paper, a complete concept for Internet Electronic Data Interchange (EDI) - a well-known buzzword in the area of logistics and supply chain management to enable the automation of the interactions between companies and their partners - using XML (eXtensible Markup Language) will be proposed. This approach is based on Internet and XML, because the implementation of traditional EDI (e.g. EDIFACT, ANSI X.12) is mostly too costly for small and medium sized enterprises, which want to integrate their suppliers and customers in a supply chain. The paper will also present the results of the implementation of a prototype for such a system, which has been developed for an industrial partner to improve the current situation of parts delivery. The main functions of this system are an early warning system to detect problems during the parts delivery process as early as possible, and a transport following system to pursuit the transportation.

1 INTRODUCTION

We write the paper to disseminate the results of our industrial project with an engine manufacturer. The aim of the project was the development of a whole new concept for Supply Chain Management to integrate their suppliers and carriers, and the realization of an easy prototype using the latest Internet technologies to evaluate the usefulness of such a system.

The starting point of the project was a problem of our industrial partner (BMW Motors Steyr in Austria). During the process of parts delivery the company has to collect manually a lot of data either from the suppliers or from the carriers. Because this method of data collection is very time-consuming, it happens only if some problems during this process are becoming visible. The system, which was developed with the engine manufacturer, has to fulfil two main functions:

- The monitoring of the suppliers and the carriers for increasing the on-time departure performance of parts delivery.
- The responsibility of parts delivery should be transferred to the suppliers. Therefore it is necessary to expose internal logistics data to the suppliers.

In both tasks there is need for a transparent, flexible and low-cost data exchange method, based on the latest Internet technology. The proposed system makes it possible that all relevant data for controlling the process of parts delivery are available automatically and electronically for the industrial partner, all suppliers and carriers of this factory, at any time.

2 SUPPLY CHAIN MANAGEMENT

Supply Chain Management (SCM) deals with the management of material, information, and financial flows in a network consisting of vendors, manufacturers, distributors and customers. Managing flows in this network is a major challenge due to the complexity (in space and time) of the network, the proliferation of products (often with short life cycles) that flow through this network, and the presence of multiple decision makers who each own and operate a piece of this network and optimise a private objective function. [11]

Material usually flows from a supplier to a buyer while information and financial flows are bi-directional.

Supply chains today are increasingly depending on effective and efficient information exchanges between the value chain partners. Over the last decade, practices such as just in time management, quick response manufacturing and lean production systems have required coordinated and reliable information exchanges between trading partners. Massive investments in information technology have been made by manufacturers, suppliers and logistics providers with the hope of achieving successful just in time implementation in their supply chains. [11]

In just in time supply chains, the coordination of material flows through information exchange is becoming increasingly important. With technologies such as Electronic Data Interchange (EDI) with suppliers, the rapid transmission of information with trading partners allows manufacturers of efficiently manage logistics activities. [11]

Figure 1 shows the market growth of Supply Chain Management.

Figure 1 SCM – Market Growth (Source: IDC 1999)

3 TRADITIONAL EDI

Over the past several decades' corporations have invested trillions of dollars in automating their internal processes. While this investment has yielded significant improvements in efficiency, that efficiency has been limited to internal processes. In effect, companies have created islands of automation, which are isolated from their vendors and customers. The interaction between companies and their trading partners remains slow and inefficient, because it is based on manual processes. [6]

Electronic Data Interchange (EDI) has been heralded as the solution to this problem. EDI is defined as the exchange of data between heterogeneous systems to support transactions. This is not simply the exportation of data from one system to another, but the actual interaction between systems. Companies that have implemented EDI rave about the various benefits. In fact, these benefits can be expanded to a chain of suppliers. [6]

There is a significant gap between the business benefits described above and the actual implementation of EDI. This is because the actual implementation of EDI is difficult and costly to implement. More importantly, however, it requires a unique solution for each pair of trading partners. Many people falsely proclaimed the Internet as the solution to this problem. By implementing EDI over a single network, our problems would be solved. Unfortunately, a network with a common protocol is still only a partial solution. This is because the systems implemented in each company are based on different platforms, applications, data formats, protocols, schemas, business rules and more. Simply "connecting" these systems over the Internet does not solve the problem. [6]

Traditional EDI is based on fixed transaction sets. These transaction sets are defined by standards bodies such as the United Nations Standard Messages Directory for Electronic Data Interchange for Administration,

Commerce and Transport (EDIFACT), and the American National Standards Institute' s (ANSI) Accredited Standards Committee (ASC) X12 sub-group. Transaction sets define the fields, the order of these fields, and the length of the fields. Along with these transaction sets are business rules, which in the lexicon of the EDI folks are referred to as "implementation guidelines". [6]

4 SYSTEM-ARCHITECTURE

The system architecture in figure 2 is based on the Internet technology and the new Internet language XML. Because the internal systems of the suppliers and carriers are not equal, it is necessary to extract the relevant data for the new system from these systems and convert the data locally into the international standardized format XML to make it available for the application, the so-called Data-Extractor, which runs on a Web server. The Data-Extractor is responsible for the whole data transmission between the different systems and should handle the requests from the different users working with a standard Web browser. The firewall between the Intranet and the global Internet is necessary to guarantee a high secure-level for the sensible industrial data.

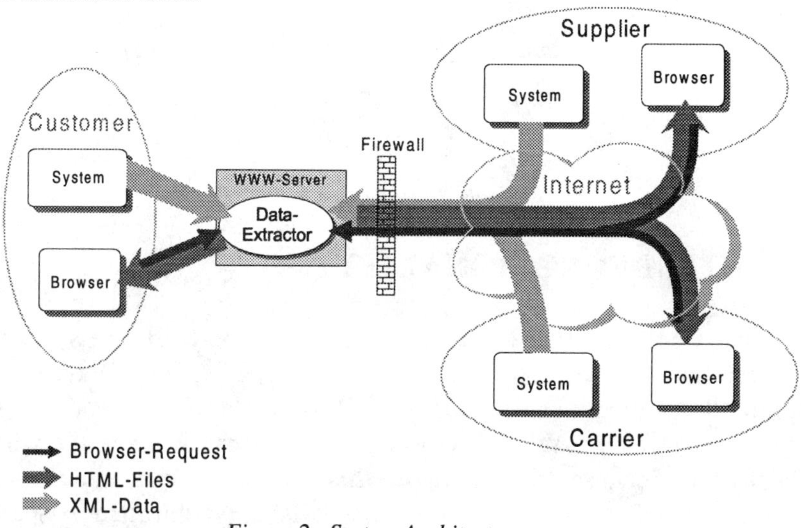

Figure 2 System Architecture

The advantages of this system architecture are:

- Platform independence of this system because the Data-Extractor is implemented with the programming language Java, exactly with Java-Servlets. The format XML is also platform independent.

- The users can work with a standard Web browser like Microsoft Explorer or Netscape Navigator.
- Using all advantages of the Internet technology like the availability of Internet or the low-cost data transmission over Internet.

The standard for Internet communication in the near future, XML (eXtensible Markup Language), was standardized in the year 1998 by the World Wide Web Consortium and is a subset of SGML (Standard Generalized Markup Language), see figure 3. SGML was standardized in the year 1986 by ISO (International Standards Organization), but it is too complicated and therefore not used in a broad range. With the easier to use language XML it is also possible to separate data and markup. XML is a universal data format that allows computers to store and transfer data that can be understood by any other computer system.

See [1, 2, 4, 6, 9] for more information about XML.

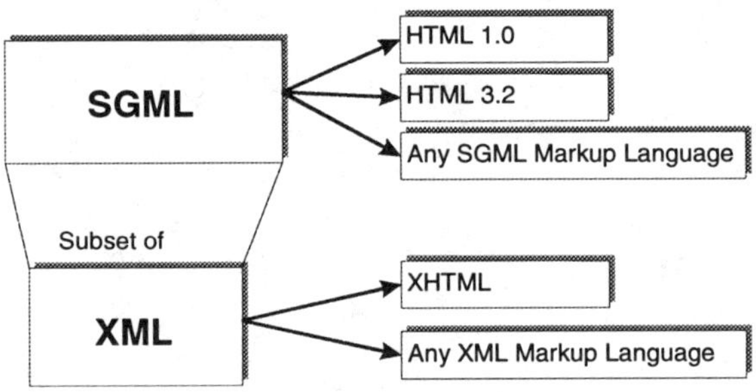

Figure 3 Difference XML - HTML

5 SYSTEM-FUNCTIONALITY

Using UML (Unified Modelling Language) the complete system model was designed. Figure 4 shows the simplified Use Case Diagram. In this diagram, the interactions between the following Use Cases are displayed:

- Warning System: With this system the user can manage the whole part delivery, because the output of this Use Case is a table, where the delivery and transport data of all parts are controlled based on the delivery order. If the data differs, the system reports this via a red sign.
- Transport Following: This Use Case reports the actual status of a part transport. The user gets all transport data from a part.
- Part Availability: The user gets information about the actual part stock at BMW and also the delivery and transport data of a part as table with time of delivery and amount.

- Calling Stock At Supplier: The user gets the actual stock at the supplier in form of a table with the part number, part name and amount.
- Report Part Delivery: It is necessary to report the system the incoming of a delivery (e. g. to update the delivery and transport data).
- Send Delivery Data: The supplier must send the delivery data to BMW.
- Get Delivery Order: The system calls the actual delivery order of a part.
- Get Delivery Data: Calling the actual delivery data of a part, which are stored locally in form of XML-files.
- Get Transport Data: The carrier calls the actual transport data of a part.
- Get Stock At BMW: The user gets the stock at BMW.

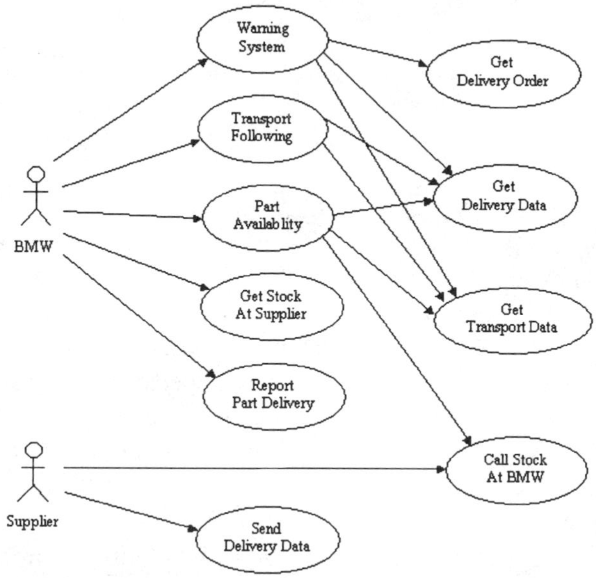

Figure 4 Use-Case-Diagram System-Functionality

6 IMPLEMENTATION AND TESTING

The whole application (Data-Extractor) is implemented in Java. Platform independence is the main goal of Java and it has been achieved. Because the application is running on a Web server, Java-Servlets, which realize the functionality of the Use Case Diagram (see figure 4), are used. In general the rise of server-side Java applications is one of the latest and most exciting trends in Java programming. A Java-Servlet is a small, pluggable extension to a server that enhances the functionality of the server and runs inside a Java Virtual Machine on the server, so unlike applets, they do not require support for Java in the Web browser.

Each servlet has the same life cycle (see figure 5): [12]

- A server loads and initializes the servlet
- The servlet handles zero or more client requests
- The server removes the servlet (some servers do this step only when they shut down)

See [3, 7] for more information about Java, and [8, 12] for more information about Java-Servlets.

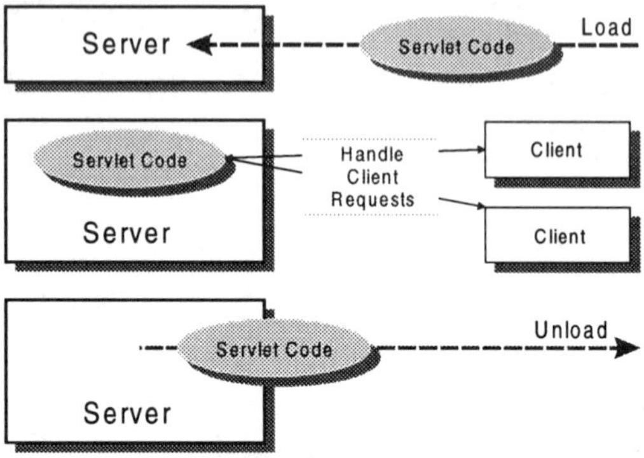

Figure 5 Java-Servlet – Life-Cycle

The Java-Servlets at the Web server handle the GET-Requests of the users (Customer, Suppliers) and work with the XML-Data. For these operations many software tools are developed since the standardization of XML. Under the developers are companies like JavaSoft, Microsoft, Oracle and IBM. This show how important the combination of Java/XML will be in the near future.

There are two major types of XML-APIs: [2, 10]

- A tree-based API compiles an XML document into an internal tree structure and allows an application to navigate that tree. The DOM (Document Object Model) is such an API, which is in the process of being standardized by W3C.
- An event-based API, on the other hand, reports parsing events (such as the start and end of elements) to an application through callbacks. The application implements handlers to deal with the various events. SAX (Simple API for XML) is such an API.

The servlets in the prototype use the DOM interface to operate with the XML data. After parsing the XML file (figure 6), you can step through the internal tree structure (figure 7) and save the necessary data into internal variables.

After working with this internal data (e.g. calculate the Warning System under using a special logic), the servlet generate a HTML-Output and send this to the client. The user can see this output using a normal Web browser like Microsoft Explorer or Netscape Navigator.

```
<SUPPLIER>
        <PART>
                <NUMBER>123456.00</NUMBER>
                ...
        </PART>
        ...
        ...
</SUPPLIER>
```

Figure 6 XML-File

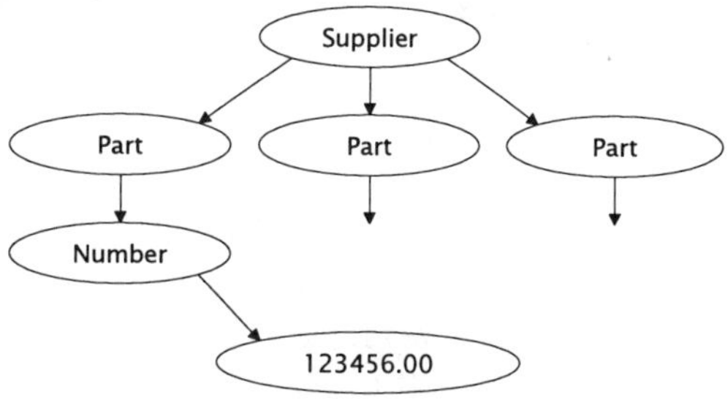

Figure 7 DOM

7 CONCLUSIONS

The results of testing the prototype give us the motivation to implement the next step of the project – the connections to the different in-house systems. Because the users of this prototype (supplier and carrier) didn't want to install additional XML-EDI software for prototype testing, we realized the necessary integration with HTML-forms, which will be filled out manually by the users. Only the interface to the internal SAP-System will be realized to work automatically. The realized prototype can work with dynamical, real-time data. After the prototype test-phase, our industrial partner can quantify the advantages of such a new, low-cost system for electronic data interchange and supply chain management using the latest Internet technologies XML and Java.

9 REFERENCES

[1] Behme H. (1999). Kunst der Stunde, *IX–Magazin für professionelle Informationstechnik*, 2/99, 36-41

[2] Chang D., Harkey D. (1998). Client/Server Data Access with Java and XML, Wiley Computer Publishing

[3] Flanagan D. (1996). Java in a nutshell, O'Reilly

[4] Goldfarb C.F., Prescod P. (1999). XML-Handbuch, Prentice-Hall

[5] Helios, http://www.wiinf.uni-wuerzburg.de/helios/startseite-main.htm (27.5.1999)

[6] Hogan M. (1998). XML and the Internet: Driving the future of EDI, POET-Software

[7] Huber G. (1996). Java – Die Referenz, Heise

[8] Hunter J. (1998). Java Servlet Programming, O'Reilly

[9] John V. (1999). XML – Weltsprache für das Internet, *OBJEKTspektrum*, 5/99, 22-28

[10] Middendorf S. (1999). XML und Java, *OBJEKTspektrum*, 5/99, 38-40

[11] Servlets-Tutorial, http://www.uni-klu.ac.at/unihome/header/tutorial/servlets/ (10.9.1999)

[12] Tayur S., Ganeshar R., Magazine M. (1999). Quantitative Models for Supply Chain Management, Kluwer Academic Publishers

A Booking Type Production System as a Collaboration Method for Virtual Enterprises

Y. Nishioka, Y. Fukuda
Hosei University, Japan
Em: nishioka@k.hosei.ac.jp

Y. Kamio
Toyo Engineering Corporation, Japan

K. Kawashima
Mitsui Engineering and Shipbuilding Co.Ltd., Japan

Keywords Advanced Planning and Scheduling, Reactive Scheduling, Supply Chain Planning, Reservation System, Virtual Enterprise, Internet

Abstract A booking-type production system (BTPS) has a coordination mechanism to synchronize manufacturing processes with customer requirements in the dynamic environment. This paper introduces BTPS, and extends it with Advanced Planning and Scheduling techniques. Some future capabilities of the system in Internet environment are also discussed.

1 INTRODUCTION

For some manufacturing companies, a booking-type production system (BTPS) can perform as an effective collaboration tool between a manufacturing division and a sales division. While a sales division knows customers' orders and puts them on an online booking chart, it is always available to promise because the booking chart describes the manufacturing division regarding the shop floor capacity and the relevant procurement situations. When BTPS was implemented in early 1990s in Japan, the booking charts were static so that there were some inefficient features. This paper deals with BTPS in dynamic environment. Furthermore, the target of the system is extended for a virtual enterprise that consists of different firms. The key technology for this purpose is re-calculating the booking chart using advanced planning and scheduling (APS) algorithms. In addition, web-based technologies such as XML/EDI are also important. This paper illustrates and discusses a framework of a collaboration architecture using BTPS and the relative Internet technologies for competitive virtual enterprises.

The remainder of this paper is organized as follows. In section 2, conventional BTPS is introduced, and then, section 3 shows an extended BTPS, that has an APS module and its capabilities. The main feature of the extended BTPS is dynamic arrangement of the booking chart. Section 4 explains this problem using a simple example. The proposed architecture of our BTPS is illustrated in section 5. Then, in section 6, an implementation example is addressed. Finally, section 7 concludes our approach as a collaboration architecture for virtual enterprises.

2 BOOKING-TYPE PRODUCTION SYSTEM

To synchronize manufacturing processes with customer's demands, BTPS uses an online booking-chart (OBC). The OBC is a user-friendly interface to reserve final products, which will be finished in the future. All sales persons can access it from different located offices. In early 1990s, some Japanese manufactures developed computer integrated manufacturing (CIM) systems successfully with BTPS. TOSHIBA Corporation, consumer electrical machine company, is the first to establish the system, in which final products on OBC are reserved [1,2]. Another industrial example can be shown by Toyoda Machine Works, Ltd., one-of-a-kind mother machine manufacturing company [3]. In this case, the target of reservation is not a final product but manufacturing resources such as machine-hours and person-hours.

Using BTPS, effective manufactures have some important advantages in the market places. First of all, customers can get a quick response of their order receipt with a concrete due date. Since the OBC shows availability to ship the required products, a sales person can enter a new order while he/she answers the due date. Secondly, from a shop floor manager's point of view, BTPS protects ineffective production orders. In other words, customers cannot select alterative dates or products that are not on the OBC. This contributes to increased productivity of the plant. Finally, long-term order reservation by customers contributes more precise demand forecasting, even if these orders might be cancelled in the future.

The business process in BTPS is similar to seat reservation systems in airline companies [4]. In accordance with the customer orders, sales persons reserve items on the OBC after a master production schedule (MPS) is fixed and a corresponding schedule is released. Up to the time a few days from the production execution date, orders are arriving one after another. When the reservation date is more than one month from the execution date, the sales person can cancel these orders with no charge. BTPS allows users to cancel their orders within one month if the modification of the order is less than fifteen percent. However, within five days, all of these modification or cancellation are denied. The business process above is a typical example, and there could be many variations for each company's characteristic.

3 SYSTEM EXTENTION WITH APS

Generally, production systems should follow each customer's order. To do this, production schedule is usually made after customer's orders arrive. In contrast, BTPS makes schedule before each order can be identified. This implies that, in BTPS, customers have to adjust to the production schedule, whereas conventional systems allow that the production schedule adjusts their customers.

Even if BPTS has many advantages in terms of synchronized manufacturing, this argument is true. Therefore, we try to modify BTPS to a more customer-oriented system. A solution to get this goal will be provided by advanced planning and scheduling (APS) technologies.

Scheduling systems were firstly developed for shop floor managers in deciding which order is started first and so on. Now, a successor of this kind of scheduling systems, so-called APS, has many capabilities. We can identify two important features for BTPS. One is rescheduling capability, which allows shop floor planners to revise the current schedule when some unexpected events occur. Another capability of APS is consideration of inventory level of raw materials, work-in-process inventories, and final products. By this function, APS can manage purchasing information, which the conventional MRP system used to provide.

Considering these capabilities, a benefit of BTPS with APS can be suggested in rearranging the OBC. There are several ways to rearrange the OBC. One is to rearrange regularly, for instance, every week. This is very simple to operate both for a production planner and a sales person. On the other hand, there are many instances of dynamic rearrangement, which is made anytime when sales persons or shop floor planners request.

In general, rearrangement should be done by combination of the two cases. However, the later case is very difficult to achieve without an integrated APS module.

4 DYNAMIC BOOKING-CHART REARRANGEMENT

The proposed system framework uses APS to rearrange the booking chart dynamically. This rearranging is affecting customer satisfaction in reserving their preference. If the OBC cannot be modified, many customers are forced to change their requirements when the OBC does not offer appropriate one. Some customer may regret their requirements and may change their mind of ordering. The rearrangement with APS preserves the OBC has most appropriate candidates for forthcoming orders. This is valuable especially in the case that a forecasting demand differs from the real trend of the orders. We found out that this modification increase customer satisfaction by numerical experiments [5].

Using a simple example, benefit of booking chart rearrangement is explained. Figure 1 shows bill of material and process (BOMP) structures of sample products. The product A is made by a process, which occupies machine X for 30 minutes per quantity. The figure also shows that, the process requires 2 of M1 and 10 of M2 as raw materials. The product B and the C are defined in the same way. In this case, the three products share the machine X and the material M1. This causes competition in BTPS in making OBC.

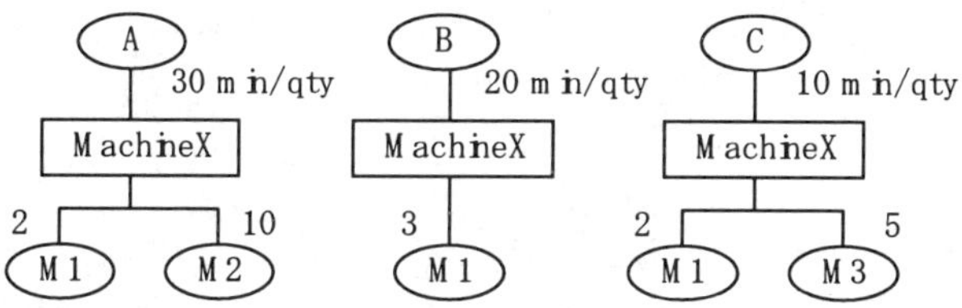

Figure 1: Bill of Material and Process (BOMP) of Sample Products

Table 1 Production availability calculation by resource and material usage.

Product	Machine X	M1	M2	M3	Minimum
A	4	10	3	---	3
B	6	6	---	---	6
C	12	10	---	12	10
Available	120	20	30	60	---

BTPS calculates each available number of the final products for OBC. The calculation procedure can be explained in Table 1. When the total availability is given for each resource and material, the number of each product is calculated for each column on the table. The total availability of each product is the minimum number of the corresponding row. The table 1 shows the results of the quantity to make A, B, and C are 3, 6, and 10 respectively.

Because there are resource and material confliction, the available numbers 3, 6 and 10 for the products are not realized at a same time. One feasible solution is 2, 2 and 2 for the product A, B and C, respectively. Obviously, if these feasible numbers are fixed on OBC, it restricts flexibility of the shop floor. BTPS with APS makes possible of dynamical rearrangement, so that, if there are particular demand such as 10 of the product C, these orders are accepted after rearranging OBC.

5 THE SYSTEM ARCHITECTURE

The system architecture of the proposed system can be shown in Figure 2. This system consists of different modules [5]. One typical module is the booking management module, which releases and revises the booking chart through the network. An order entry system in the booking management module and the OBC accessible from the different places perform cooperatively. These orders are divided into long-term orders and short-term orders, and then stored in the database.

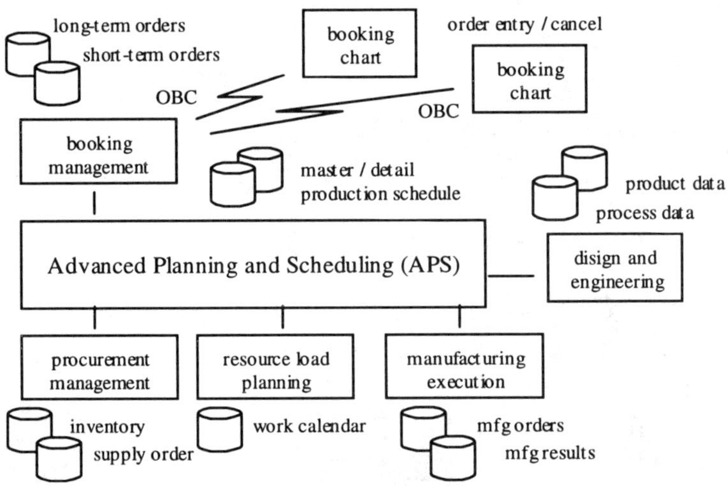

Figure 2: System framework of BTPS [5].

The APS module is located in the central part of the system to calculate an initial schedule using MPS. After that, the APS revises the schedule when predicting gaps on the OBC appear. The result of the scheduling is stored as a detail production schedule, and frequently used to rearrange the ineffective OBC. Around the APS module, there are several modules such as procurement management, resource load planning, manufacturing execution, and design and engineering modules. The advantages to integrate BTPS with APS come from these close connections of the modules.

In the current Internet infrastructure, BTPS should change its framework to more effective one. In this situation, sales persons are not only the persons who enter the orders. Various kinds of customers, who might be procurement staffs of a different company, can operate the BTPS through Internet. An example of use cases can be illustrated in Figure 3.

Regarding Internet, OBC should be upgraded to a web-based booking chart (WBC). Using a web browser with some security protection, WBC can be shared by all relevant customers without any investment. As shown in Figure 3, an enterprise needs its suppliers' BTPS to reserve some parts or materials as its planning/scheduling preconditions. Of course, these reservations might be changed. From supplier's point of view, they can prepare the future demand. And more importantly, it is clear that the business speed and occasion are obviously increased by the web-based BTPS.

A virtual enterprise, which consists of different companies, needs a collaboration method against their different interests. BTPS has a capability to coordinate stakeholders using a reservation system. Consequently, a dynamic supply chain will be successfully managed by the extended BTPS.

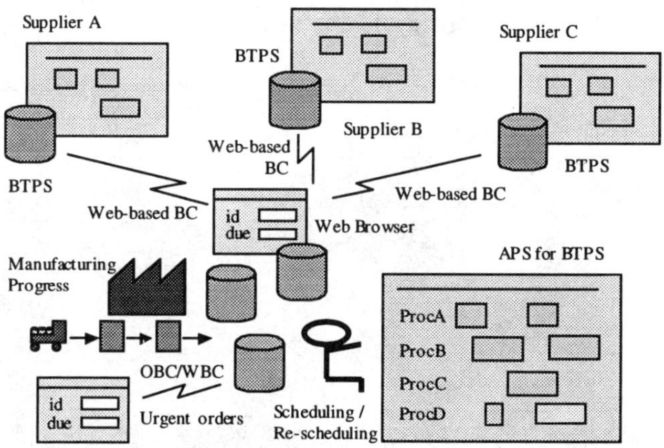

Figure 3: BTPS in Internet environment for VE

6 IMPLEMENTATION

We have an APS system named APSTOMIZER [6]. Our APS has a simulation based scheduling logic and penalty propagation re-scheduling logic with a flexible user interface. In order to deal with practical-level constraints, the system keeps tight relationship to Planning and Scheduling

Language (PSL), which might be a standard of a specification language in all kind of manufactures in Japan [7]. Using APSTOMIZER, we developed an extended BTPS in the Internet infrastructure. Figure 4 shows a Gantt chart of a typical schedule used in the BTPS. The sample manufacture has two product series A and B. Each product series has same instances of A1, A2, B1X, B1Y, B2X and B2Y. This scheduling result is used to make OBC/WBC information for the BTPS.

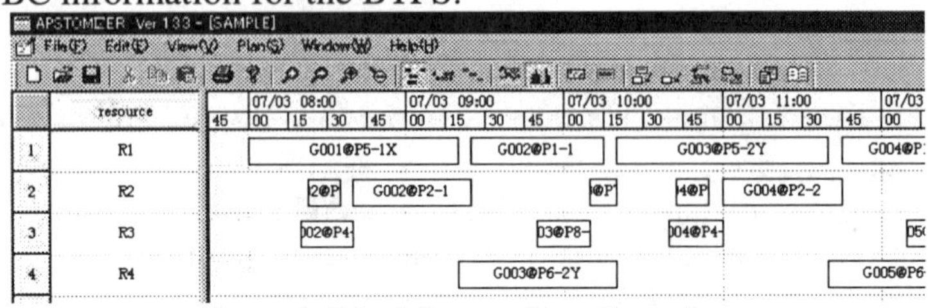

Figure 4: Scheduling result on Gantt chart

When the forecasted demand and the actual demand have a gap, and rearrangement of OBC/WBC is required, APSTOMIZER, first, removes all the tasks that are not reserved. Then, APSTOMIZER regenerates new orders according to an up-to-date forecasting. After that, APSTOMIZER reschedules the additional orders, regarding the pre-allocated orders. Finally, OBC/WBC is modified according to the new production schedule.

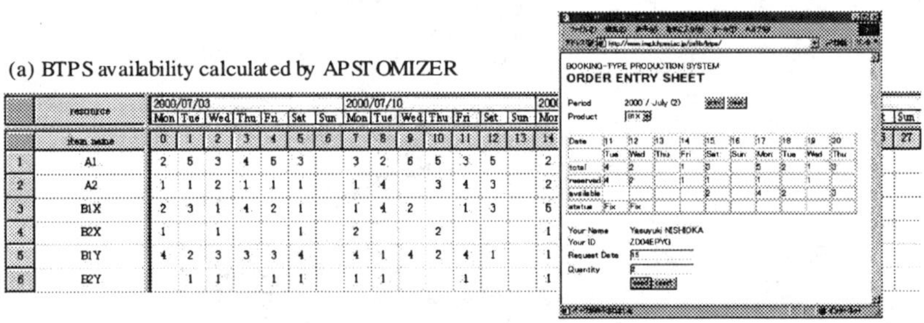

(a) BTPS availability calculated by APSTOMIZER

(b) Order entry page for BTPS on the Web

Figure 5: A table of booking chart and a web-based BC

Figure 5 (a) is a table of the available numbers on the booking chart. This also calculated by APSTOMIZER. In this chart, the time bucket is a day, so that, each day has the number of products that will be available to ship. Using this information, the system publishes OBC/WBC to all the sales persons or customers. If BTPS is used only inside of an enterprise, client and server type OBC might be developed on a local network. However, considering supply chain in a virtual enterprise, WBC is necessary. Figure 5

(b) is an example of web pages provided by the web-based BTPS. In our system, these contents of web-browsers and appropriate web-applications are developed using XML/EDI technologies on a World Wide Web (WWW) server.

7 CONCLUSION

A booking-type production system (BTPS) has coordination architecture between a sales division and a manufacturing division. This feature is also valuable between different enterprises, when the system is established on Internet technologies. Using APS capabilities, this paper shows a system framework by modifying a conventional BTPS, in order to make it more customer-oriented. While APS makes rearrangement of the booking chart dynamically, customer satisfaction is obviously increased. Furthermore, a web-based BTPS will provide an effective supply chain in virtual enterprises.

We developed a prototype system of BTPS with APS. The system can be extended for the Internet infrastructure. This feature has a huge impact on virtual enterprise collaboration, especially in the case that members of the VE don't have any close relationship in advance. Even if the user of web-based BTPS belongs to different organizations and different cultures, the system performs well according to the pre-agreed reservation rules.

For future works, we apply this framework to one-of-a-kind production on a virtual enterprise such as ship building manufacturers and a plant engineering companies. In this production, partnership between enterprises is much important, so that, a web-based BTPS will take an important role in the virtual enterprise. In addition, many research topics on BTPS such as price discount in reservation, and lot sizing for the booking chart are also our future target.

8 ACKNOWLEGEMENT

This research has been carried out under the GLOBEMEN project, one of the IMS International Programs. The authors would like to acknowledge all the members of GLOBEMEN project for variable support.

9 REFERENCES

[1] Matsuura J. (1990). CIM in assembly process of information processing machinery industry: A case of Toshiba Ome plant, *Automation*, **37**(8), 29-43 (in Japanese)

[2] Kuga T. (1994). Production control system for individual order: Load levelling by Toyoda's seta assignment system. Management Systems, **4**(1), 20-24 (in Japanese)
[3] Tamura T. and Fujita S. (1995). Designing customer oriented production planning system (COPPS), *Int. J. of Production economics*, **41**, 377-385.
[4] Tamura T., Fujita S. and Kuga T. (1997). The concept and practice of the production seat system, *Managerial and decision economics*, **18**, 101-112.
[5] Nishioka Y. (2000). Coordination Architecture for Supply Chain Management Using A Booking-Type Production System, *POM 2000 Proceedings,* Sevila Spain.
[6] Nishioka Y. (2000). Production Scheduling Library, http://www.img.k.hosei.ac.jp/pslib/
[7] Nishioka Y. (1999). An Generalized Language of Production Scheduling Problems as a Basis of Supply Chain Management, *Proceedings of APIEMS'99*, 681-684.

Collaborative Design Procedure for Supply Chain Process Integration Using UML

Francis E. Plonka, Mohammed S. Ahmed
Wayne State University, Detroit, USA
fplonka@mie.eng.wayne.edu, msa1@roogna.eng.wayne.edu

Dan Carnahan
Rockwell Automation, USA
Dlcarnahan@Ra.Rockwell.Com

Keywords UML, Collaborative Design, Supply Chain, Process Integration

Abstract The auto industry has embraced B2B e-commerce as a way of assembling their suppliers at one place so as to purchase parts and process equipment over the Internet. Suppliers who want to be successful in this environment will need to use Internet capabilities not only to quote projects but also to collaborate with other suppliers to execute the design and build activities at the quoted price, delivery and functionality requirements at an appropriate quality level. This will require interaction between product, process and tool engineers. Collaboration and interactions do not happen by accident. They require a common vocabulary and language along with explicit planning and management framework. The Unified Modelling Language (UML), is on its way to being adopted by the computer industry as the standard modelling language. This paper suggests that this object modelling language can be one component of this communication framework. In order to successfully use the Internet to accomplish these tasks, a framework/ procedure is needed to capture needed interaction and assignment of responsibilities that become the basis of the bid. These agreements can also be used to manage the interactions needed during design, construction and tryout stages. The proposed procedure utilizes a fundamental objectives tree to define the voice of the customer, the IDEF0 model language to define the life cycle of the design and construction activities and the UML to document and guide the needed interactions and communication. The process can also be used to manage the program in order to achieve the performance objectives. This paper describes the procedure and provides a case study for the design of a press material handling system. The equipment manufacturer relies on the die process and design functions as well as press manufacturer to interact at appropriate time to ensure successful operation of equipment.

1. INTRODUCTION

E-commerce is about building better relationships between customers, producers, suppliers and making communications that flow more efficiently while lowering cost. E-commerce can be divided into three categories; business-to-business, business-to-consumer and consumer-to-consumer [1]. When discussing e-commerce in a business-to-business framework, it is often associated with supply chain management and e-commerce's ability to improve the supply chain process.

Supply chain collaboration occurs when two or more companies share the responsibilities of exchanging common planning, management, execution and performance measurement information. Collaborative relationships transform how information is shared between companies and drive changes to the underlying business process, [2].

Supply chain strategies are undergoing tremendous changes. Outsourcing and partnering with other local or global enterprises are becoming more commonplace as companies seek to share the burden of demand for more complex products and more responsive services, [2]. These changes require greater collaboration and interaction with individual at diverse locations.

This paper will outline a framework for documenting the interactions and communication that take place between participants in the supply chain during a collaborative design process. Before we explain the methodology we will define three concepts that are being used in the paper. A more complete description of these concepts can be found in an article by the authors, [3].

1.1 Function Objective Tree

The Function Objective Tree (FOT) helps us to identify the Voice of the Customer that processes in the enterprise work to satisfy.

1.2 IDEF0

IDEF0 is a structured design and analysis technique that has been used extensively for the modelling of industrial and computer systems. The methodology allows the user to describe functions and activities and to decompose them into sub functions. The suite of IDEF tools, IDEFx, are used to model "as is" and "to be" scenarios, to create data models, to develop sequence logic and to simulate the throughput performance of the system.

1.3 UML

UML provides a visual representation using object-modelling notation. The modelling of system elements using UML is recommended by the ISO TC 184/SC5/WG5 technical committee in their draft standard for Application Framework Requirements for Industrial Automation.

2 METHODOLOGY - SUPPLY CHAIN COLLABORATIVE DESIGN PROCESS

In this paper we are outlining a methodology that provides a systematic means of capturing and documenting the interaction that takes place in a supply chain collaborative design process. The aim is to outline the process and make it web-enabled so as to bring about a new dimension to e-commerce supply chain management. We name the methodology the Supply Chain Collaborative Design Process, (SCCD).

The SCCD identifies customer requirements by using the FOT. Once objectives are outlined and linked with the activities, IDEF0 is used to describe the life cycle of the design process. IDEF0 is then used to model each activity that makes up the process. The actors that are associated with the mechanisms are identified on the IDEF0 model. These actors and their interactions with the mechanisms are modeled as a Use Case model in UML. The final step is to develop various UML diagrams to document the process and also to outline the interactions that are needed.

3 CASE STUDY: PRESS MATERIAL HANDLING SYSTEM

The case study presented here is an extension of the research conducted under a NIST ATP grant to the Auto Body Consortium entitled the Near Zero Stamping, (NZS), Project. One NZS task outlined a process that documented the interaction and collaboration in developing a material handling system to move parts between stations in a progressive die. Next, we apply our methodology to capture and document the interaction that takes place between various parties while developing a material handling system for fabricating stamped sheet metal automotive parts.

In order to apply the SCCD process, we outline five basic customer requirements for an auto body and its associated parts. These primary customer requirements are set so as to achieve a near zero dimensional variation on the panels. These primary customer requirements are Precision Fit, Function, Finish, Fast (to market) and High Value to the customer. For each requirement a detailed FOT is developed to identify the role of the original equipment manufacturer (OEM) and their process equipment suppliers as well as the collaboration need to achieve it. Figure 4 shows the FOT for Precision Fit and Fast (to market).

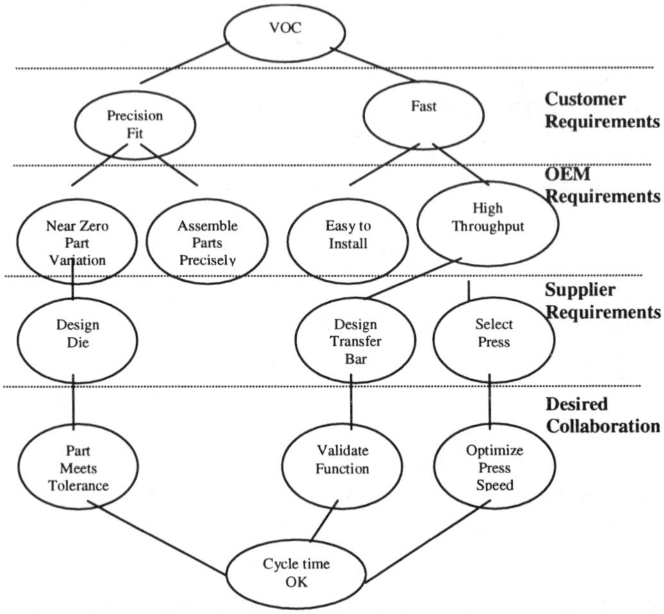

Figure 4 Function Objective Tree defining collaborative requirements

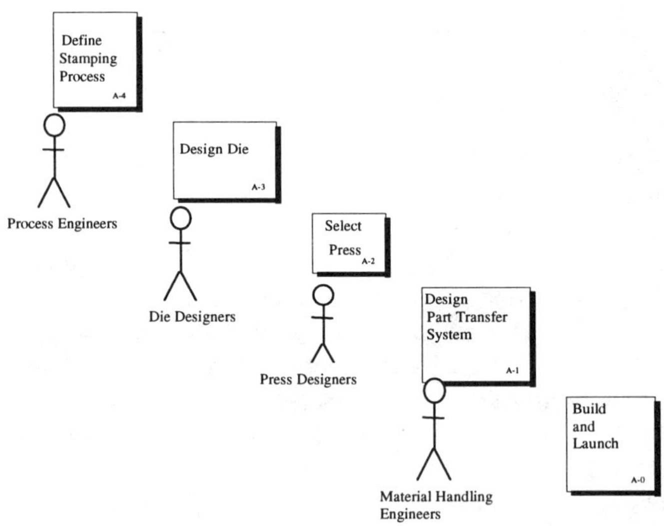

Figure 5 IDEF0 Process for Press Material Handling System Design

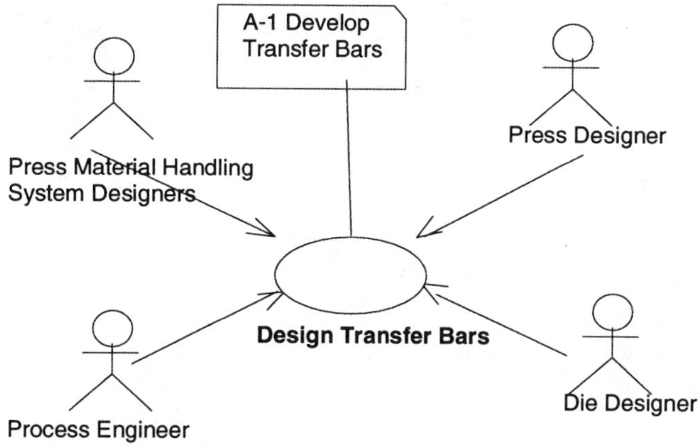

Figure 6 Use Case Diagram for "Develop Transfer Bar" activity

Using these requirements, activities are defined and a process can be outlined as shown in Figure 5. The IDEF0 model was developed based on interviews conducted with the following actors that make up the process. The interviews captured, in detail, the process of designing the press material handling system in a collaborative environment.

The next step is to develop a use case diagram using the UML as shown in figure 6. The diagram shows the interaction between use cases and their associated actors for the activity "Develop Transfer Bar" which is a sub-activity of "Design Part Transfer System".

For each Use Case, a sequence diagram, is developed that outlines the sequence of operations for the system or the flow of functionality through the use case. A UML sequence diagram, depicted in Figure 7 shows how various actors exchange information to iterate through the design process. The diagram also shows the sequence of each transfer of information as well as, the process step performed by each actor. It is a complete representation of the collaboration that must take place defining when something must be done and when information must be transferred to support decision making for the task.

Focusing on one of the actors, we develop specific design requirements for various sub-systems based on the process needs. The requirements for each sub-system, such as the material handling, define the specific values for the various attributes for the sub-systems entities or "objects". Figure 8 illustrates the objects that define the task of the process engineer using a collaboration diagram. This provides a definition of his or her responsibilities in executing the process elements. The UML provides a rich set of diagrams that can adequately document individual roles and responsibilities as well as the interactions with others in carrying out each others job functions that ultimately satisfy the customer.

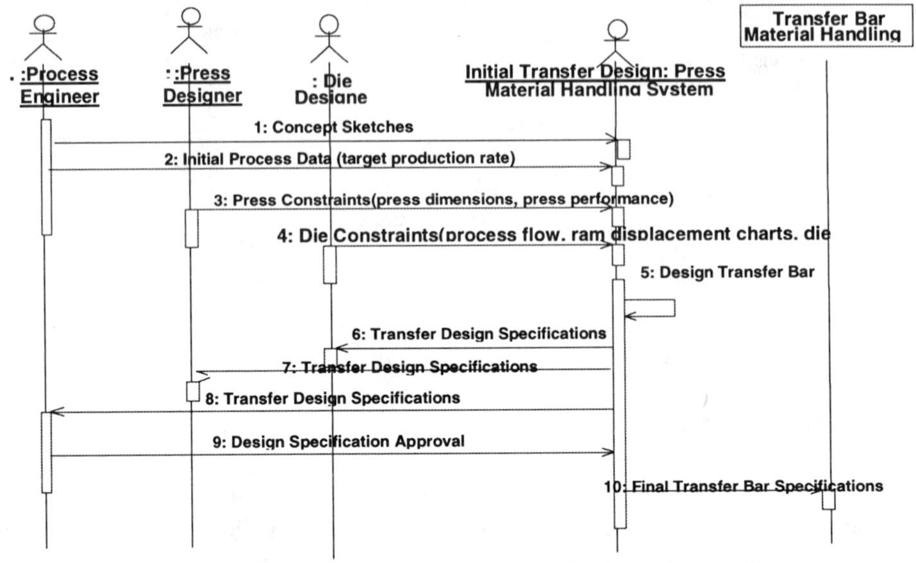

Figure 7 Sequence Diagram for "Develop Transfer Bars"

A collaboration diagram view of the requirements, as shown in figure 8, provides the first step in abstraction of the requirements to the various sub-system objects. While this figure is fairly simple, as more sub-system interaction become involved, the collaboration diagram provides a good view of the all of the interactions with the sub-systems.

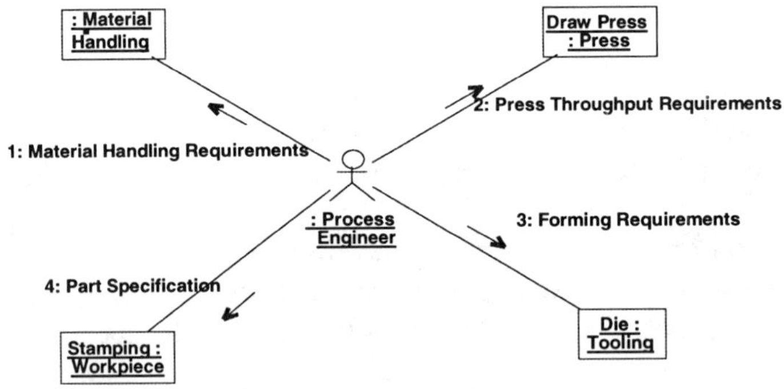

Figure 8 Collaboration Diagram for the Process Engineer

Figure 9 shows the roles and responsibilites of a process engineer using a sequence diagram. This provides a representation of when things must be done relative to each other. An indication of the time it takes to perform each specification and decision is shown as a small rectangular box on each vertical line.

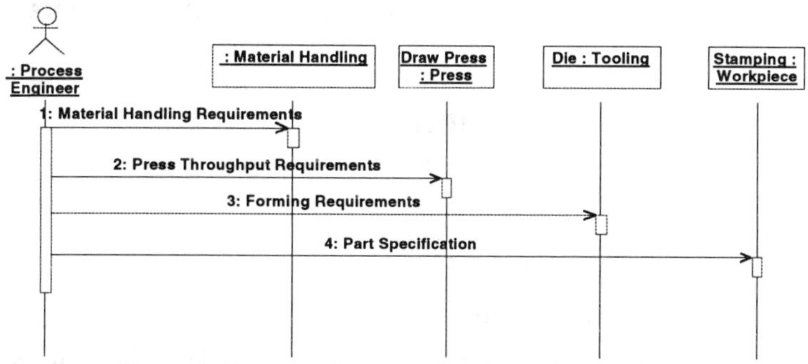

Figure 9 Sequence Diagram for the Process Engineer

Figure 10 is a representation of a class diagram that defines the end state desciption of each design artifact. It describes the information that must be defined and stored. The modelling language facilitates the design of screens, data bases and enabling programs that can support the web based collaboration. The class diagram depicts the various attributes and operations that the sub-system designer defines. By interacting with the users of the sub-systems (in this case study, the process engineer), the values of the attributes, and the timing of the operations can be specified based on the system design requirements. For instance, in the case of the material handling sub-system, the motion profile could be specified indicating a "cam profile," or it could be more generally specified using a mathematical expressions.

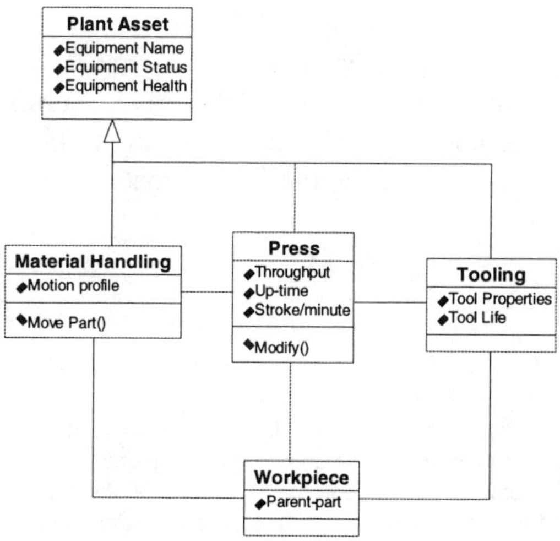

Figure 10 Class Diagram defining the end state description of each design artifact.

Based on the UML diagram we can either implement the documented interaction or develop web-based tools that can facilitate collaboration. The web-based tool, Computer Aided Decision Support System, CADSS, was developed under the Near Zero Stamping Program that provides an effective interaction and communication between the Automobile Body supply chain participants.

4 CONCLUSION

The proposed methodology based on the Voice of the Customer, IDEF0 and UML permits the documentation of a process to be followed during the collaboration between individuals in a supply chain. The application of the procedure discussed in this paper involved the design, construction and installation of manufacturing equipment. A case study was provided regarding the design and construction of a material handling system for stamped sheet metal parts fabricated using a progressive die in a high-speed press.

The methodology provides several advantages for users. The first advantage permits the documentation of the interaction between individuals that can be used for collaborative design and training purposes. This demonstrates the power of the UML language in the training application domain. A second advantage of the methodology is that it provides interactive collaboration requirements for system developers of web based tools. A web-based system based on this methodology will enable and facilitate needed collaborative outcomes. This methodology can be easily extended and elaborated to develop web based tools to execute the desired collaborative interactions. A third advantage is that interactive information flows among the various actors in a collaborative environment are clearly delineated. The information exchange requirements provide the basis for developing simulations that can be used to validate the collaborative design process.

5 REFERENCES

[1] Robert Aaron; IEEE Communications Magazine September 1999, vol. 37, No 9; Electronic Commerce; Enablers and Implementation.

[2] Sherman Bai; NSF workshop on Supply Chain Management in Electronic Commerce, Department of Industrial and Systems Engineering University of Florida.

[3] Plonka, F. E., Carnahan, D., Ahmed, M. S.; Voice of the Customer ; An elaborative design technique for design synthesis and generation; Computer Technology Solutions for the Manufacturing Enterprise, SAE September 1999.

Agent-based Architecture for Flexible Lean Cell Design, Analysis and Evaluation

T.E. Potok, N.D. Ivezic, N.F. Samatova
Collaborative Technologies Research Center, Computer Science and Mathematics Division, Oak Ridge National Laboratory, Oak Ridge,TN, USA
E-mail: potokte@ornl.gov

Keywords supply chain information infrastructure, lean cell formation, agent-based systems

Abstract We present an agent architecture in support of lean operations across manufacturing supply chains. Our architecture addresses the issue of heterogeneous information management, the need for analyzing mixed manufacturing approaches, and efficiency and accuracy tradeoffs of analysis methods. We propose five distinct layers to address these requirements: an XML/RDF/DAML Information Access Layer, a Resource Vector Representation Layer, an Analysis Layer, a Coordination Layer, and a Visualization Layer. Our successful results in developing agent systems in support of lean manufacturing cell and supply chain management validate our layered approach.

1 INTRODUCTION

In today's manufacturing environment, the concept of producing the right product, at the right time, for the right price is a driving goal. To achieve this goal, a manufacturer must have great insight into the supply chain that feeds the manufacturing process. Many would advocate the use of MRP or ERP software systems to achieve this objective. The drawbacks with this approach range from an enormous initial investment to regimenting dated practices. Once such a system is put into place, it can stifle the move to new manufacturing technologies. We believe that the flexibility of agent-based systems provides a great benefit over existing manufacturing control software.

We propose a flexible agent framework that is based on rich representation of manufacturing resources, and advanced analysis techniques. With this framework, we can address two key problems within the manufacturing supply chain. The first being the formation of lean or efficient manufacturing cells, and the second is the analysis of material and part flows through the supply chain. The first problem looks at optimal ways of organizing a manufacturing environment, or multiple environments. The second looks at how to understand and analyze the flow of activities that are required to build products.

We have recently developed and deployed two agent-based systems for optimizing manufacturing processes and analyzing supply chain logistics [1-3] that provide great insight into these problems. From our experience with building agent-based systems for supply chain analysis, we believe that agent technology is ideally suited to be the foundation for a supply chain information infrastructure. In this paper we present an adaptive framework and some preliminary findings for such systems.

2 BACKGROUND

The idea of software agents interacting with a manufacturing environment is not a new one. The typical approach reported in the literature is to have agents represent various components of manufacturing, such as parts, machines, cells, and people [4]. Through negotiations among these agents, work can be efficiently scheduled based on a variety of possible constraints. However, the scheduling of work is not performed from a systematic engineering approach, but from efficiencies gained from advanced negotiations techniques [5]. These negotiation-based techniques have proven very effective. However, rather than generalizing these concepts over the supply chain, we believe there may be a more effective approach. Rather than letting agents simulate real-world human interactions, we propose that agents operate based on manufacturing efficiency principles. By doing so, these agents can help build, analyze, optimize manufacturing cells and supply chains from engineering principles. We develop our framework based on two aspects of supply chain analysis, 1) the formation of lean or efficient manufacturing cells, and 2) the analysis of material and part flows through the supply chain.

3 AGENT FRAMEWORK

Several unique challenges arise when building agent-based systems.

1. The agent framework must be able to access and understand heterogeneous and distributed information within the supply chain system. Typically, each supplier will represent and store information in

different ways than other suppliers. This information may include metrics on part inventories, machine utilizations, or production goals.

2. The agent framework must be flexible enough to be able to adapt to different manufacturing methods. One supplier may produce products using traditional batch production methods, with a focus on machine utilization, while another supplier may use agile production processes, where a part value stream analysis is important.
3. The agents must be able to adapt to different analysis methods. There are times when a quick, ballpark estimate may be required, and other times when an in-depth analysis is necessary. The system must be capable of producing both types of analysis.

The typical way of addressing these requirements is to employ an MRP or ERP system that integrates a vast amount of information, and makes this information available to a broader range of participants. However, unless the suppliers in the supply chain adopt the same system, the information will be available only from within a single manufacturing environment, and not the entire supply chain. Other approaches, such as distributed databases, or distributed object brokers can provide a solution, however, a great deal of specialty software must be written and maintained.

A software agent approach provides a number of attractive features. One is the flexibility of communication among agents. Given a common framework, agents throughout the supply chain can communicate to determine the best course of action. Another is the distributed nature of agents, where the information can literally be anywhere.

Looking at the above requirements in terms of agent frameworks provides a more complete justification for a supply chain information framework. Much has been written about the first requirement. The challenge of this requirement is quite pervasive, and unfortunately, current technology is not able to provide much help.

In order for an agent to be able to gather and understand manufacturing information, the agent must have a priori knowledge of the information, such as how the information is organized, what it means, how it is formatted, and what units it is in. Our proposal is to build an encapsulation layer from the current and emerging standards: XML, RDF, and DAML. The Extensible Markup Language (XML) is the universal format for structured documents and data on the Web; The Resource Description Framework (RDF) is a foundation for processing metadata; it provides interoperability between applications that exchange machine-understandable information on the Web; DARPA Agent Markup Language (DAML) is an emerging facility built on top of RDF to enable creation of a Semantic Web and fully interoperable agents.

The second requirement is critical to the manufacturing environment. If the representation of information is limited by the current manufacturing technology, then changing to a new technology will require a dramatic change to the existing information infrastructure as well. The resources of the manufacturing environment must be represented in a way so that improvement in the future will not require changes to the resource representation. This representation must be flexible and capture essential information about a resource, and critical relationships among resources. We believe that a vector-based representation for manufacturing resources would be ideal for the resource description, cell analysis, and visualization.

The third requirement is to be able to direct the level of analysis needed. The general impact of a change for forecasting purposes will require much different analysis than that needed for short-term production planning. This requirement calls for the user to be able to choose between the duration of analysis versus the accuracy of the analysis. Again, a key aspect of this challenge is to represent this information in such a way that various levels and types of analysis can be performed.

The key to our framework is how the supply chain is represented, and how agents are used to analyze this information. We propose five layers as shown in Figure 1: 1) an XML/RDF/DAML access layer that provides a means of searching, accessing, and interpreting manufacturing information. 2) a resource vector representation that relates resources, such as parts and machines, so that each process step for a part is represented as a dimension in space. 3) an analysis layer where this part/machine information can be clustered or grouped to form cells, 4) a coordination layer to enhance agent communication; and 5) a visualization layer where the results of this analysis can be displayed.

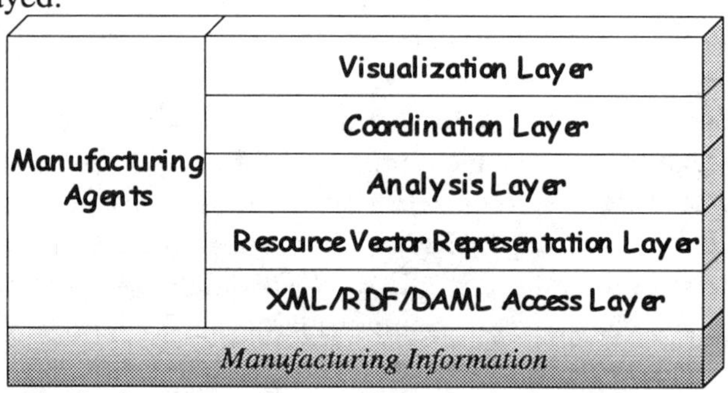

Figure 1 The five layer agent framework for supply chain information analysis.

3.1 XML/RDF/DAML Access Layer

This layer provides access and definition to the information available from the manufacturing environment. The information can be gathered, converted or transformed to feed into the resource representation layer.

We have used the extended markup language (XML) as a means of describing and integrating distributed information. XML can provide an encapsulation layer that provides the ability to translate and manipulate representations of information. This XML representation can form a basis for an ontology for a broader range of agents to access and understand the information sources. Furthermore, this XML representation can be used as a method to define uniform syntax of a language for communication among agents.

To achieve a full ontological foundation of communicating agents, additional components of the layer are needed. We have experimented with RDF, a simple, but extensible and potentially powerful representational layer delivered on top of XML. In addition, we have begun to investigate the impact of DAML to provide much more expressive semantics on top of the RDF layer using formal logic to express concepts, relationships, rules, and functions. The XML/RDF/DAML progression of increasingly expressive languages allows one to capture the meaning of manufacturing concepts for their use by higher layers that perform manufacturing-specific activities.

3.2 Resource Vector Representation Layer

The second layer calls for a solid representational structure for a manufacturing environment. These systems are typically represented by traditional means, such as a bill of material, or a part-machine matrix. These are usually one or two-dimensional structures, and fairly simple to manipulate. Unfortunately, they are not well suited for representing other factors, such as the direction of the flow of a part through a cell, or the similarity of resources. We believe that with a stronger representational scheme, the agents will be able to provide better information and improved analysis.

We propose a representational structure called a *Vector Space Model (VSM)*. This model helps us to capture all of the relevant features of a resource in relationship to other resources. The basic premise in the VSM is that manufacturing resources, such as machines and parts are represented as elements of a multi-dimensional vector space. Specifically, parts and machines are all vectors in the vector space. The mathematical framework is composed of a k-dimensional vector space and standard linear algebra operations on vectors. In general, the use of a vector space model requires the specification of several aspects such as the set of basis vectors, the dimensionality of the space, an interpretation of the values for vector coordinates, and a similarity measure or distance between two vectors.

One of the main virtues of the VSM is the ease with which individual vectors can be modified, making it possible to adapt resource vectors to a dynamic environment without reconsidering the entire data set. Moreover, the vector representational structure provides the desirable flexibility to the Analysis and Visualization Layers.

3.3 Analysis Layer

Once we have a strong representation of the manufacturing information, we then are able to quickly analyze the information, and make recommendations. There are two key ways of analyzing the VSM representation. The first is by grouping vectors that are similar into clusters. These clusters can be parts that are manufactured in a common way, machine capabilities, or personnel skills. The other way to analyze this information is by looking at the projection of the multidimensional vectors onto two or three dimensions. This provides a meaningful way of viewing information.

The Vector Perturbation Approach [6] integrated into the Analysis Layer is capable of directing the level of analysis needed. Different aspects specific to a manufacturing environment can be emphasized or obscured due to the variations of the vector perturbation parameter. For instance, it can provide a desirable compromise between the ease of resource assignment and the increase in production throughput by taking advantage of interleaving various parts through a common set of sequenced operations.

This method was applied to three large sets of military aircraft part data. The results, though preliminary, are very promising. Our conclusion is that the VSM representational model coupled with the Vector Perturbation Approach are well suited to address the supply chain information analysis problem, and can be used to enhance the efficiency and utilization of most supply chains.

3.4 Coordination Layer

The coordination layer supports interactions within multi-agent systems. We have experimented with skeleton-based approaches such as described in [7]. This approach captures constraints on, and specifications for agent behaviour, specifically, within the context of an agent's participation in dialogs with other agents. Such dialogs may arise through coordination with other agents and collaborative problem solving, to name a few.

The basis for an agent behavior description using skeletons is messages sent or received by the agent. Each message has a corresponding event specified by a skeleton. A skeleton is a directed, possibly cyclic, graph that consists of states (nodes) and events (links). Each graph has a unique start node and one or more end nodes. Each event (link) represents a transition from its start state to its end state. Each state and event has an associated name, action, and type. The actions associated with states and events are executables. The types of states and events are defined based on the semantics of the intended use of skeletons. Temporal logics can be defined over the skeletons to capture time-dependency among events and states in groups of agents.

3.5 Visualization Layer

The visualization layer must be very versatile to represent different types of information in an appropriate way for a human user. We have developed several components of this layer. First, we developed a capability to visualize statistical properties of a distribution of resources or their measurements over some manufacturing space. Second, we developed a capability to render results of statistical clustering methods that we used extensively within some part/resource grouping algorithms. Finally, we developed animation component to visualize interactions/communications among agents, movements of materials and products, and the evolution of important manufacturing concepts such as critical path analysis.

5 SUMMARY

An agent-based architecture has been proposed for supply chain information analysis based on manufacturing efficiency principles. Such an architecture facilitates heterogeneous and distributed information processing, flexibility, and adaptability. Flexibility is facilitated through a solid representational structure for a manufacturing environment. Adaptation is supported through the layered structure of the architecture and through multi-level analysis mechanisms.

The architecture is built on the information integration and agents communication standards. The proposed architecture also addresses other specific requirements for next generation manufacturing systems, including scalability, interoperability, cooperation, and visualization.

The experimental results from applying the architecture to the formation of lean manufacturing cells, and the analysis of material and part flows through the supply chain have demonstrated the potential of the agent-based approach for advanced supply chain analysis systems. The architecture is generic and can be applied to heterogeneous and distributed information analysis systems in other domains in addition to that of supply chains.

REFERENCES

[1] Potok T.E., Ivezic N.A. (1999). Multi-agent Spare Part Grouping System for Logistics Optimization, *Proceedings of the 5th International Conference on Information Systems Analysis and Synthesis*, 582-585.

[2] Ivezic N.A., Potok T.E. (1999). An Agent-Based Approach for Supply Chain Management, *Proceedings of the 5th International Conference on Information Systems Analysis and Synthesis*, 570-573.

[3] Ivezic N.A., Potok T.E., Pouchard L. (1999). Manufacturing Multiagent Framework for Transitional Operations. *IEEE Internet Computing*, 3 (5), 58-59.

[4] Cantamessa M. (1997). Agent-based modelling and management of manufacturing systems, *Computers in Industry*, 34, 173-186.

[5] Baker A.F., Parunal H.V.D., Erol K. (1999). Agents and Internet: Infrastructure for Mass Customization, *IEEE Internet Computing*, 9, 62-69.

[6] Samatova N.F., Potok T.E., Leuze M.R. (2000). A Vector Perturbation Approach to the Generalized Aircraft Spare Parts Grouping Problem, *Proceedings of the 10th International Conference on Flexible Automation and Intelligent Manufacturing,* 1017-1026.

[7] Singh M.P. (2000). Synthesizing Coordination Requirements for Heterogeneous Autonomous Agents, *Autonomous Agents and Multi-Agent Systems*, 3(2), 107-132.

Supply Chain Business System Reference Model: A Business Process Description Using IDEF0

Shigeki Umeda, Hu Bin
Musashi University, Tokyo Japan
shigeki@cc.musashiu.ac.jp

Key words Supply chain system, Enterprise model, Business Process Re-engineering

Abstract This paper proposes a generic business process model for supply chain management. Effective operation of a supply-chain system requires frequent communication, information exchanges, among the member companies to achieve cooperative goals. Such business communication should include various kinds of data, such as ordering processes, manufacturing processes, transportation, product specification, quality, consumers' data, and so on. The objective of the model is to specify these process functions and their related information flows.

1. INTRODUCTION

Supply chain management is one of the keys to success in today's global manufacturing environment. Firms will need not only to cooperate with their vendors but with their vendor's vendors or supplier's suppliers, which are often firms in different countries. They must develop global strategies to coordinate operations at all phases of the value chain. Coordination of the supply chain will become strategically important as new forms of organizations - such as virtual enterprises, global manufacturing and logistic networks, and different company-to-company alliances - come into existence. Supply chain structures for these organizations will range from intra-firm stable networks to inter-firm dynamic networks.

Effective operation of a supply chain requires frequent information exchanges among the member companies in the chain to perform cooperative operations. Such business data communication should cover various kinds of operational data, such as product design, Bills Of Materials (BOM), operations directions, and process performance. Furthermore, the communication should include lots of planning data on operations. In some particular cases, these communications can be very complex because they contain confidential data from the chain members. Consequently, well-structured communication protocols and security mechanisms are needed in ensure successful collaboration among the chain member companies.

A well-organized business process model is the basis for successful supply chain systems. Such a model must describe the planning and control of all major business operations in the view of entire chain. Coordination of these functions across the whole chain will be an important factor in the implementation of a well-structured supply chain.

This paper proposes a generic business-process model that describes the core business activities of a supply chain, information communications in the chain, and specifications of business data of the individual chain-member companies. This paper builds on some of the concepts described in [1]. One of its objectives is the description of business functions in the company organizations. The other is the identification of the communication interfaces among them for the definition of software application interfaces.

2. MODELLING OF PLANNING AND CONTROL IN THE SUPPLY CHAIN

Several models have been proposed to capture features and activities of manufacturing enterprises. CAM-I, ESPRIT/IMPACT, PERA, etc. are typical enterprise models that have been used to implement such computer-integrated environment [2]. ARIS is a framework that provides a methodology to model business processes and to configure enterprise information systems [3]. SEMATECH CIM-framework is an application framework based on object-oriented methods [4]. The scope of specifications in this model is restricted to shop floor operations within the semiconductor factories.

SCOR (Supply Chain Operations Reference model) is a framework for modelling the business and management processes in a supply chain [5]. SCOR defines the four core business processes, "PLAN", "SOURCE", "MAKE", and "DELIVER", and it defines twenty-four core process categories, which are used to assess the business process implementation. However, it does not provide descriptions of information exchange rules or information communication protocols.

Each of these models describes planning and control either within a single enterprise or across multiple enterprises. The coordination of these functions across member companies is critical to the construction of a powerful value chain. When a firm wants to exploit the potential of its manufacturing network to satisfy an order from a client, the firm has to configure that network and schedule the needed operations through that network to deliver the order at minimal cost and on time. The resulting network is a virtual supply chain composed of manufacturing plants, warehouses, and transportation firms.

Coordinated, optimal planning and control require efficient information exchanges among firms. These exchanges are critical to the development of plans and their subsequent implementation. Consequently, the new exchange protocols should be built to ensure the timely and accurate exchange of that information. Supply chains should be built on a strategic understanding of the topology of the network of member firms and on the capabilities of their business and manufacturing processes. Relations are established through formal and informal information exchanges between the different firms in the network.

The Japanese manufacturing companies are well known for the way they use information sharing to get their supply chains to be competitive. They have made information exchange a prime part of most of their manufacturing strategies, especially in the case of just-in-time implementation [6]. For example, Srinivason et al.[7] showed that in a just-in-time environment, information exchange enhances shipment performance. Nishiguchi and Brookfiled [9] have described Japanese subcontracting practices. These are strongly based on collaborative, deep, true and real-time information exchange. Based on above discussions, this paper discusses a model for the planning and control operations for a supply chain. The model includes the information exchange protocols among the members of the chain.

3. IMPLEMENTATION OF THE MODEL

3.1 IDEF0 modelling method

The IDEF0 Function Modelling method is widely used to model the decisions, actions, and activities of an organization or system. It is also the most commonly used functional modelling method for analysing and communicating the functional perspective of a system. Effective IDEF0 models assist in organizing system analysis and promoting effective communication between the analyst and customer. Furthermore, the IDEF0 modelling method establishes the scope of analysis either for a current functional analysis or for future analyses from another system perspective. As a communication tool, IDEF0 enhances domain expert involvement and consensus decision-making through simplified graphical devices. As an analysis tool, IDEF0 assists the modeller in identifying the functions performed and the resources, in particular information, needed to perform them.

The basic activity element of an IDEF0 model diagram is represented by the simple syntax illustrated in Figure 1. A verb-based label placed in a box describes each activity. Inputs are shown as arrows entering the left side of the activity box while the outputs are shown as exiting arrows on the right side of the box. Controls are displayed as arrows entering the top of the box and mechanisms are displayed as arrows entering from the bottom of the box. Inputs, Controls, Outputs, and Mechanism are all referred to as concepts.

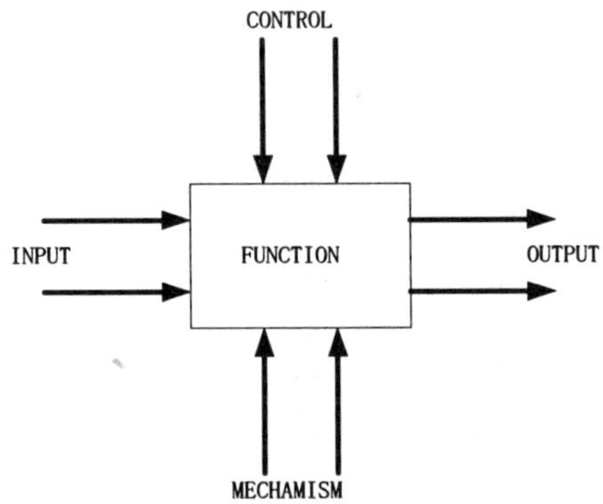

Figure.1 IDEF0 diagram

An IDEF0 model diagram is then composed of several activity boxes and related concepts to capture the overall function. IDEF0 not only captures the individual activities but also reveals the relationships between and among activities through the activities' related concepts. For example, the output of one activity may in turn become the input, control, or even a mechanism of another activity within the same model.

IDEF0 includes both a procedure and a language for constructing a layered model of the decisions, actions, and activities in an organization. Applying the IDEF0 method results in an organized representation of activities and important relations among them in a non-temporal, non-departmentalised fashion. IDEF0 is designed to allow the user to "tell the story" of what an enterprise does.

3.2 Supply chain business model

The proposed model covers operations of discrete-typed supply chain firms, which pose the material inventory. The typical examples are automobile and electric companies. The core business processes of the whole of chain can be identified in the model. These are "Planning", "Execution", and "Monitoring". Each core process owns hierarchical structure, which is explored to the detail level. The "Planning" includes all of the activities, which make plans of operations in the chain. These include demands analysis, capacity assessment, purchasing planning, production planning, inventory planning, and transportation planning. The "Execution" includes all of the activities, which make products and deliver to customer. The "Monitoring" supervises the executed activities (*Table.1* and *Table.2*)

Table.1 Definition of the functions in the model (1)

A0 Provides products to market in JIT			
	A1 Define the collaboration rules among the chain members		
		A11 Define business process rules among the chain members	
			A111 Define manufacturing rules
			A112 Define distribution rules
			A113 Define delivery rules
			A114 Define quality rules of deliverance
			A115 Define delivery stock rules
		A12 Define information exchange rules among the chain members	
			A121 Define data formats
			A122 Define data interface formats
			A123 Define data exchange rules
			A124 Define data security rules
	A2 Plans production and distribution activities in the chain		
		A21 Demand Planning	
			A211 Make MTO Planning
			A212 Make MTS Planning
			A213 Make Demand Plans
		A22 Supply Planning	
			A221 Supply capacity assessment
			A222 Material requirement planning
			A223 Supply capacity optimisation
			A224 Make supply plans
		A23 Manufacturing Planning	
			A231 Make manufacturing plans of products
			A232 Make manufacturing plans of supplied parts
			A233 Make procurement plan
		A24 Distribution Planning	
			A241 Make distribution plans of supplied parts
			A242 Make inventory plans of supplied parts
			A243 Make distribution plans of products

The model is composed of five sub-models. A1 is the function that defines rules of both business processes and information exchanges. This function provides collaboration rules of the chain member companies. Once this function would be executed, the defined rules will be used till the system configuration or environment is changed. A2 is the planning function of operational activities in the chain.

Table.2 Definition of the functions in the model (2)

A0 Provides products to market in JIT			
	A3 Manufactures products		
		A31 Procure materials	
			A311 Scheduling procurement
			A312 Ordering purchase
			A313 Receive materials and tests
			A314 Records receipt of material
			A315 Procurement accounting
		A32 Manufacture products	
			A321 Scheduling manufacturing
			A322 Manufacturing operations
			A323 Maintenance manufacturing tools and equipment
		A33 Transport products	
			A331 Scheduling transportation
			A332 Withdraws products
			A333 Transport operations
			A334 Payment process
	A4 Distributes products		
		A41 Package products	
			A411 Scheduling packaging
			A412 Packaging operations
		A42 Deliver products	
			A421 Scheduling delivery
			A422 Delivery operations
			A423 Delivery/Inventory record processing
		A43 Provide services	
			A431 Scheduling service activities
			A432 Activate services
	A5 Monitor chain processes		
		A51 Monitor distribution of products	
		A52 Monitor manufacturing products	
		A53 Monitor service activities	

The plans built in this function denote what procedures should be done in the chain. The demand planning (A21) is to make plans what items of products the members make, and how many volumes of products they make. The supply planning (A22) is to make plans how many volumes of parts they require, and when those parts should be supplied. The manufacturing planning (A23) is to make plans when the parts or products should be produced. Finally, the distribution planning (A24) is to make plans when the parts or products should be delivered, and how many volumes of parts or products should be stored. The "Manufactures products"(A3) defines manufacturing operations, which should be done in the suppliers, and the "Distribute products"(A4) defines distribution operations, respectively.

3.3 Characteristics of the model

This model is based on the decomposition of the "Chain" as a virtual enterprise, which is a group of companies. Individual business function is represented as the inputs, outputs, constraints (controls), and conditions (mechanism). The relations among the business functions are represented as a serious of block diagrams, which is straightforward to understand the process flow.

The model represents enterprise functions as a combination of core element models, which are planning, procurement, manufacture, and delivery. This model also represents the details of the business planning and control operations, which supervises all of processes in the whole of the chain. Individual communication protocol specifies the senders, the receivers, and the condition of them. The proposed model defined schedule-driven (push) system and buffer-driven (pull) system as the communication mechanism among the supply chain members. These specifications will support system integrators to define the communication interfaces among the business sections of the individual chain member company.

4. Conclusion

The proposed model is applicable as a reference model to represent company's organizations to perform BPR. The model can be used as a reference model to redesign business flows among the chain. The uniqueness is a definition of planning function whose scope the whole of the chain. Each process is breakdown of not the member firm but the chain. The model defines individual process of the business in the chain and information flows among them. These descriptions will especially help supply chain planners to understand chain-level planning functions.

The model will be also used as a reference model to build information system that support supply chain operation. Each element of the model (function) is hierarchical decomposition of activities. This hierarchical structure helps the practitioner keep the scope of the model within the boundaries represented by the decomposition of the activity. Therefore, the feature specified in individual model is applicable to specification of the module, when the practitioner designs software to support supply chain operation. This feature will be also applicable to define the communication protocols in the chain. The defined elements of the individual function such as input, output will be the definition of feature of interfaces among the modules.

This model will be also applicable as a framework to benchmarking of the individual process. Individual function specifies its inputs/outputs of materials and data. Accordingly, the indices can be defined to measure performance of the individual function. The practitioners can use the indices in order to review and improve their business processes continuously.

Supply chain system is an implementation of manufacturing enterprise integration. It needs generic frameworks to design and implement business processes and information systems. The main scope of current version is the representation of planning and controlling activities in the whole of the chain. We are currently expanding the models to describe the further detail levels. And we are also developing representation of the data structure definition by using IDEF1X modelling method. These will be also reference models to support supply chain implementation as well as the activities models.

4. REFERENCES

[1] Umeda, S. Jones, A. (1999) A Simulation-Based BPR Support Systems for Supply Chain Management, Reengineering in Action: The Quest for World-Class Excellence (Koong, C.M. eds.), 95-117

[2] Bernus, P., Nemes, L. eds. (1997) Modelling and Methodologies for Enterprise Integration, Chapman & Hall

[3] Scheer, A.W. (1998) ARIS – Business process Frameworks, Springer

[4] SEMATECH, CIM Application Framework Specification 1.3, Tech. Transfer #93061697F-ENG, 1996

[5] http://www.supply-chain.org/

[6] J.H. Dyer, W.G. Ouchi, Japanese-style partnerships: giving companies a competitive edge, Sloan Management Review, Fall 1993 (1993) 51-63.

[7] K. Srinivasan, K. Sunder, T. Mukhopadhyay, Impact of electronic data interchange technology on JIT shipments, Management Science 40 (10) (1994) 1291-1304.

[8] T. Nishiguchi, J. Brookfield, The evolution of Japanese subcontracting, Sloan Management Review Fall, 1998, pp. 89-101.

PART FIVE

e-Commerce and e-Service

B2B E-Commerce Infrastructure Using Agents and Standards – A Potential Impact Analysis and Architecture

N. Ivezic
National Institute of Standards and Technologie, Oak Ridge National Laboratory, USA
Em: Nenad.Ivezic@nist.gov

L. Fong, T. Rhodes
National Institute of Standards and Technologie, USA

Y. Peng
University of Maryland at Baltimore County, USA

Keywords business-to-business, electronic commerce, agents, standards, business case

Abstract Multi-agent systems are typically being developed independent of industrial standards. In this paper, we present an agent architecture for business-to-business electronic commerce in the domain of circuit board manufacturing that builds on a number of standards. Our architecture builds on the GenCAM standard from the circuit board manufacturing sector and RosettaNet from the B2B community.

1 INTRODUCTION

The Internet Commerce for Manufacturing (ICM) project is working with the printed circuit (PC) board and PC assembly (PCB/PCA) industries to extend and apply emerging Internet-based technologies in support of manufacturing. In this paper, we focus on the ICM project component that develops and evaluates agent technologies to address two PCB/PCA industry issues: the PC board manufacturing information exchange and the supply chain information integration. Differently from other approaches in developing agent systems, our agent approach attempts to use existing PCA/PCB industry standardization efforts. First, we rely on the emerging

GenCAM standardization effort in support of PC board manufacturing information exchange [1]. Second, we use RosettaNet standardization specifications in support of supply chain information integration in the electronics component industry [2].

In the rest of the paper, first, we provide an assessment of potential impact of agent technologies within the realm of the Internet-based Commerce for Manufacturing (ICM) project. Following the assessment, we identify two areas of agent technology development that we pursue and describe the agent architecture for this work. Finally, in the last section, we summarize the value-added contribution of the agent technologies within the ICM domain of interest.

2 AN IMPACT ANALYSIS FOR ICM PROJECT

To date, the ICM project has looked closely at the manufacturing Business-to-Business (B2B) area as the primary domain of investigation. The rationale for this focus was the potential impact of the new Internet-based solutions in the emerging electronic markets of manufacturing goods. The team identified several alternative manufacturing functions and B2B use cases where agent solutions could take place. We have taken the identified opportunities for the agent-based technologies from B2B area and performed an evaluation of these opportunities, as they apply to the ICM domain. The metrics used to perform the evaluation were the potential impact, complexity, and relevance to the ICM area of interest. Table 1 gives a summary of the outcome of the evaluation process:

First, *Flexible Customer-to-Supplier Interfaces*, points at the opportunity for agent approaches to 'wrap around' or completely circumvent the existing form-based interfaces on the Web that have pre-defined syntax, implicit semantics, unpublished interaction protocols and, instead, enable automated, on-demand B2B interface construction.

Second, *Optimized Negotiation of Service Cost and Terms,* points to the fact that an expert human is solely responsible to negotiate 'optimal' set of terms and costs of service. This decision making approach requires an expert human, who, often has only a limited view of the business situation and cannot react immediately to new business situations. Agent approaches carry promise of embedding significant decision making capabilities and supporting the human in identifying the optimal choices.

Third, *Efficient Intra-Enterprise Technology Adoption and Adaptation*, points at the issue that currently, few means exist to accelerate adoption and adaptation of new e-commerce technologies within an enterprise. An opportunity exists for agent approaches to provide for easier integration with legacy systems through usage of shared languages and ontologies and efficient updates of interaction protocols.

Fourth, *Efficient Engineering Change Order (ECO) Processing,* indicates that, currently, processing ECOs require each human participant in the engineering and manufacturing process to manually sign-off on the change and to make sure that the change is appropriately reflected in the part of the process for which the human is responsible. Agent technologies hold potential to make this process much less tedious and error-prone.

Fourth, *Efficient Inter-Enterprise Interaction Technology Support,* indicates that currently, very few means exist to accelerate adoption of new B2B communication technologies across enterprises so that the enterprises can quickly engage in new inter-enterprise interactions. Agent solutions promise to automate such inter-enterprise interaction setup and support.

Table 1 Opportunities for Agent-based Internet Commerce for Manufacturing.

	Impact	Complexity	Relevance
Flexible Customer-to-Supplier Interfaces	Med-High	Med	Med-High
Optimized Negotiation of Service Cost and Terms	Med	Med-High	Low
Efficient Intra-Enterprise Technology Adoption and Adaptation	Med-High	Med	Med-High
Efficient Engineering Change Order (ECO) Processing	Med-High	Med-High	Med-High
Efficient Inter-Enterprise Interaction Technology Support	Med-High	Med	Med-High

3 AGENTS-BASED SOLUTIONS IN ICM PROJECT

Based on the impact analysis in Section 3, we selected two areas of development for the agent technologies. The first area was the Flexible Customer-to-Supplier Interfaces where we focused on development of GenCAM information retriever agent to enable automated B2B interface construction. The second area was the Efficient Inter-Enterprise Interaction Technology Support to develop a supply chain integration agent technology in support of emerging B2B standards such as RosettaNet. The rationale for this selection is demonstrated in Table 1 – it was determined that we could achieve a significant impact through the technology development in the area that is very relevant to the ICM domain while managing the development complexity. Moreover, the GenCAM and RosettaNet standards are becoming mature to the degree that basing the agent technology on these standards seemed to be warranted. In the following, we use an example business-to-business situation (i.e., an RFQ scenario) to describe architectures of the agent technology developments.

3.1 An Example Problem: RFQ Scenario

In this RFQ scenario, an original equipment manufacturer (OEM), responsible for developing a PC board, issues a request for quotes to its suppliers – electronics manufacturing service (EMS) providers – to perform a specific service to build the PC board (e.g., bare board production, part supply, part preparation). The EMS provider requests additional information that is not provided within the technical package. The OEM searches for the additional information in the technical database, identifies the requested data, and provides the data back to the EMS provider.

Some of the EMS providers may provide a counter-offer with a portion of the original specification altered (e.g., time of delivery, quantity) and traded off (e.g., cost versus time of delivery). The OEM, than, may enter negotiation with one or more suppliers to determine completion of the service. A series of counter-offers may be exchanged until the process is completed at the time OEM selects the winning bids.

The scenario serves to demonstrate some of the principal traits of the following agent based solutions and is not meant to capture all peculiarities of the RFQ process.

3.2 GenCAM Agent-based System Architecture

Figure 1 shows the proposed architecture for an intra-enterprise information retrieval agent system supporting OEM RFQ functionality. The system is composed of four agent types. Inter-Enterprise Broker Agent (BA) is responsible for collecting advertisements from all system agents regarding their individual capabilities. Human-Computer Interaction Agent (HCIA) facilitates interaction between the human user and the agent system. Web Assistant Agent (WA) supports interaction between the agent system and the EMS client that consists of a human user with a Web browser. Finally, GenCAM Specialist Agent is responsible for retrieving information from the GenCAM technical database.

The proposed architecture addresses several issues in intra-enterprise information retrieval and in support of RFQ functionality. First, the GenCAM technical data standard for PC board manufacturing is becoming increasingly complex. By providing a wrapper agent capable of searching, retrieving, and reasoning about PC board data, the standard interface is localized and isolated from the rest of the system. Changes in the GenCAM standard and technical databases may be handled in a local fashion without affecting the remaining system. Second, the interface between the EMS user and the OEM agent system becomes simpler and more general by isolating the 'content' of transactions in the GenCAM agent. Only the GenCAM agent on the OEM side and a plug-in, a client within the Web browser, or an application program on the EMS side needs to understand the GenCAM content. The remaining agents of the OEM agent system need not to be concerned with the content.

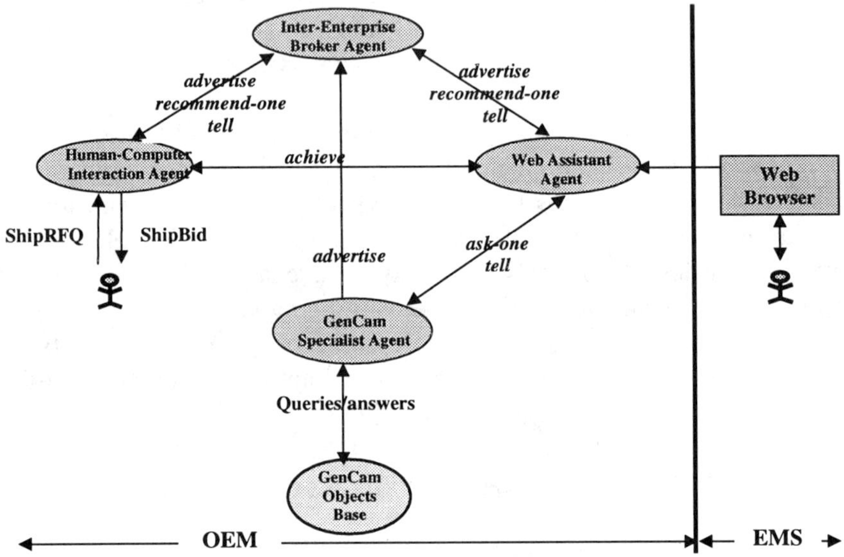

Figure 1 Agent architecture for intra-enterprise information retrieval agent system

While this agent architecture addresses the intra-enterprise information retrieval and interaction with a human on the side of the EMS client, a complementary supply chain integration system is described next that allows automated agent-based inter-enterprise interactions.

3.3 RosettaNet Agent-based System Architecture

Figure 2 illustrates the proposed architecture for an inter-enterprise supply chain integration system in support of RFQ scenario. The system is composed of two new agent types. Supply Chain Interaction Agent (SCIA) takes place of the Web Assistant Agent (WA) from the previous architecture to enable agent-based interaction among enterprises. The Inter-Enterprise Broker Agent (IEB) is responsible for matching advertised capabilities and requirements on part of agents in the supply chain. Within the EMS provider, there is an opportunity to introduce a Bidding Specialist Agent to automate the bidding process. A human user interacts with the system through the HCIA to check and approve the bids.

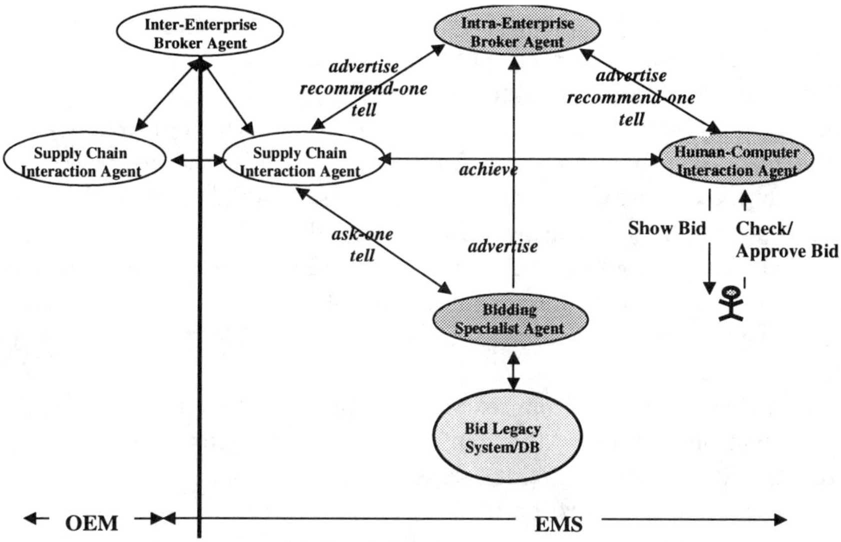

Figure 2 Agent architecture for inter-enterprise supply chain integration system

The supply chain integration architecture further provides for automation of interactions among OEMs and EMS providers. Our intent is to use the proposed architecture to map its coordination protocols onto emerging standards such as RosettaNet and provide approach to capture business-level interactions among customers and suppliers. Another goal is to merge the inter-enterprise and the intra-enterprise informational retrieval architecture to explore issues in supply chain integration standards development and harmonization.

4 EXPECTED IMPACT OF ICM AGENTS SOLUTION

There are four distinct value-added opportunities in adopting agent-based technologies within the ICM project:

First, an agent-based approach for distributed systems is potentially superior when there is a considerable autonomy of participating entities: The common ontology for the application domain allows efficient building of interoperable systems with shared semantics of the communication. In addition, the agent coordination modelling approaches allow for specification and management of interactions among the participating entities (e.g., organizations or people). Moreover, the agent-based architectures provide for a common high-level schema on which to build future systems that are highly maintainable and manageable. Also, the autonomy of agents provides for a fundamentally new and potentially significant capability to have in the new electronic markets. The capability to automatically initiate actions may change the way that the manufacturing businesses collaborate and search for opportunities.

Second, the standards for information sharing and exchange in PCB/PCA industries (e.g., GenCAM) may benefit significantly from providing an agent-based service that would use the standard dictionary of the industry when providing data to suppliers. Different than the current established manner of sharing data among the customers and suppliers (which is based on complete file transfers), a GenCAM-based agent service would be able to provide information about the product on on-demand, as-needed basis. In this manner, a more manageable and agile way of information sharing is feasible.

Third, the agent-based approaches provide for modelling and synthesis of coordination policies for interacting autonomous entities. Such coordination policies provide basis for modelling interactions of collaborative, autonomous entities in business processes. An obvious example of the need that these agent-based approaches can fulfil is in the case of the RosettaNet standard. RosettaNet provides for Dictionaries, Framework, and Partner Interchange Process (PIP) definitions which, in turn, allow 'Dialog' among the partners (see Fig 3 taken from [1]). However, RosettaNet does not provide for capture of eBusiness processes. This is where the agent-based approaches can provide value to both an industry and standards making body as they provide high-level modelling and synthesis methodologies.

Last, the initial feedback from the workers in the manufacturing industry indicates that the agent-based approach could provide a desirable capability for flexible product information retrieval and supply chain information integration. Although preliminary, this feedback begins to validate our intuition in focusing on the two important areas for agent-based information systems: product information retrieval and supply chain information integration.

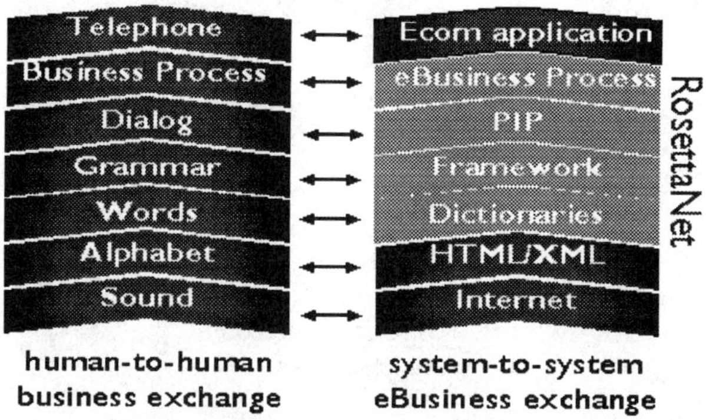

Figure 3 Parallel between RosettaNet deliverables and entities required for human-to-human business exchange.

5 CONCLUSION

A need, an impact, and systems architecture analysis have been presented to address concerns within the Internet Commerce for Manufacturing (ICM) project. ICM is a National Institute of Standards and Technologies (NIST) sponsored project working with industry to create an environment where small manufacturers of mechanical and electronic components may participate competitively in virtual enterprises that manufacture printed circuit assemblies. The motivation and current work in adopting agent-oriented approaches for *intra-enterprise information retrieval* and *inter-enterprise supply chain integration have* been discussed in the context of emerging industrial standards.

6 REFERENCES

[1] GenCAM Standard Specification (2000). An Overview Document, http://www.gencam.org/overview/fact/fact.htm/

[2] RosettaNet Consortium (1999). An Overview Document, http://rosettanet.org/general/index_general.html/

A Web-based Bidding Workbench for Global Manufacturing

Mingwei Zhou, Jeffrey Zheng, Angela Williams
CSIRO Manufacturing Science and Technology, Australia

Bob Alexander
Hawker de Havilland Pty Ltd, Australia

Key words Virtual Enterprise, Architecture, Global Manufacturing, Information Infrastructure, Bidding, Web-based Teamwork

Abstract Global manufacturing business requires a platform to support dispersed partners working together to bid for a contract. The bidding process usually involves rapid product definition in order to prepare the technical proposal that gives a competitive advantage. This paper presents our findings from the Viewbid Project of the IMS Globeman'21 Consortium, in which a web-based bidding workbench was developed and a proof-of-concept system has also been implemented.

INTRODUCTION

With the globalisation of manufacturing industry, the core competence of a manufacturing enterprise is increasingly relies on its intellectual resources, its global and regional partner networks, and more importantly, its capability of making best use of its resources and partner networks in critical missions such as bidding. A manufacturing enterprise involved in the global manufacturing business therefore needs a platform that can support its collaboration with partners to bid for contracts. This kind of bidding processes usually involves rapid product definition and business development in order to prepare proposals that gives reliable and competitive total value solution that makes a difference for the customer in all aspects of their operations.

The IMS Globeman'21 was an international collaboration Consortium operated for 3 years since 1996 and involved over 40 industrial and academic organizations from Australia, Europe, Japan, Canada and USA. It aimed to investigate the business practices and techniques required to operate globally distributed enterprises in the environment and under the conditions anticipated for the 21st century. The VIEWBID (*VI*rtual *E*nterprise *W*orkbench for Worldwide *B*usiness *I*ntegration and *D*evelopment) project is one of the 14 Globeman'21 demonstrator projects. The VIEWBID project aims to develop and demonstrate the tools and methods for the design and operation of a virtual manufacturing enterprise to compete in the distributed global manufacturing environment. The requirements for such kind of bidding platform have been identified and analysed, and the architecture of a web-based bidding workbench has been proposed. A proof-of-concept system of the web-based bidding workbench has also been implemented.

THE BIDDING PROCESS

In a globalized manufacturing environment, a typical bidding process can be described in a 5-stage development, as shown in Figure 1. People from different parts of the world are involved in the bidding. These include core bidding team members, sales, finance, customers and suppliers.

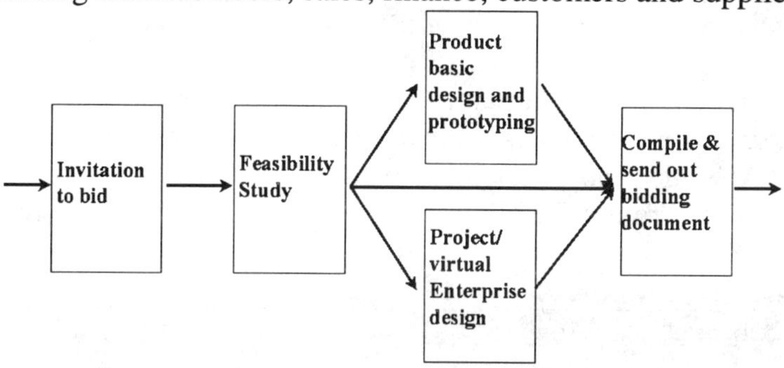

Figure 1: The bidding process

The system requirement for such a global collaboration environment is inevitably based on the internet. At the core of the system, a database comprising information of products and production, information of business partners and capability of risk assessment is required. Design of the database structure requires thorough understanding of the company's bidding process by using the PERA methodology [1,2].

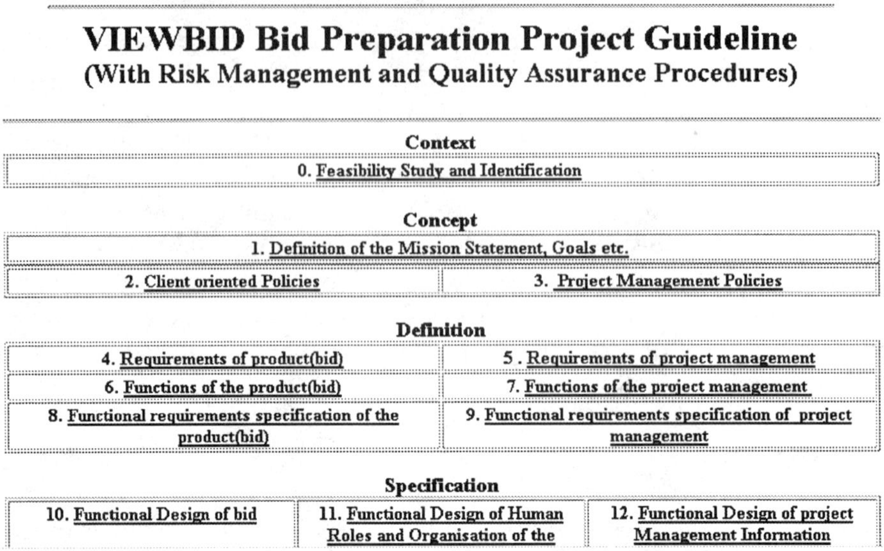
VIEWBID Bid Preparation Project Guideline
(With Risk Management and Quality Assurance Procedures)

Context

0. Feasibility Study and Identification	

Concept

1. Definition of the Mission Statement, Goals etc.	
2. Client oriented Policies	3. Project Management Policies

Definition

4. Requirements of product(bid)	5 . Requirements of project management
6. Functions of the product(bid)	7. Functions of the project management
8. Functional requirements specification of the product(bid)	9. Functional requirements specification of project management

Specification

10. Functional Design of bid	11. Functional Design of Human Roles and Organisation of the	12. Functional Design of project Management Information

Figure 2: Bidding project guideline using PERA methodology

The PERA (Purdue Enterprise Reference Architecture) methodology provides a guideline for the analysis of the bidding process in product design and development projects. The formalism in the PERA methodology provides a generic listing of these tasks that must be carried out in order to achieve enterprise integration. It allows views in market, plant, product, resources, stakeholder and technology to co-exist. Figure 2 shows the top part of the Bidding Project Guideline developed using PERA Methodology.

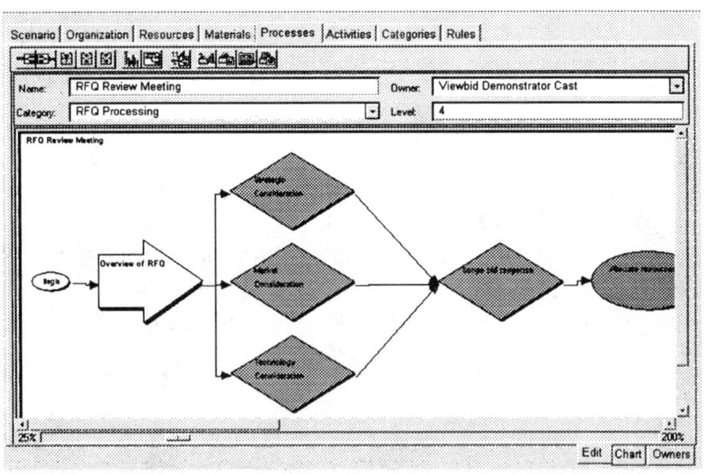

Figure 3: The Bidding process modelling using FirstSTEP Designer

Based on the Project Guideline, we modelled, simulated, and analysed the bidding process, and identified the information and functional requirements for a bidding workbench. As shown in Figure 3, an enterprise modelling tool FirstSTEP Designer from *Interfacing Technology Corp.* was used to model the bidding process.

SYSTEM ARCHITECTURE

Based on the requirements identified in the bidding process modelling, A web-based bidding workbench architecture was developed, as depicted in Figure 4. This architecture is based on the Virtual Enterprise Workbench Architecture [3], which is jointly developed with VRIDGE Project -- another Globeman'21 Demonstrator.

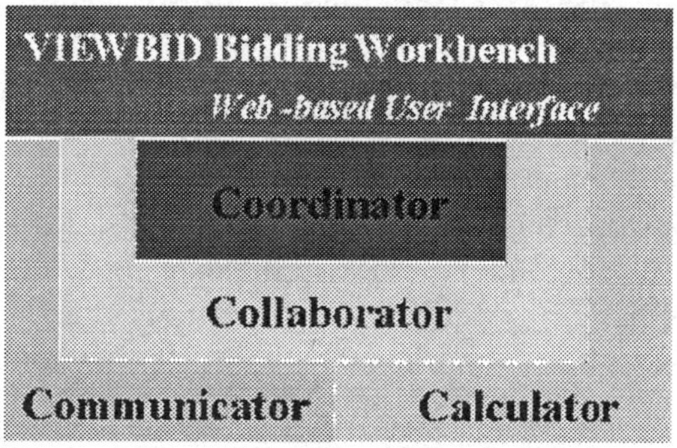

Figure 4: Viewbid bidding workbench architecture

The VIEWBID bidding workbench provides two layers of functionality: the virtual enterprise workbench for supporting the design and operation of a virtual enterprise, and the distributed bid preparation platform.

At the first layer, the workbench provides four major function modules: the Coordinator, the Collaborator, the Communicator, and the Calculator.

The Coordinator provides the tools for bidding process management:

- team design and operation support;
- corporate knowledge management facilities such as partner profiles, contacts, capabilities, etc.;
- daily project supervising, monitoring, and supporting tools; and
- risk management and quality assurance tools.

The Collaborator provides the tools supporting dispersed teamwork during bid preparation:

- sharing and exchange of product and production information;
- review and release of product and production information;

- revision and change management of product and production information;
- configuration and co-authoring of the bidding document.

The Communicator provides the essential facilities for bidding team communication:

- secure IT infrastructure for dispersed teamwork;
- central information repository; and
- access control and session management.

The Calculator provides the tools for calculation during bidding:

- cost estimation;
- risk analysis and assessment;
- simulation of the best, worst, and most-likely cases; and
- formula-based calculations.

At the second layer, the workbench provides a bidding document co-authoring platform. This layer is tailored to suit the specific requirements of a bidding team. It includes three major modules: the bidding document configuration tool, the bidding document component editor, and the bidding document synthesising tool.

The bidding document configuration tool provides facilities for defining a framework or structure of the bidding documents, which specifies what kind of information should be included in the bidding document, the possible source of the information, and the bidding decisions or costing guidelines.

The bidding document component editing tool is mainly a web-based word processor for editing the sections and chapters of the bidding document by re-using or modifying the component from the previous bids or creating a new component.

The bidding document Synthesising tool is used to compile the final bidding document based on the specified bidding document configuration, and all the necessary components, and finally delivers the complete set of the bidding document.

IMPLEMENTATION AND CONCLUSION

A proof-of-concept system of the VIEWBID bidding workbench has been implemented. Figure 5 shows a snapshot of the bidding document component editing tool.

The system was implemented using Java, and made use of the Lotus eSuite Devpack Java applet library to develop the web-based editing tools.

By demonstrating the proof-of-concept bidding workbench to our industrial partners, we get a better understanding of user's requirements for the bidding workbench. It also helped us to formulate a new project to further the research to develop a more intelligent bidding workbench with enhanced corporate knowledge management functionality. This new project is currently continued as part of the IMS Globemen Project.

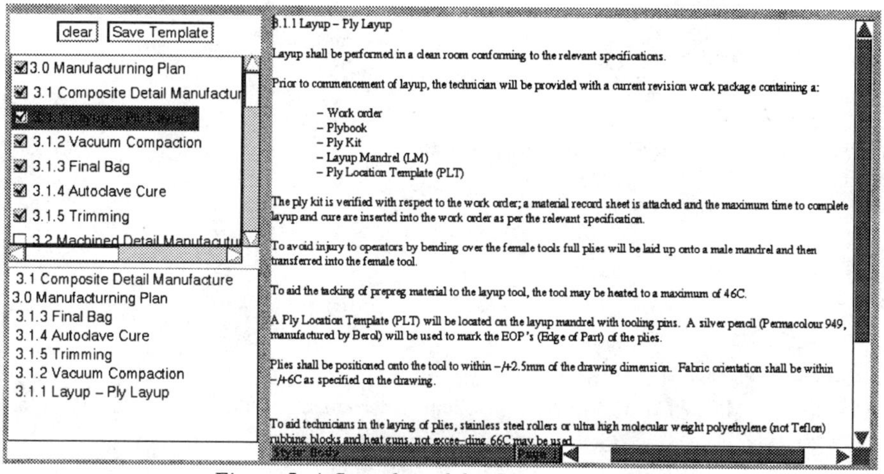

Figure 5: A Snapshot of the Component Editor

ACKNOWLEDGEMENT

The authors wish to thank all project partners for their contributions, especially Dr. Laszlo Nemes and Dr. John Mo from CSIRO, Mr. Ron Beckett from Hawker de Havilland, and Mr. Masashi Shinonome from Toyo Engineering.

REFERENCES

[1] Williams T.J. , The Purdue Enterprise Reference Architecture, Instrument Society of America, North Carolina, 1991

[2] Rathwell G.A. and Williams T.J., Use of the Purdue Enterprise Reference Architecture and Methodology in industry (the Fluor Daniel example), in *Modelling and Methodologies for Enterprise Integration*, edited by Burnes, P. and Nemes L. Chapman and Hall, 1996

[3] Zhou M. et al, A framework for design a virtual manufacturing enterprise and its implementation as a workbench, in *Information Infrastructure System for Manufacturing*, edited by Miles J.J. and Kimura F., Kluwer Academic Publishers, 1998

Web-based maintenance manual with three-dimensional simulation model

K. Morita, K. Kawashima
Mitsui Engineering & Shipbuilding Co., Ltd, Japan
kmorita@mes.co.jp, kkawashi@mes.co.jp

Y. Fukuda
Hosei University, Japan

Keywords IETMs, WWW, simulation, object oriented technology, product model, INTERNET

Abstract Currently manufacture industry increasingly needs to make their service engineer to understand for maintenance process easily. In order to use the maintenance information effectively, maintenance documents are necessary to apply the multimedia technology and telecommunication technology. This research is to develop the Web-based multimedia manuals using three-dimensional simulation model applying for the maintenance. The goal of this research is to implement to industries the multimedia maintenance manual that integrates a virtual equipment model with assembly process and disassembly process through INTERNET. We developed Web based simulation environments named World Wide Simulation Environments (WISE). The WISE is a remote simulation system on the Web. The multimedia maintenance manual is applied the WISE and other information communication technologies.

1 INTRODUCTION

Under increasingly competitive conditions in manufacturing sector, process engineering requires drastic improvements in all process. Simulation-based design is a new approach, which, aims at creation and operation of virtual and real products and process, in order to increase real product value and to reduce life cycle cost.

The main object of simulation-based design is to generate a virtual environment that can provide support, information and control during all phase of the product life cycle. Furthermore, this environment has to support internationalisation and decentralization of the design, production activities and maintenance processes. To realize the environment, a simulation technology is very effective tool. However, a simulation technology is not good enough to apply to maintenance process. Therefore, to construct the next simulation-based design, the following items must be considered; (1) the environment can support the all phases in the product life cycle, ranging from the planning to production, maintenance and reuse, (2) the environment can use a multimedia technology with Web browser, and (3) the environment has a simulation ability to predict the future behaviours.

The final goal of this study is proposed to construct the simulation-based design environment that simulates a virtual manufacturing on a global communication network. This concept of the simulation environment is named the World Wide Simulation Environment (WISE).

This paper features an application of the WISE in the maintenance phase. The WISE has physically meaningful models or product models with behaviours for maintenance activities. The models are built in a distributed environment and distributed throughout a network. The models are made to represent the function, behaviour, and performance of real equipments and human beings by giving behaviours programs to three-dimensional CAD models that represent virtual equipment with motions and correspond to the "behavioural multimedia manual" of real equipment.

Users can thereby collect these multimedia manual through a wide-area network, confirm and evaluate these maintenance activities. The virtual environments that these activities are considered constrains among them is easy and effective for service engineer to understand and capture these disassemble activities and assemble activities in the maintenance phase. The virtual environment drastically improves the maintenance service effectively among distributed company and virtual enterprise.

2 WISE ARCHITECTURE

The WISE architecture is shown in Fig. 1. The WISE model can be transfer to the client side to the designated URL address in the WWW browser based on the HTTP protocol. The model is an encapsulated components file of three-dimensional model, HTML multimedia-documents, inputting menus for GUI and object oriented programs showing its behaviours. An engineer can execute the simulation by communicating via HTML multimedia-documents after acquiring the digital manual through the INTERNET.

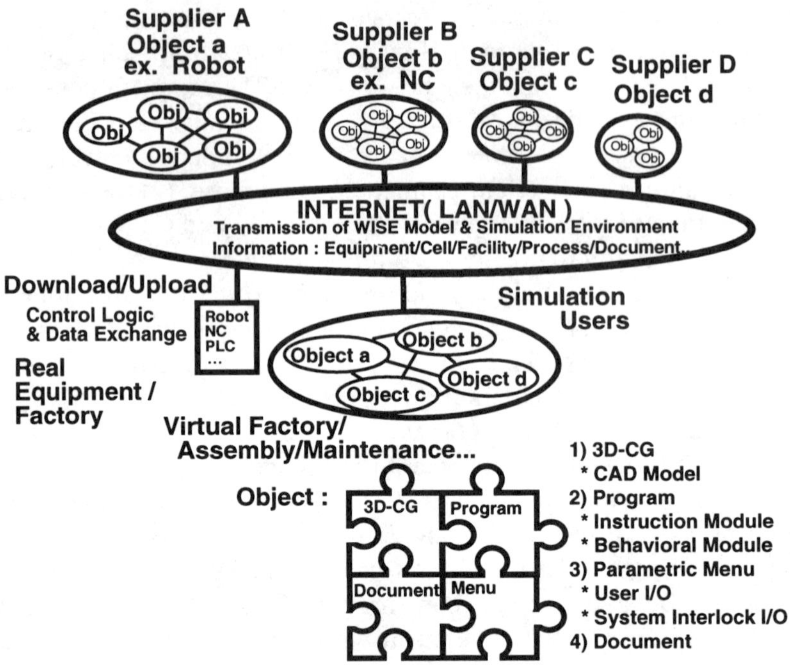

Figure1 World Wide Simulation Environment (WISE) Architecture

3 REQUIREMENTS FOR MAINTENANCE MANUALS

Currently, maintenance manual is publishing as the digital documents. However, it only includes the static two-dimensional drawing data. It's not effective only static data to train maintenance process from any viewpoint. Considering the Information Technology, the various method are possible to applied to the maintenance manual for service engineer to confirm and to train the disassemble process and assemble process, For Example, multimedia technology, three-dimensional model from CAD systems and simulation technology for three-dimensional model are effective in visualizing documents to understand easily. VTR have a reality to visualize, however, the viewpoint is limited. VRML technology or Java 3D, is the effective in visualizing motion form any angle. However, it is not suitable interaction between the instruction sentences in HTML and three-dimensional model with motions like maintenance manual. Integration of multimedia technology, three-dimensional model from CAD systems and simulation technology still not used. We redefine the requirements from view point of Integration among multimedia technologies, three-dimensional model from CAD systems and simulation technology. Therefore WISE technology is applied to the maintenance manual. Additionally reuse of three-dimensional model from CAD system is important from all life cycle cost. In order to realize the maintenance manual on the proposed WISE architecture, there are the language level requirements. The requirements are follows:

- The creation of a three dimensional model in order to represent each product model.
- As shown the object-encapsulation specification described in the reference [1], contents data and program components treat as an encapsulated component.
- Hierarchical programming, which is handled as a unique unit, and creating a new program combining with the existing program and the object oriented language that can create a reusable object encapsulating data and function.
- Cooperated operation with a scheduler function, a control function for an autonomous behaviours of each machine object, and a definition and communication function for controlling a signal to control behaviour.
- HTML file and programs linking each other, if an icon or an instruction on documents is clicked, related behaviour or activities are executed.
- Interpreter and compiler language for creating the program flexibly and quickly.
- As shown the electronic addressing specification described in the reference [1], transmission of the digital manual via the INTERNET by the designated URL address, accessing to the component on the Web, which implement the programs in HTML function.
- An animation function that operates while simulation is running to enable the engineer to easily understand the behaviour of the virtual machine and activity of the virtual factory.

4 SYSTEM ARCHITECURE OF WEB-MAITENANCE

Virtual enterprise has to consdsider the change of maintenance in global decentralization environments. Therefore, they require needs for maitenace manual via global communication and managed the contents in centralization or decentralization environments. Our company have to support maintenance-training system through global branch. Considering the maintenance manual via INTERNET, the contents should be to manage in server. The update contents should be reflected to the client. Most new information is distributed to the clients. Browser environments are suitable documents interaction. Therefore, to realize WISE architecture, Web-based client and server environments are selected. Behaviour Engine is selected to implement the application programs in order to satisfy the requirements of maintenance manuals. It has integrated capability of documents, three-dimensional model from CAD systems and simulation technology.

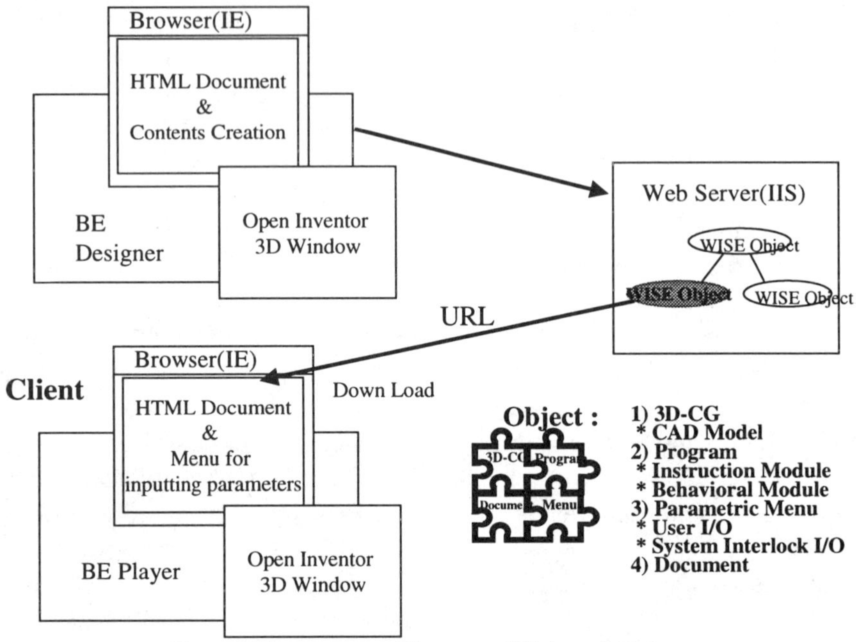

Figure2 The system architecture of Web-maintenance

The system architecture of Web-maintenance is shown in Fig. 2. This system uses the client and sever environment on the WISE architecture. The client side has a browser, the Open Inventor windows and the helper software called Behaviour Engine Player. Sever side is installed the Web-server like Windows IIS. The WISE objects are created by Behaviour Engine Designer and place on the Web-server. The WISE object is downloaded to the client side by URL address in other object or main WISE object. The instruction and inputting menu written in HTML and Active X is called a motion program in WISE object via the JavaScript. The motion programs in WISE object written by Umbel like Pascal Syntax represent the motion of three-dimensional model by simulation clock. This simulation function has a capability to do the concurrent simulation of motion for three-dimensional model. Therefore, the geometry transfer matrix of three-dimensional model and motion class is used to move the motion.

Clicked on instruction text or button by HTML, motion programs are executed concurrently. Since the simulation has a functionality of signal control of motion, Interlocks of motion are implemented.

5 WEB-BASED DEMONSTRATION SYSTEM

The Web-based demonstration system for maintenance activity is developed in order to evaluate the proposed architecture and a possibility of usage to a maintenance field. The system is applied to two kinds of product. One is the AGV for heavy weighted part and second is the turbocharger for diesel engine. These products are produced in Mitsui Engineering & Shipbuilding Co., Ltd. The multimedia manual consists of equipment class, behaviour programs, three-dimensional models by Open Inventor, and an input panel for teaching and executing behaviours by HTML. It is able to communicate through INTERNET.

5.1 AGV for Heavy Weighted parts

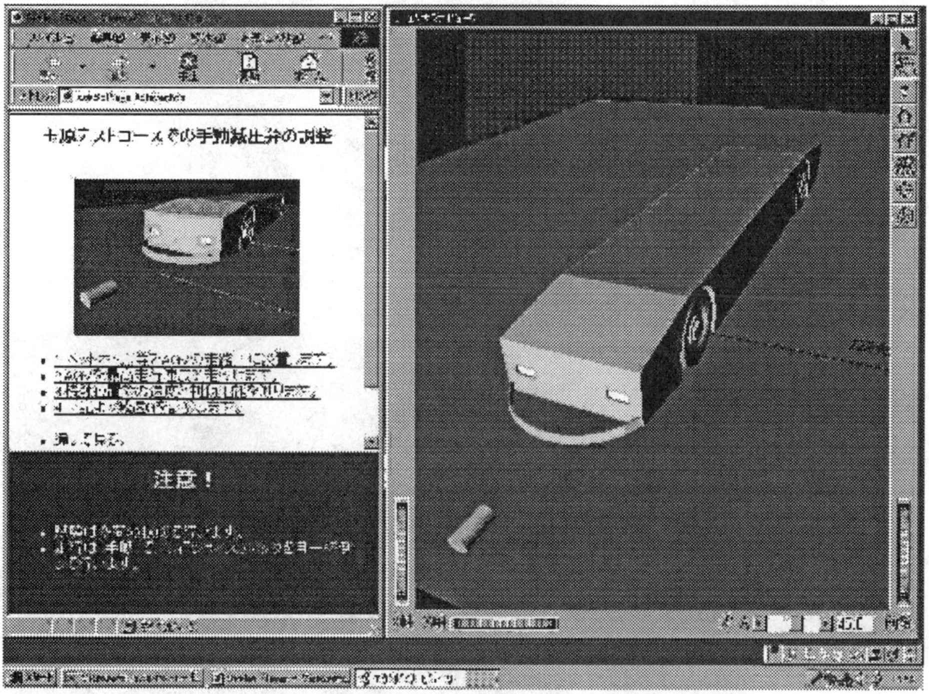

Figure 3. A manual of Braking test instruction for AGV

The AGV for heavy weighted parts are used in the steel manufacture fields and container terminal of ports. It has a few dozen of adjustment items. All of them adjust to motion parts. The purpose of this demonstration is linkage between an adjustment instruction and a three-dimensional motion in the virtual environment. The developed Web-manual of the rubber tire in the AGV is shown in Fig. 3. The instructions of adjustment activities are written in HTML. Each instruction links the corresponding adjustment motion. The figures show experiment tests for braking.

5.2 Vessel diesel engine with turbo-charger

The diesel engine is the main machine for the vessel. They require the periodical maintenance. Therefore, these training processes are important for the technical crews. To acquire effectively the maintenance activities from crews, the manual has to improve serviceability for maintenance. Figure 4 is an example of web-manual for turbo-charger of diesel engine. The Figure shows the disassembling process a turbo-charger for overhaul.

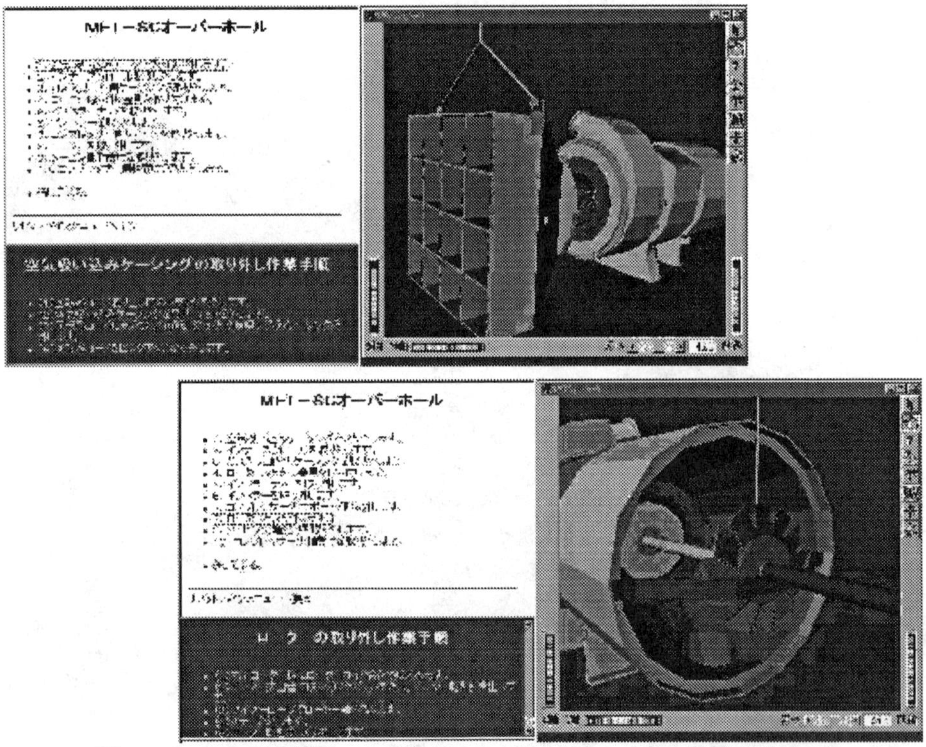

Figure 4 a manual of disassemble process maintenance for a turbo-charger

6 CONCLUSIONS

Web-based simulation architecture and its configuration for a communication model (WISE model) that is able to represent exactly real equipment in the software are presented. A WWW browser-based maintenance system is then developed to meet the requirements. The effectively of the principle demonstrated is validated.

The purpose of the study is to create such an environment that can execute a virtual maintenance with simulation in accordance with instructions in HTML through INTERNET. These web-based maintenance manuals satisfy with the requirements of DoD IETMs's Class 5 by Web-Server. This web-based multimedia manual confirms that they are very effective to the service engineer to train, evaluate the maintenance activities. The selection of tools is very good for maintenance manual, especially interaction sentence in HTML and Thee dimensional model with motion. The service engineers can effectively learn and confirm the maintenance process from these documents with motion. The size of the contents for turbocharger is 5Mbyte. The fidelity of three-dimensional model is suitable for service engineer to understand the shape of part itself. These maintenance manuals offer the effective computer aided training ability for service engineer. It took 1 week to add the motion to the three-dimensional model from CAD system. This is the tolerable time to develop the contents. Furthermore, the virtual maintenance itself provides through INTERNET, This system offers wide-area collaborative engineering environment in order to support the virtual enterprise.

Next Step of our study is the creation of system that satisfies intelligence of Class 5 in DoD IETMs specification by XML and linking procurements system and diagnosis system.

Finally, this research and development is implemented as a part of GLOBEMEN project under the umbrella of International IMS program.

7 REFERENCES

[1] Jorgensen E.J, et.al. "A Web-Based Architecture for Interactive Electronic Technical; Manuals (IETM)", http://navycals.dt.navy.mil/

[2] Jorgensen E.J, et.al. . "DoD CLASSES OF ELECTNIC TECHNICAL MANUALS", http://navycals.dt.navy.mil/

[3] Morita K., et.al. (1997)."World Wide Simulation Environment (WISE)", Proceedings of CAPE '97, pp.330-339, IFIP

e-Service for Complex Technical Products - a New Approach for Supporting Life-cycle Services

S. Bürkner, H.-P. Wiendahl
Institute of Production Systems, Hanover, Germany
Em: Buerkner@ifa.uni-hannover.de

Keywords Life-cycle, e-Service, Internet, Re-manufacturing

Abstract Service is an essential element of capital goods today. Sometimes it is even the most profitable business field for the manufacturer. This paper presents an investigation of the re-manufacturing industry in Germany and its competitiveness against new products. Requirements for making service more effective are adduced from this study, leading to the description of an internet-based life-cycle service system. This service system supports all phases of the life of a complex product from configuration to operation, repair and re-manufacturing.

1 INTRODUCTION

Companies producing capital goods have for a long time focused on technological leadership or low prices, but users increasingly ask for the benefit a product generates and not just the product itself. They demand not only more service but customized service. Service is widely seen as the central post-industrial product. According to Sihn, however, original manufacturers of equipment provide only 20% of the service of the capital goods they produce [1]. This is only a small fraction of the money they could earn through service. The delivery of service parts is an especially important business in this respect. In good years, the car manufacturer Volkswagen alone generates a profit of 2 billion Euros just by selling spare parts [2].

The potential for profit from service parts has been investigated by Cincom, a supplier of Enterprise Resource Planning systems (ERP) for life-cycles. Cincom estimates that the profit to be gained from products in this field can be much higher than the profit from selling the original products if the number of products sold annually does not exceed 2-6% of the total sold to date. The profit which a capital good generates throughout a product's life-cycle can be considerably higher than the profit from selling the product itself [3].

In Eastern Europe, machines have been running for more than 50 years due to cheap labor costs and a lack of new machines, thus creating a large service market. Similar machines in the West have been replaced by new ones much earlier. If the processes of repair and re-manufacturing in countries with high salaries are also to compete better against new products, then ways have to be found to make the practice more efficient and to reduce the amount of work that does not add value.

2 USAGE: A POST-INDUSTRIAL PRODUCT

The usage or benefit of products is becoming more important for customers than the product itself. This is due to the fact that companies no longer regard the ability to use and maintain a machine as a core competence. An airline, for example, understands its mission as being to transport people and not to repair aircraft, which can be done more efficiently by other companies. This is leading to a sort of dematerialization, where customers buy the use of the product rather than the ownership of the product. Producers of photocopying machine are pioneers of this approach. They mainly lease machines to offices and then exchange them when they become too old for the customer. As a result, the office always has up-to-date and reliable copy machines. The manufacturer takes back the old machines and re-manufactures them as new. Producing new machines is getting to be the exception [5].

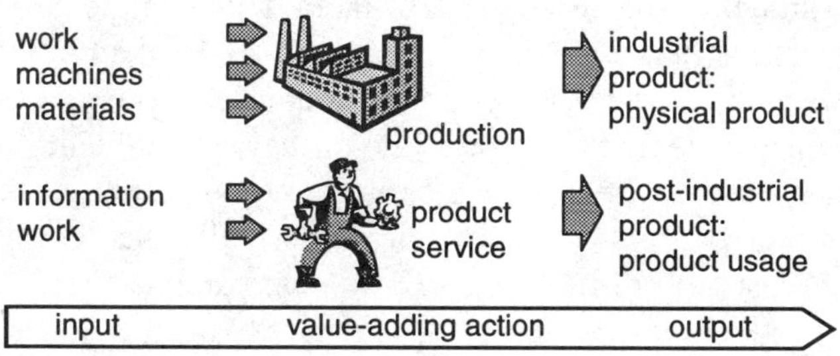

Figure 3: industrial and post-industrial products

Not only companies such as IBM, Hewlett Packard but also traditional industrial companies such as General Electric are also starting to focus on service. IBM already earns more money from service than from selling new computer systems [6]. General Electric Aircraft Engines already generates 57% of its income from the after-sales market. Selling service has a great impact on the value-adding process, because the input factors, the process of creating value, and the output change. The traditional input factors of production (i.e. work, machines and materials) are gradually being substituted by information and skilled labor. The result of the service process is not a physical product (which already existed) but merely a benefit generated by the service (Figure 1). Another difference between the industrial and the post-industrial product is that the latter is often created on the customer's premises or in local workshops.

3 PRODUCING USAGE: RE-MANUFACTURING COMPANIES IN GERMANY

The re-manufacturing industry is the biggest sector to undertake service throughout the life of the product. A survey of 102 re-manufacturing companies was conducted in order to understand their problems. These companies do not usually produce new products, but restore old products to an as-new condition. They attract little notice despite being a large industrial sector. 73,000 re-manufacturing companies are registered in the USA alone. They employ 480,000 people, which is as much as the US steel industry employs in total [7].

Most of the companies in the study were engaged in the re-manufacturing of tool machines (27%), plastic processing machines (8%), office technology (9%), and construction machines (8%), but the survey also covered companies that re-manufacture car components, cranes, motors and aircraft. The authors can provide a detailed analysis on request.

Re-manufacturers usually sell re-manufactured products in as-new condition at a much lower price. 60% of them, for example, are able to offer the products at less than half the price. Only 10% of the companies charge more than two third of the price of a new counterpart. The re-manufacturers are also competitive against producers of new equipment in respect of delivery times. While the producers of new equipment have to build a product completely and produce or order all the components, re-manufacturers usually re-use most of the non-standardized components. Savings in production and ordering times mean that 60% of the re-manufacturers are able to deliver a complex overhauled product within the first month. Only 5% require more than three months.

Since these companies usually work closely with their customers and since they often carry out maintenance on the customer's site, one might expect the service to be supported by advanced information systems. In practice, however, most of what they offer on the internet, for example, consists of general product information and a presentation of the company. It is very rare that spare parts can be ordered through the internet. In response to the survey's question about the importance of internet, many companies (35%) said that the internet would be an important additional medium. 15% felt the need to provide more content on the internet, but admitted to not having a strategy.

If the quality of the re-manufactured products is as-new and the price is lower, the shortcomings of these companies can be identified in the provision of information and the insufficient automation of processes by means of computerized systems. The information supplied both to the service employees and to the customer is mostly inadequate.

4 MAINTAINING USAGE: PRODUCT SERVICE THROUGHOUT A PRODUCT'S LIFE-CYCLE

The most common medium used in product service is the telephone. It connects service employees with customers all around the world and can be used for a direct exchange of information. This is often problematic, however.

If a customer cannot identify the required exchange part, for example, he either sends the part to the manufacturer, who then does the identification, or he describes the part verbally, which can be time-consuming and may cause mistakes. If complex information has to be exchanged or if sources such as drawings or pictures are needed, communication via email can be time-consuming. International service can also entail language problems.

Internet technology can provide information in the right context, i.e. customized to the product, to the user and to the particular phase of the life-cycle. Such a system can be easily kept up to date and can deliver information in various languages without having to change its structure. Throughout the life-cycle of capital goods, two major phases can be supported by internet-based product service.

During the period of use, services are needed to keep a product functioning and to pre-empt breakdowns. At the end of usage life, a recovery service can restore the product to an up-to-date technological state. Figure 2 depicts a typical product life with its remaining usage.

If and when parts have to be replaced, spare parts need to be ordered. This is usually done by selecting them from drawings or parts lists, which are often out of date or unavailable. For the identification and ordering of the required parts, electronic systems using interactive drawings can help, making this process faster and avoiding such mistakes as the ordering of the wrong part.

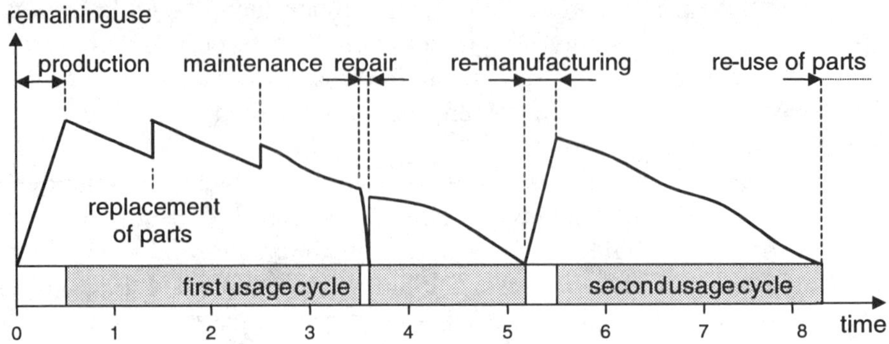

Figure 4: Services in the life-cycle of a product

During the period of usage of a product, the documentation, checklists and guidelines have to be available. Having this documentation electronically accessible helps keep it both up-to-date and in the right context and language.

For maintenance, specific information is required about the condition of a product, its history, the parts already replaced, modifications, and the current operating cycles of parts that wear out. This information is dynamic and cannot be generated in advance. It needs to be recorded during the period of usage. If it is available (e.g. if the manufacturer is the service provider), valuable information and assessment can be provided electronically. It is possible, for example, to generate a current list of parts which are due for replacement. In this case, the ordering could be triggered instantaneously. It is also valuable to provide an analysis of the running costs of a product within a particular time period, including all the parts, service and energy consumption.

If re-manufacturing is to be undertaken, information is required so as to be able to plan the process. Even before the product is disassembled it is often necessary to predict which parts will have to be exchanged and to order them. Assessments based on the age of specific parts and their typical failure rate improve these predictions. In searching for parts that could be salvaged from discarded products, product-specific electronic parts lists can help in finding suitable ones. In this connection, a part with a remaining life-expectancy similar to that of the overall product should be chosen.

5 SUPPORTING USAGE: INTERNET-BASED PRODUCT SERVICE

The Institute of Production Systems, in cooperation with EyeProm GmbH, has developed a web-based system which takes all these considerations of product service into account. The aim of the system is to support all service needs during and at the end of a usage cycle. The Life-Cycle Service Solution (LSS) consists basically of functions that can be combined into modules which work as portals for user-groups with different information needs

The product structure and the documentation structure are defined in a relational database that has to be created for the LSS. This database contains individual items and their connections.

Items in the product structure (product groups, product types, products, components and parts) are linked together forming products. The items can be part of different product variants, by defining different connections. This enables a rapid definition of customer-specific variants.

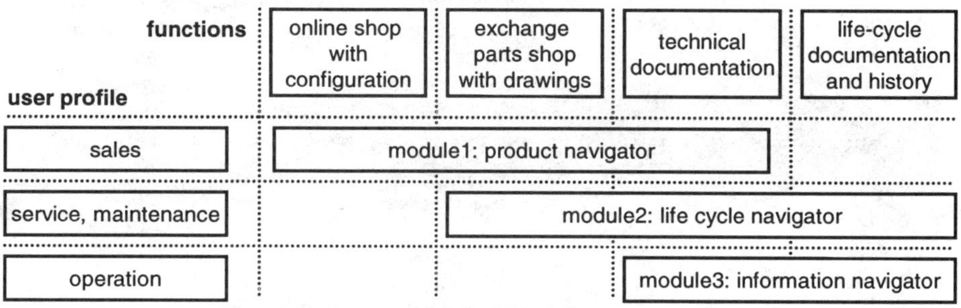

Figure 5: Structure of the life-cycle service system

Items in the documentation structure are also organized hierarchically. Information items can thus be combined to create complex manuals by flexibly linking these items together. This enables the automatic generation of customer-specific manuals by linking the general product information with variant-specific information modules (e.g. for extra accessories).
Other data sources for the information content are Product Data Management (PDM) systems for the manuals, pictures and drawings, and ERP systems for delivering commercial information about products and parts (e.g. availability and price). As the LSS can extract information from existing data sources, an application is possible without much additional data handling. This information is used by various service functions of the LSS (compare Figure 3):

- the online shop with configuration allows the definition of complex products individually for a customer. It is possible to check beforehand in the ERP system to see if the required components are available in the stock;

- the replacement parts shop allows parts to be identified from drawings and to be bought directly. As in the online shop, it is possible to generate an order using email, fax or ERP systems;
- the technical documentation contains general information about product types, such as operating manuals, drawings, diagrams, maintenance information and diagnosis advice;
- the life-cycle documentation records the history of serialized products with all modifications, part exchanges, repairs and specific configurations. Various assessments with regard to wearing parts, availability and costs can be performed on the system.

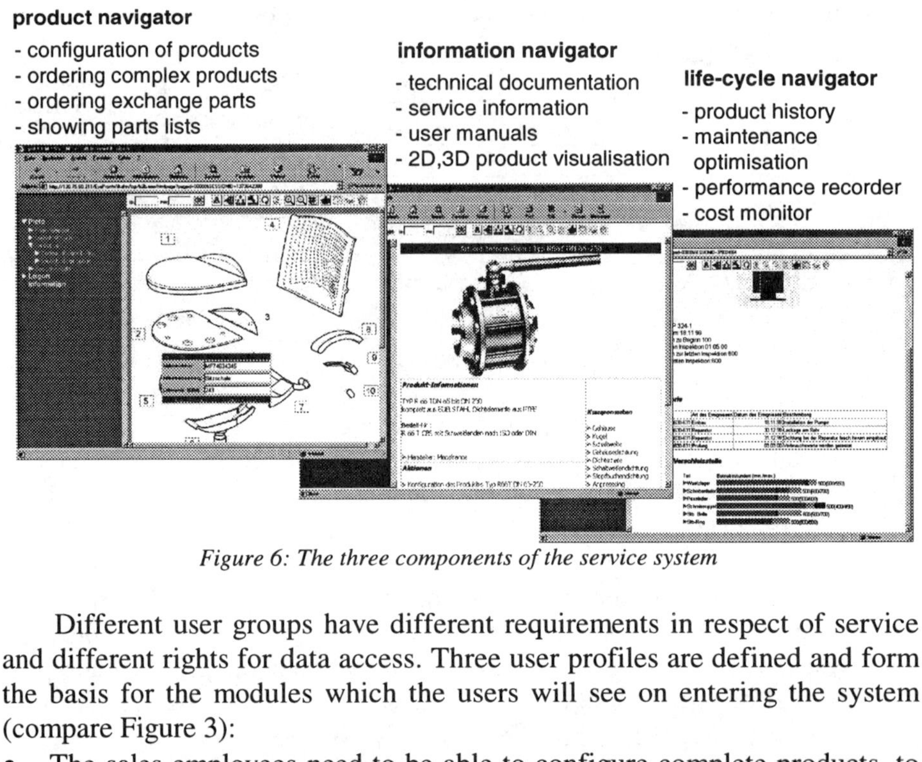

Figure 6: The three components of the service system

Different user groups have different requirements in respect of service and different rights for data access. Three user profiles are defined and form the basis for the modules which the users will see on entering the system (compare Figure 3):

- The sales employees need to be able to configure complete products, to order them, and to select and buy exchange parts. They also require general product information.
- The product service and the maintenance departments require information about the life and the service history of a product as well as the facility to order exchange parts.
- The operators only need access to the general and the product-specific documentation. They cannot make use of shop functions. The product

scope to which they have access can be limited to one product (Figure 5).

When the user logs into the system, a language adaptation is performed. The system then identifies those products for which he has rights and which user profile he has. He then gets access to the modules, which is configured for his specific requirements. Three basic modules can be dynamically generated: the product navigator, the life-cycle navigator and the information navigator. These modules combine different functions into integrated product-, user- and profile-specific service pages as depicted in Figure 6.

Security is an important feature of the LSS. Users need a password and a login (to load the user profile, language and product scope). A transaction number is then created for the rest of the session. Such an access can be generated customer-specifically and directly printed on the tag of the product or into the manual.

7 CONCLUSIONS

Information is a key element of today's complex technical products and the main resource for keeping them functioning. This paper has presented an approach for delivering information quickly and safely. Different functions are essential for product service systems. However, these functions need to be assigned user profiles and geared to product-, problem- and user-specific contexts. The result of this is a user interface, the so-called navigator. A user logging into the system receives the right information for the products serviced by him, and in the right language. Three different navigators have been implemented. Password-protected transponders make possible the physical combination of product information and the product itself.

8 REFERENCES

[1] Sihn, W. (1999). Deutsche Firmen brauchen mehr Innovationskultur, VDI Nachrichten, **34**, 2

[2] Wüst, C. (2000), Nirvana der Radkappen, Spiegel, **2**, 16

[3] Cincom: (1998), OEM Service Parts Management - Supporting on Out-of-Product Line, Brochure of Cincom, **8**, 2

[4] von Weizsäcker, E. U., Lovins, A.B., Lovins, L.H. (1997). Faktor 4 - Doppelter Wohlstand - halbierter Naturverbrauch, München, Kaur

[5] IEEE Expert (1996), Green engineering: AI pioneer cutting a trail, in: IEEE Expert Intelligent Systems, **6**, 4-6

[6] Stewart, T.A. Der Vierte Produktionsfaktor, Hanser Verlag, 1999

[7] Mühlemann, P. R. (1997), Up and down: Cycling auf jeden Fall: Technische Rundschau, **3**, 24-28

Automated Management of Quality Control System for Network Enterprise

J.P.T. Mo
CSIRO Manufacturing Science and Technology, Australia
Em: John.Mo@cmst.csiro.au

Keywords Quality Control System, Remote Operations Support, Web Technologies, Process Modelling

Abstract One of the driving forces in manufacturing in the last decade is the demand of customers on the quality of the product they buy. However, as products become more complex and involve a lot of other subcontractors, inter-organisational quality control functions require strong support from the organisations involved as well as an effective IT infrastructure. This paper describes a project involving a large transport equipment manufacturer using a methodology known as FIRST to develop an automated quality management process. The FIRST process provides a framework for the company to capture quality control information into a process model which can then be converted to the final user-end system automatically.

1 INTRODUCTION

It is generally accepted that quality is a keyword for winning current global competition in every business. Total Quality Management is one of prominent management strategies to win this competition [1]. In a company, the way to manage quality is by establishing a quality system. This comprises a systematic approach to plan, to do, to monitor and to control company's operations in achieving customer satisfaction [2].

In aviation industries, requirements for quality system are common practices. They are usually tied with safety requirements such as the airworthiness issues. Therefore, an aircraft manufacturing company must establish a quality system in its operation. Both quality manual and procedures must be in place.

Along with internal efforts on quality achievement, a company needs to acquire quality materials, parts, or components from its suppliers. This is crucial because inconsistent quality of incoming resources may cause difficulties in achieving quality products.

There is an obvious conflict in buyer-supplier relationships. Suppliers may perceive that high quality products will increase manufacturing cost. On the contrary, buyers will always try to find a lower price for every material they purchase. The most common approach is to establish a supplier development program.

With the above imperatives, this project is motivated by the need of an aircraft manufacturing company to obtain such commitment on quality from its widely distributed suppliers.

2 ROLES OF QUALITY SYSTEM DOCUMENTATION

In quality management, there are two major roles of documentation. Firstly, system documentation demonstrates that the quality system is under control, clear and convincing. Management should know how the system is regulated. Employees need to comply with the system. Internal auditors should understand the system before they check the actual condition. Secondly, documentation on quality records is of authentic significance. To evaluate achievement level of quality parameters, records are needed. The evaluation process can then be used to determine corrective actions for process capability and product quality improvements.

A typical quality system in a company comprises a hierarchy of documents. This hierarchy forms a pyramid structure, which places quality policy at the top of the pyramid (Figure 1). Other quality documents must be developed based on the top level policy.

Figure 1 Hierarchy of Quality System Document

Like other documentation systems, quality system documentation traditionally uses paper as its media. Up to now, it is still a popular media that keeps the consistency of information in a reasonable period of time.

There are several alternatives of documentation media, for example microfilms and optical disks. These media have considerably reduced the use of paper. However, for daily operation, they need high cost equipment, materials and processes. Moreover, by using these media, the documentation control system remains the same and it is difficult to propagate changes to lower levels.

Nowadays, efforts to reduce paper in human activities is growing. Companies may gain potential advantages of using a paperless documentation, especially in quality system. Several efforts have been started since the wide use of computer, but the developments are still in the initial stage [3].

3 WEB-BASED QUALITY SYSTEM

The use of computer is an important consideration in achieving paperless documentation system. The latest research has involved the Internet and Intranet technologies. These technologies promise a great benefit in distribution problems, because they will eliminate delays in delivery even though the users are located in a long distance.

A quality recording system may as well take advantage of web-based technologies by using capabilities of Java programming in its pages, thus making the system truely interactive and uniform. This makes it easier to enter and retrieve data.

An interesting result that comes out is when the recording system is combined with the web-based documentation. With an interactive interface, a seamless and consistent look-and-feel quality system may be produced. Last but not least, companies can benefit from the independence of web system on the users' platform. The investment for new computing systems is minimised.

4 FIRST METHODOLOGY

FIRST stands for Framework for Integration of Remote Support Technologies developed by CSIRO Manufacturing Science and Technology. FIRST methodology was initially developed to provide automatic creation of remote operation support system [4]. The main concept of the framework is to integrate the World Wide Web technology, hypertext markup language (HTML), interactive forms, multimedia, artificial intelligent, and supported by a knowledge based system. Besides easing remote customer support, the framework may benefit in reducing customer support related costs and minimise service delay time when used over long distances.

An important aspect of the framework is the use of an integrated knowledge-based system [5]. To enable this function, the system must understand the entire customer support process including the capability to collect facts and to navigate customers within the process. In traditional approaches, off line support is provided by experts performing such tasks in a direct mode.

In the FIRST methodology, this function is carried out using a knowledge repository, which is built through IDEF3 process modelling [6]. IDEF3 process modelling has capabilities to capture processes including logical and temporal relationships between activities in the processes. A fundamental treatment of IDEF3 to drive knowledge representation has been written by Mo and Menzel [7]. From users' view, operational cost for accessing remote system is quite low, because it uses web browsers, which is a free software.

5 REMOTE QUALITY SYSTEM IMPLEMENTATION

Using the same framework, a remote quality system capturing the quality system manuals, procedures and work instructions can be developed. Depending on the system design, additional scripts to access quality control database can also be attached to the remote support system.

The importance of knowledge repository is not only for initial publication of the quality system documents, but also for modifications that may apply during their life cycle. When change is required, the model is modified and the web-based support system is regenerated. Thus, this approach accommodates the documentation control function.

As stated before, the object of this project is to develop a candidate quality system for suppliers of an aircraft manufacturing company. Currently, this company has implemented a similar system with the same content, but in non-electronic media.

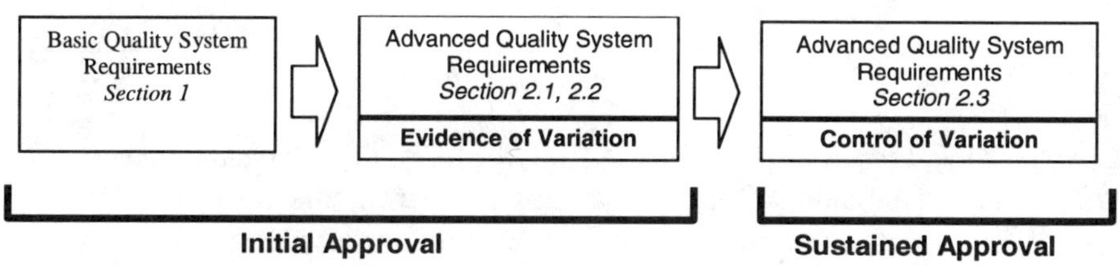

Figure 2 Quality System Approval Requirements

To make the system easier to understand, an overall Quality System workflow is provided by the company (Figure 2). It illustrates the step-by step process and documentation requirements.

The project started with the creation of a system model. The system is divided into four different subsystem and function (Table 1).

Table 1 : Main Subsystems and Their Function

Subsystem	Functions
AQS Concepts	Overview and basic understanding of the newly introduced system
Basic Quality Requirement	Descriptions of basic quality requirements that has been in place
Advanced Quality Requirements	Description of new quality requirements for suppliers, including recommended actions and flows to be accomplished.
AQS Tools	Explanation of various tools those can be utilised in fulfilling AQS requirements.

This documentation allows the main branch of the system model to be created as illustrated in Figure 3 and guides the users to select one of different subsystems they are interested to explore. A new user will easily decide to explore the Quality System Concept, while an expert will directly explore the Quality System Requirements to get the information he/she wants. Similarly, executives who will decide to implement the system can easily read the summary in the Quality System Concept section, while their technical staffs or subordinates will explore the details in the requirement sections.

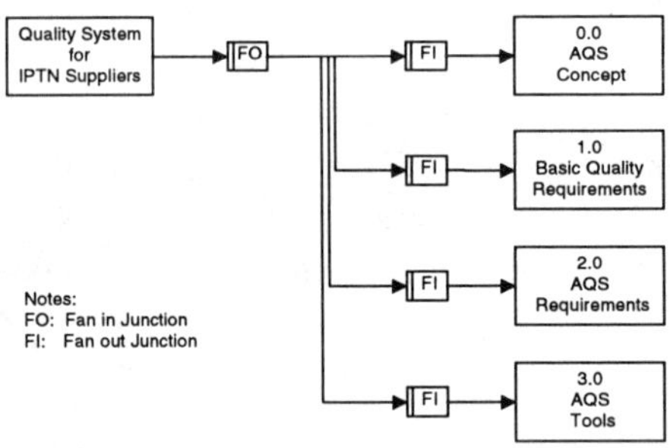

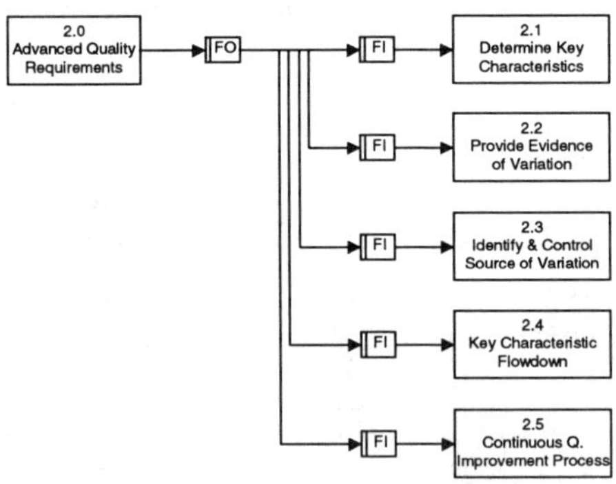

Figure 3 : Main Branch in IDEF3 Model *Figure 4 : Core Workflow Main Branch*

The AQS (Advance Quality System) Requirements section is the core message in this documentation. So, the system is developed intensively in this section. The main objective is to guide the user to explore the requirements in a sequential basis. The users are guided to follow the information interactively with the system. After exploring a certain process, users will be linked to some possible pages that relate to that process. There are two kinds of links, viz. links to the next process that follows the current process and links to referred pages that contain related material and tools. With this arrangement, it is expected that users can be easily and systematically navigated through the system (Figure 4).

In the complete model, apart from the quality related processes, one can also see two additional processes attached to the model (Figure 5). They are *Welcome* and *Thank You* processes. The function of *Welcome* process is not only to give a welcome speech to users, but also is the starting process providing a fan-in junction leading to one of the next level of processes. This is important, because the next process after the *Thank You* process is a *Home* base of the model.

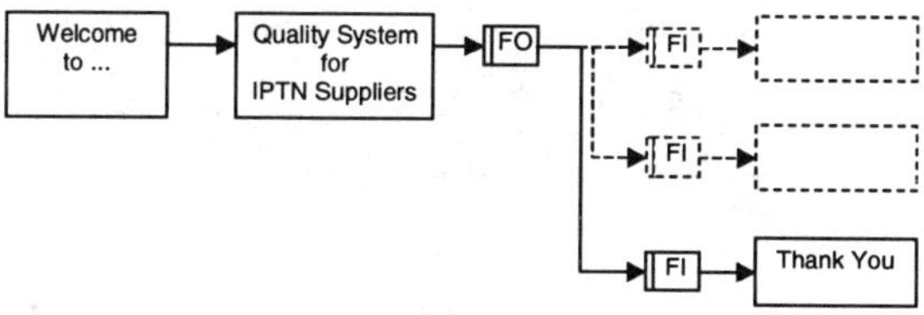

Figure 5 : Additional Processes

FIRST generates 101 web pages from the model. These pages build a web site and link one another providing trials of the quality system information and procedures. The FIRST methodology produces a start page in every site it develops. This page is always named as *start.html*. It represents the first process from which the model starts. For this project, the first process is *Welcome* page (Figure 6). A single click in the *Next Page* link word will start surfing the site.

Figure 6 : Welcome Page

It is impractical to explain the model step by step. However, to give an overall picture of how the model works, one can find four typical patterns of relationship among the processes in the model. They are *simple continue* pattern, *yes - no* pattern, *multiple choice* pattern, and *reference* pattern.

A *simple-continue* pattern is created in the model when the current process being modeled only have one possible link to the next one. Sometimes, this pattern can be used when we need to divide a long process into two or more process (Figure 6).

A *yes-no* pattern is needed when users are required to answer a yes-no question after evaluating the current process. If the user can identify the special cause of variation, then he/she will choose a 'yes' answer. This will cause the system to guide him/her to remove that special cause of variation, otherwise, the system will ask whether gauge variation study has been performed or not. For the example given in Figure 7, the site will bring the users to section 2.3.4 if they answer 'Yes' and to section 2.3.6 if they answer 'No'.

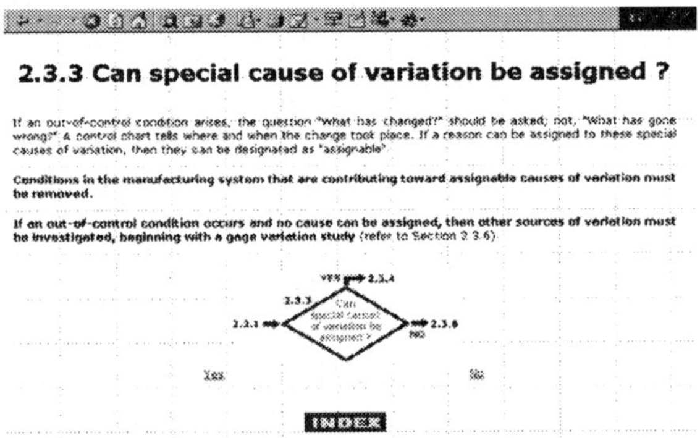

Figure 7 Yes-No Pattern Results

A *multiple-choice* pattern is created when users are advised to choose more than one possible answer. It is almost like *yes-no* pattern, but the answers to the question are more than simple *yes* or *no*. FIRST methodology has enabled this possibility by expressing all possible links in the current process/page. The link names are the possible page titles or ones those have been defined in *'process facts'* when modelling. An example of the result is shown in Figure 8.

6 CONCLUSION

Sharing information over long distance is a potential benefit of the emergence of the internet/intranet. To exploit this benefit, CSIRO has developed a methodology called FIRST. This methodology enables systems for remote customer support through the World Wide Web to be created with an integrated process model driven knowledge based system.

This project has implemented a Quality System using FIRST methodology. The main users of this system are suppliers of an aircraft manufacturing company to enable easy quality system access and maintenance. It is predicted that many areas in quality system may likely take advantages of this development, for example, supplier control, quality audit or even manufacturing control like measurement equipment control, facility and personnel certification and non-conformance material control.

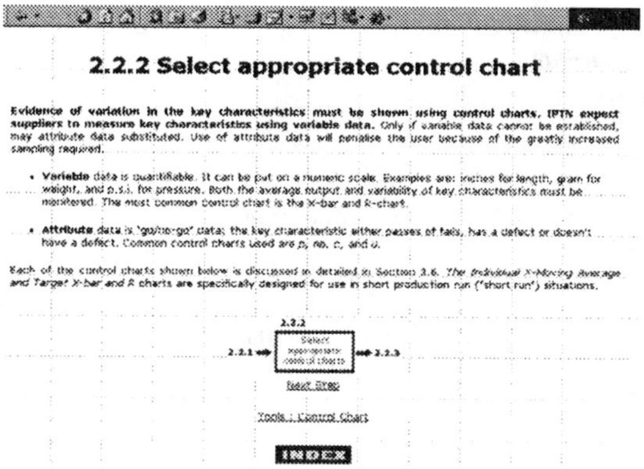

Figure 8 : Multiple-Choice Pattern Results

7 REFERENCES

[1] Mo J.P.T., Chan A.M.S. (1997). "Strategy for the Successful Implementation of ISO 9000 for Small and Medium Manufacturers", *TQM Magazine*, Vol.9, No.2, pp.135-145

[2] Scholtes, P.R. (1991). "The Team Handbook", pub. Joiner

[3] Koshy T.T., Gramopadhye A.K., Kennedy W.J., Ramu N.V. (1996) "Application of Hypertext Technology to Assist Maintenance on the Shop Floor", *Computers and Industrial Engineering*, Vol.30, No.2, pp.283-295

[4] CSIRO Manufacturing Science and Technology (1998). "FIRST-Framework for Integration of Remote Support Technologies", Development Manual, Ver.1.1, January

[5] Mo, J.P.T, and Nemes, L. (1997). "A Knowledge Based Remote Diagnostics System", IFAC Symposium on Artificial Intelligence in Real Time Control, 22-25 September, Kuala Lumpur , Malaysia, pp.541-546

[6] Mo J.P.T., Nemes L. (1998). "Framework for automatic creation of remote operations support systems", Third International Conference on the Design of Information Infrastructure Systems for Manufacturing (DIISM '98), May 18-20, Fort Worth, Texas, U.S.A., pp.93-104

[7] Mo, J.P.T, Menzel, C. (1998). "An Integrated Process Model Knowledge Based System for Remote Customer Support", *Computers in Industry*, Vol. 37, pp171-183.

Neo-kaizen Applications on the Generic Operations Support and Renewal

K. Mori, N. Yoshikawa
Omron Corporation, Japan
Em: kenichiro_mori@omron.co.jp

K. Morita
Mitsui Engineering & Shipbuilding Co., Ltd, Japan

T. Kimura, H. Goto
Japan Society for the Promotion of Machine Industry

M. Asamori, Y. Kamio
Toyo Engineering Corporation, Japan

Y. Fukuda
Hosei University, Japan

Keywords Model, Improvement, Operation support, Renewal, KAIZEN

Abstract Conventional TQC (Total Quality Control) activities do not use information technologies adequately. The TQC activities adapted to a global manufacturing environment have been limited. This research deals with using information processing systems for the improvement of manufacturing systems based on a TQC method for a global manufacturing environment. We are developing the G-OSR (General Operation Support and Renewal) in the IMS program. The G-OSR is a generic information architecture for operations support and renewal in the manufacturing. This paper describes to realize the KAIZEN demo scenarios based on the G-OSR and to show three prototype systems for applications. The application systems include KAIZEN support and remote monitoring support by information technologies.

1 INTRODUCTION

Continuous improvement activities for manufacturing systems are recognized more important than ever. Almost excellent companies carry out TQC (Total Quality Control) activities for the improvement of their processes to maintain the productivity for operating a manufacturing system. The activity is called KAIZEN activity in general.

However, conventional TQC activities have not been supported by information technologies compared with operation and maintenance [1][2][3]. The reasons are (1) lack of information models for the KAIZEN activity, (2) the cost of building an information system for the activity is higher than the benefit from the KAIZEN activity, and (3) operators or managers in the shop floor do not focus on their information systems as an improvement target.

In this paper, we propose the Neo-KAIZEN system to support continuous improvement based on an enhanced model and generic support system architecture. The system consists of data monitoring and logging from plants, statistical quality control assisting tools, simulation systems for evaluating the system performance, and operator support systems. The Internet and local area networks link all sub-systems.

Firstly, the KAIZEN activity and its extended model for a support system are proposed. Secondly, Generic Operations Support and Renewal (G-OSR) information architecture is described for a generic system design. Thirdly, Neo-KAIZEN systems based on the G-OSR are explained as application demonstrations. We discuss the performance and the improvement agility from the applications.

This research and development project has been supported by *Global manufacturing toward 21century* (*Globeman-21*) project and *GLOBEMEN* project in IMS International Research & Development Program. The G-OSR is developed in the project based on the discussion with Australia, Europe and Japan partners.

2 THE KAIZEN ACTIVITY MODEL AND ITS SUPPORT SYSTEM

The basic method of continuous improvement (KAIZEN) system is TQC activity [4]. The purpose of TQC activity is to solve problems relating Q(Quality) C(Cost) D(date of delivery) S(Safety) M(Morale) on operating plants. All members in an enterprise from president to operators should join to TQC activity. The activities are carried out in all product life cycle such as market research, research and development, design, operation, and maintenance. The TQC activity consists of four processes, P(Plan) D(DO) C(Check) A(Action).

To apply information technologies to the KAIZEN activity, PDCA process is extended to a new process model. As a result of analysis of basic models, we propose a spiral approach model for the KAIZEN activity support system (Neo-KAIZEN system) which has six phases instead of PDCA. Figure 1 shows the spiral approach model and the relation of PDCA process. Then the system is able to improve the QCDSM concerned with operation in the life cycle of manufacturing supported by information technologies. The six phases in Figure 1 are followings,

(1) Get Operation Status -The information concerned to operate the manufacturing systems is collected from the real shop floor.
(2) Analyse the Operation Status -The operation status of the manufacturing system is analysed based on the collected information.
(3) Find the Problems for the Operation -The major problems of the manufacturing system are found out from the result of analyses.
(4) Plan to Solve the Problem -The improvement proposal to solve the problems of the manufacturing system is planed.
(5) Evaluate the Plan -The proposal to solve the problems is evaluated in advance by simulation.
(6) Implement to the Process -The evaluated plan is implemented to the manufacturing system to improve the process.

Executing these six phases is repeated continuously on the Neo-KAIZEN system.

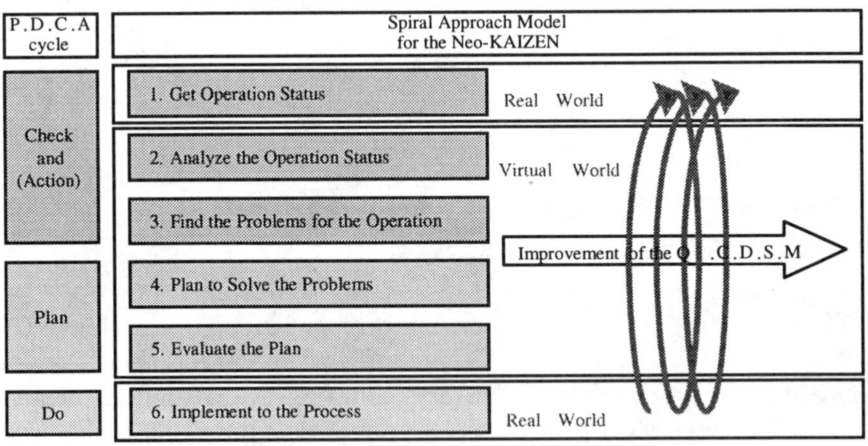

Figure 1 Spiral Approach Model for the Neo-KAIZEN system

3 G-OSR INFORMATION ARCHITECTURE

G-OSR (Generic Operations Support and Renewal) proposed here is a trial to realize a systematic approach to develop support systems of the processes of operations, maintenance, and renewal using the information technology. The objective of G-OSR is to develop a generic methodology for the common and cost effective design of the support systems [5].

The G-OSR architecture is defined based on the analysis of similar systems that supports remote operation and the KAIZEN activity. Additionally, the information infrastructure required to support systems is defined in the G-OSR architecture. The hierarchy of architecture and information technology corresponding to common applications and common tools are defined as the information infrastructure of G-OSR. G-OSR is visualized as a model that can make the use of the group of tools developed in this study, thus facilitating simplified integration with future systems. In addition, the groups of tools developed in the Neo-KAIZEN system are mapped to the G-OSR architecture.

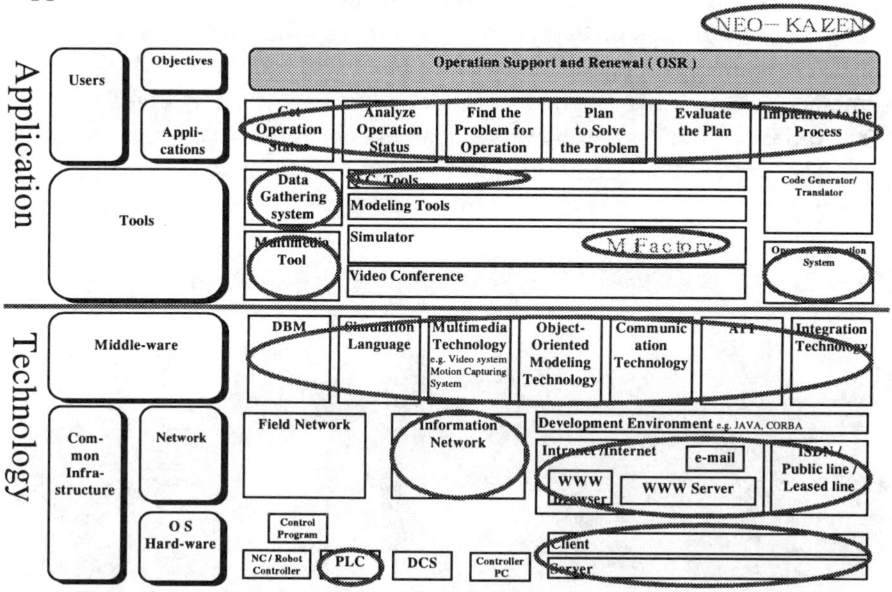

Figure 2 Mapping of elements of Neo-KAIZEN on the Architecture of G-OSR

Primarily, the G-OSR architecture is a matrix, within which groups of tools are classified as 'Technology' consisting of Middle-ware and Common Infrastructure on the vertical axis and 'Application' consisting of Users and Tools on the horizontal axis. On the horizontal axis, for instance, groups are classified into applications equivalent to each phase of the spiral approach of Neo-KAIZEN. Figure 2 shows the mapping of the groups of tools developed in Neo-KAIZEN systems to the G-OSR architecture. Common use of the information infrastructure and middle-ware enables to design and build the KAIZEN systems easily in the global environment. Low cost and rapid design of the KAIZEN systems can be realized by the G-OSR architecture.

4 EXPERIMENTAL DEMONSTRATION SYSTEMS

The Neo-KAIZEN spiral approach model was applied to the three experimental demonstrations of (1) operation improvement of an electronics product assembly, (2) quality improvement on an automated assembly and (3) improvement on a continuous process plant.

4.1 Operation improvement

The proposed Neo-KAIZEN system has been evaluated by the data from an U-shape assembly line in OMRON Corp. The assembly line of electronic equipment is shown in Figure 3. Over two hundred kinds of products are produced by the groups of two or three workers in the line. Their operations change dynamically depending on the model type or the batch size to be assembled. Usually the change for operation balancing is required. The system monitors the operation status, gathers the status data and stores the data to the database as time synchronized data [6]. Then, an improvement item is quickly proposed by the system.

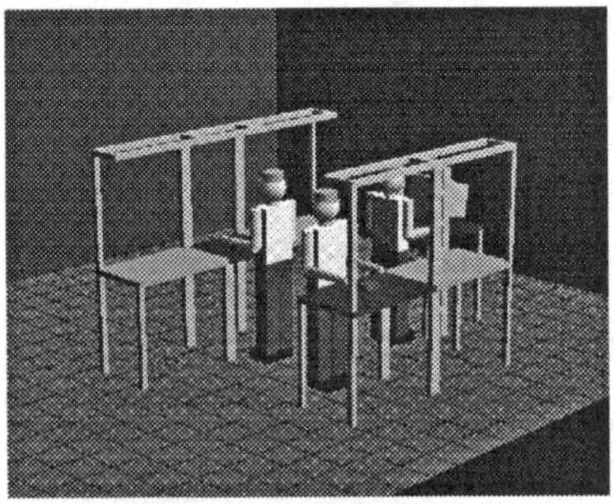

Virtual simulation model

Real Shop floor

Figure 3 U-shape assembly line

In this system, POP (Point of production) system and video camera monitor the workers for getting status. The system stored the logging data in the Web-based Synchronous Data Logger. The logged data is analysed by QCAS (a QC assistant tool). Then, a small group of engineers and workers make a solution for improving their operations. The solution can be evaluated by MiFactory (a simulation tool) before implementing to the real shop floor. After the evaluation, the Operator Instruction System informs the operation manual to workers using multi-media. We could get the successful result from this application.

An improvement scenario paying attention to worker's performance is described as a demonstration system according to the following steps.

(1) Record video images of the process if deviation from the pitch time exceeds 30%.
(2) Analyse causes of referring to the image.
(3) Enter data obtained from the analysis into the simulator model and make simulation to see if a relationship with worker's skill exists.
(4) Through simulation find out such work combinations and number of workers that promise higher work efficiency.
(5) Modify the work procedure.
(6) The result: As to the work that is analysed, time including preparatory time is reduced. After the KAIZEN activity is performed, the system performance is improved.

4.2 Quality improvement

The second application is improvement of yield of PCB (Printed Circuit Board) assembly operations in an automated assembly line.

(i) Target

In the inspection process of the PCB assembly line, an automatic machine solders parts accurately at a specified position, and the inspection equipment checks the soldering work.

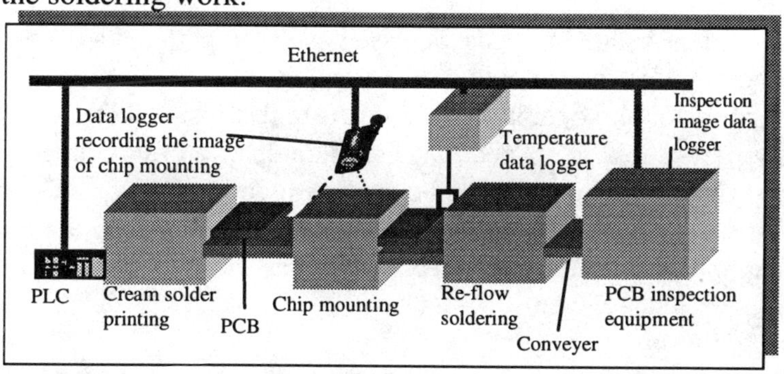

Figure 4 Configuration of the information system of PCB assembly

(ii) Tasks for improvement

The goals are to reduce the number of defective soldering when the line is in operation and to increase the assemblies yield.

(iii) Required information for improvement

- Visually check the soldered appearance.

- Check the shape, the height and the quantity of solder.
- Deviation in the quantity of solder can be determined from checks of the soldered area.
- Low Solder temperature.
- Defect of timing of chip mounting and wrong parts.

(iv) Improvement steps

An improvement demonstration scenario is described according to the following steps.

(1) Collection of primary information mainly obtained as data from the inspection equipment ascertains the real cause. (2) To establish the cause, collection of detailed information is realized using a data logger. When the inspection result is NG, solder temperature, video image of chip mounting and surface image of solder are recorded. (Collection of secondary information)

(3) Soldering process specialists are employed for the analysis of causal factors and solve the problems easily using the information from the automated assembly line.

(4) The result: The information gathering and editing time is reduced to half by the support system.

4.3 Demonstration in Continuous Process Plant

A Neo-KAIZEN system was implemented to a continuous process plant for a demonstration.

(i) Objectives

A chemical plant located in Asia was used as the target, and a demonstration was carried out with the following points as objectives.

- To support for site operation troubles from the design side.
- To support checking, inspection, and maintenance of equipment.
- Actual plant operation data to be analysed, and the data used in the next design such as process improvement or equipment specification review.
- To support for management works such as production management or equipment management.

(ii) Configuration

To accomplish above objectives (i), the system was reviewed using the configuration shown in Figure 5. The study concentrated on (a) data collection, (b) data transfer, and (c) some data analysis and evaluation.

(iii) The result

Experiment with the demonstration system configuration was applied at an existing overseas plant. It was confirmed that the collection and transfer of the data through the Internet could be carried out automatically. The system has now entered the stage of operational test.

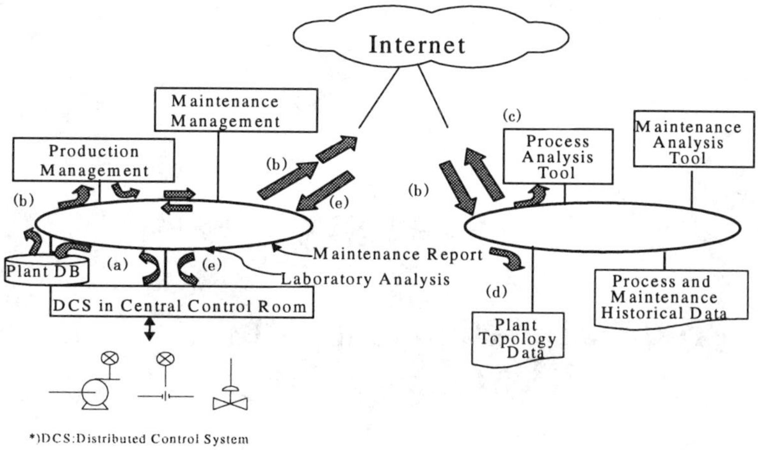

Figure 5 System Configuration of a Continuous Process Plant for Demonstration

5 CONCLUSIONS

The Neo-KAIZEN systems based on the spiral approach model were proposed. Three applications of demonstration systems using the data from some plants were designed. It has been verified that the Neo-KAIZEN systems on the G-OSR information architecture are feasible and can efficiently perform the KAIZEN activity on operating electronics assembly lines and a chemical plant. These systems will be easily designed on the G-OSR information architecture in the future. We will continue to study the methodology of design for the G-OSR systems.

6 REFERENCES

[1] Goto, H., et.al.,(1997). "Remote Renewal By Aggregating Real and Virtual Models In Manufacturing Systems", Proceedings of IMS '97, pp.77-81, IFAC.

[2] Mitsuishi, M., et.al.,(1995). Proceedings of IROS'95, pp.13-20, IEEE.

[3] Iwata, K., et.al.,(1995). Proceedings of EI'95, pp.154-167.

[4] Hosotani, K.,(1992). The QC Problem Solving Approach - Solving Workplace Problems the Japanese Way, ISBN4-906224-91-1 C0034.

[5] Ollus,M.,et.al.(1998). "Systems for Support of Process Operations", Proceedings of IiM98.

[6] Mori, K.,(1998). "Computer Aided Continuous Improvement (KAIZEN) System on Operating the Manufacturing System", Proceedings of PTK98, pp.375-380, IPK.

PART SIX

Product Development and Life Cycle Management

Towards Information and Knowledge in Product Realization Infrastructures

John J. Mills, Jan Goossenaerts
Group Information Technology, Faculty Technology Management, The Technical University Eindhoven, Eindhoven, The Netherlands.

Keywords Product Realization, Knowledge, Information Infrastructures

Abstract For many decades now, the focus in information technology to support the product realization process has been on data: its creation, management, and use. A general sense is emerging that data alone is no longer sufficient to support the people involved in this process. In this paper we suggest that it is time to start understanding what Knowledge is necessary to the product realization process and to move towards thinking of and implementing knowledge infrastructures and product knowledge managers. The authors provide some definitions of data, information, and knowledge in general and show how it relates to information and data. Specifically, we highlight the importance of (a) relationships or patterns in data and information, and (b) context to understanding knowledge. As part of our definitions, we provide some brief examples of relationships and contexts in the product realization process and show that the manufacturing community has already started to manage knowledge in the product realization process.

INTRODUCTION

As products become more sophisticated and complex, and computers are used to perform more and more tasks during the product realization process, it is becoming clear that current approaches to managing the representation of the product do not support the realization team well. Product data managers (PDM's), for instance, do not provide adequate support for either designers or production planners.

While there is work on defining what the next generation of PDM's should look like [1], there is also a growing realization that managing the product data alone is not sufficient. As Thoben et al have said: "...a learning organization requires the management of product data and knowledge." Several authors have started to address the issue of managing the knowledge created during the realization of the product and specific methodologies are beginning to emerge [1,2]. Further, there is increasing attention being paid to managing knowledge as a corporate asset [3,4]. However, it is not at all clear what is meant by "product knowledge" or by "knowledge" in the Product Realization Process (PRP). Discussion of "knowledge" in the literature ranges from the philosophical (i.e. epistemology [5]) to vague general descriptions. In this paper, we provide a perhaps more pragmatic approach to defining and understanding what knowledge is and what product knowledge might be. In particular, we show that context and interrelationships are a neglected but very important aspect of knowledge in general and product knowledge in specific. In this discussion we draw on examples of data, information and knowledge from the product realization process and show that, as a community, we have already started along the path of creating knowledge infrastructures and product knowledge managers.

DATA, INFORMATION AND KNOWLEDGE

Data

There appears to be general consensus in the literature on the definition of data but that the definitions for information and knowledge vary considerably. ***Data*** are simply symbols (e.g. numbers) with no context and no relationships [6,7,8]. For instance, the separate characters, "1.2,""3.4", "in.," "C," "A," and "B" are just data. In the product realization process, data is just the numbers and symbols used in describing, for instance, a line, a vertex, the material used, the machine capacity. Other data is required and relationships must be identified before these numbers and symbols can be interpreted as lines, vertices, etc..

Information

Information has been variously defined as: 1) a quantitative measure of the content of information [6]; 2) consisting of processed data, the processing directed at increasing its usefulness [7]; 3) the result of comparing data that are structured to provide a meaning within a given context [8].

Information exists when we recognize relationships in data within a specific context. The symbols "1.2", "in.", and "B" taken independently, do not mean much to anyone. They are just data. Stated as "1.2 in. from B," most people would understand this relationship as a distance from a location, "B". Written this way, these symbols impart meaning to the reader. Looked at another way "1.2 in. from B" is a pattern that the mind recognizes and associates with other patterns stored in long term memory.

In general terms, therefore, when patterns or relationships can be discerned in data, we have information and meaning is given to data: the brain is able to (a) perceive the patterns, (b) match them to patterns which it has stored and (c) connect them to other patterns. This perception of the pattern and its matching and connection to existing patterns we call *recognition* and *understanding*.

As another example of information, an ASCII view of a STEP or IGES file is just a list of numbers or just data, but when this data is read by a software program than can interpret each number or symbol according to a data model which determines the relationship between the numbers, and display this relationship on a computer screen, the user recognizes the relationship as a geometric shape. The recognized relationship between the numbers makes them information. Given a different data model or context in the relationship might end up meaning something else. This is where information begins to become knowledge.

Knowledge

Knowledge is much more difficult to define because it has so many possible interpretations as illustrated by the entry in Webster's Dictionary [6]. In what follows we avoid delving into epistemological arguments about knowledge and attempt to stay on a more practical level. First, there is the definition (in Webster's) in terms of "the sum of what is known" or "the body of truth, information and principles acquired by mankind." [6] It is the collective knowledge of mankind. We call this "***Total Knowledge***" and it exists mostly in a very diffuse, not very useful state. Part of this Total Knowledge can be codified or made explicit while the rest will remain locked up in the brains of humanity until ways can be found for extracting it from individual human minds. This event is not likely in our lifetime.

The individual contributions to Total Knowledge are what Polanyi has termed ***Personal Knowledge*** [9]. As Polanyi has suggested, part of ***Personal Knowledge*** is ***Tacit*** and part is ***Focal*** [9]. While Tacit Knowledge can play a substantial role in decision making and problem solving in the product realization process , it is hard to provide information technology support for it. Nonaka and Takeuchi have suggested ways in which tacit knowledge can be usefully tapped in product realization teams [3].

Tacit knowledge is the sum total of a person's experience and training and includes parts which cannot be articulated by the person. As Polanyi has put it " We know more than we can say." [9]

According to Polanyi [9] *and Weggeman* [8], ***Focal Knowledge*** is knowledge that is applied to a specific task. Focal knowledge encompasses codified or explicit knowledge. The codified part of focal knowledge is the knowledge we can usually articulate in some form. From these definitions, it should be noted that "codified knowledge" and "explicit knowledge" are the same. This is the personal knowledge which can be written down in some form or other. The rest of our focal knowledge, we cannot make explicit, at least at that point in time. Figure 1 illustrates our suggested relationships between these different types of knowledge.

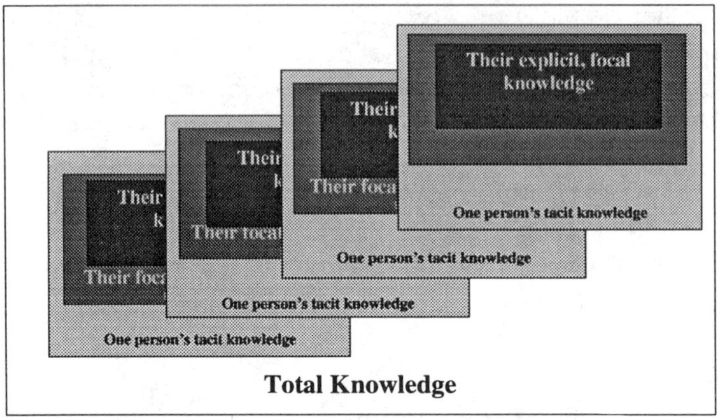

Figure1. Proposed relationship of the different kinds of knowledge to each other

To view knowledge from a different perspective, we examine the mechanism by which data becomes information becomes knowledge. In our view, the mechanism by which a string of symbols becomes information is recognition and understanding or comprehension. Understanding and comprehension are used interchangeably here. First, a pattern is recognized within a particular context, then internalized (or indwelling as Polanyi calls it [9]). In this recognition process, the pattern in the data is matched with an existing pattern within its context creating information.

When a person recognizes and understands complex patterns in data (i.e more complex than the relationships which create information) and in information itself, knowledge is created. In the understanding (Polanyi's indwelling) process, connections are made in the mind to other, related patterns (i.e. personal information and knowledge) in the same context. When appropriate connections are made, the person "understands" the information, and it has meaning for them: it becomes knowledge.

In another method of creating knowledge, a person connects patterns, recognized in information within a specific context, to similar patterns in other contexts and then, recognizing and internalizing that pattern across contexts, creates knowledge in their mind. They also create knowledge when that recognized pattern is used in other contexts. The pattern "F=ma" by itself is simple information. The understanding that the pattern "F=ma" can be applied in predicting the motion of sub atomic particles, atomic particles, human sized bodies (e.g. a car), space craft, planets, star systems, galaxies, each of which is a different context, creates knowledge.

Moreover, knowledge can be also created from other knowledge. For instance, Newton had access to all the codified knowledge of his peers and predecessors on the motion of bodies but only he was able to see a pattern which resulted in his equations and theory of motion thereby creating new knowledge. Similarly, Planck was able to see patterns in available codified scientific knowledge that allowed him to formulate his Quantum Theory.

Hence, knowledge is created or exists when (a) complex patterns are recognized and understood in data, (b) patterns are recognized in the simple relationships of the information and these patterns extend across several contexts; and (c) patterns are recognized in existing knowledge itself.

This leads naturally to the concept of Re-ification in which the mind collects all the knowledge about a concept under a single "thing" or Res [10]. As suggested by Wilson, reification involves taking a complex set of knowledge and making it into a single entity. This “Res” can then be incorporated into other concepts of greater complexity, leading to a layering of knowledge as suggested by Feynman and Polanyi. Feynman's layering seems to be mostly hierarchical [11], but both Polanyi and Wilson indicate that the connections between layers is much more complex than a simple hierarchy with connections within layers and multiple connections between layers [9,10]. This leads to the idea of what we call connectedness or the degree of connectivity among concepts or Res. Our current hypothesis is that the degree of connectivity is related to the complexity of the relationships existing within knowledge and hence to the level of knowledge.

Knowledge, therefore, comes into existence with the recognition and understanding of patterns and how they can be used in tasks like problem solving and predicting phenomena across a wide variety of contexts at increasing levels of complexity. As Polanyi has said; the true knowledge of a theory *"lies in our ability to use it."* [9] Since context independency and connectedness appear to be two parameters that define knowledge it seems natural to use a graph to provide a framework for describing knowledge, Figure 2. The large arrow indicates the transformation of data into information and information into knowledge etc. This figure is derived from a similar figure by Bellinger [12]. The difference is discussed below.

Since all knowledge is ultimately personal, there must be some human factor in this graph. What makes knowledge personal is a person's abilities. It seems to us that the greater the person's capacity for identifying, storing, matching and connecting patterns, the greater their ability to create knowledge of greater connectedness and context independence. This view recognizes the ability of some people to recognize significant patterns in a lot of noise and to be able to make good decisions in the presence of uncertain information. We call this ability "*Innate Intelligence*" for want of a better phrase at this time.

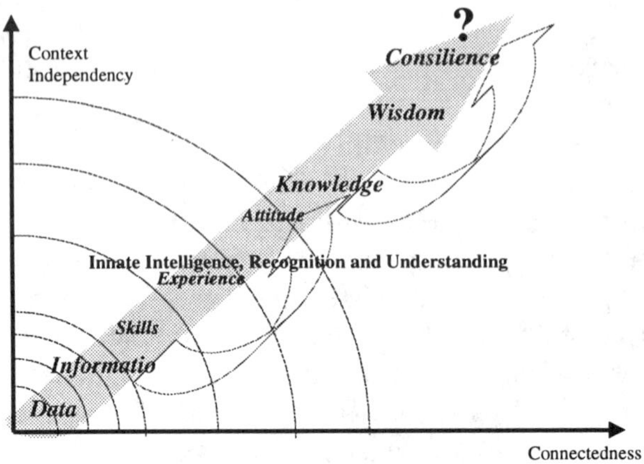

Figure 2. Concept of knowledge as a function of context independency and connectedness

Consider the example of Planck given above. Although the information and knowledge that Planck used to create Quantum Theory was available to a wide variety of scientists, only he had the "*Innate Intelligence*" to recognize the pattern in all the existing information and knowledge, much of which was not relevant and hence was probably noise from his viewpoint [9]. Innate intelligence must be applied to create higher levels of knowledge. Thus, these words appear in the mechanism by which data , information and knowledge move to higher levels as indicated by the curved arrows.

In this diagram, a natural issue arises: when does information become knowledge and what role does context play in this transition? It is not clear to us that this is a useful debate. It seems to us that there is a broad grey area in which information might be data and knowledge might be information.

The data-information-knowledge space is in one sense continuous with no abrupt barriers between them. Data merges into information into knowledge etc and one person's knowledge may be another's information. It seems to us that, because of the way the human mind works, there is no sharp dividing line between data and information and between information and knowledge. However, this space is structured into various levels, in which each level can merge into the next by means of the application of a human's innate intelligence. In Figure 2, therefore, data occupies a very small circle – the lowest level - around the origin. Information occupies a larger circle encompassing the data circle, but the actual boundary to data is uncertain. Knowledge occupies a much larger circle but its boundary to the information also extends over a broad range.

As discussed earlier, we think that the space illustrated in Figure 1 can either represent Total Knowledge for mankind or a single person's focal or tacit knowledge. However, none of the above really discusses what Knowledge is and while helpful in understanding what product knowledge might be, it does not move us very far forward in determining how we can help engineering teams make better products. Another perspective of Knowledge is that explicit or codified knowledge has been described as facts, procedures, rules, images, etc. However, facts, procedures, and rules can also be formatted as patterns or relationships. Hence, knowledge, as far as we can tell, is all about relationships and contexts.

KNOWLEDGE WITHIN THE PRODUCT REALIZATION PROCESS

However, in this paper we do not discuss the management of tacit knowledge or even focal knowledge, but only the codified part of focal knowledge because that is what is created, transferred and used in the product realization process using computer systems. Further, space precludes our discussing the myriad relationships and contexts in any detail.

Modern products are complex, sophisticated artifacts with manifold relationships. The idea of managing the relationships among product data is not new. As long ago as 1988, Wasserman was highlighting the importance of this topic [13]. The first and most obvious of all product system relationships is that between the geometric entities of the physical embodiment of the product structure. Modern CAD systems are already managing these relationships quite well and provide tools to the users to identify and understand them. For instance, most CAD systems provide visualization of the geometry in a variety of forms (e.g. line, hidden line, surface. 3-dimensional, etc). Modern CAD systems also allow users to manage another form of relationship, namely constraints between geometrical entities, as in Parametric or Variational Design. They also typically allow the user to manipulate a third kind of relationship: the bill of materials or product structure. Product data managers (PDMs) also manage the bill of materials (BOM) and are starting to provide multiple contexts for it: engineering, vs. production vs. procurement. PDM's typically also allow relationships among versions, and alternatives to be managed. Some PDM's are also beginning to allow users to work with a fourth kind of product relationship, that of the product family. To manage the product relationships along the product realization process some companies are introducing workflow management systems and PDM vendors, responding to the trend are incorporating such systems into their products. Workflow management tackles the management of the life cycle relationships in product realization.

As an alternative to this approach to life cycle relationship management, several researchers are suggesting that we need to define all the product information and the interrelationships of this information in a single product schema. This purpose-function-behavior-embodiment relationship, first suggested by Gero [14], expanded by Rosenman and Gero [15] and then further by Wang and Mills, can be called abstraction-concretization relationship because, as one moves from purpose to function and on to the physical embodiment, the level of abstraction decreases and the amount of concrete detail increases. Others have provided slightly different versions of the representation of a product schema [16,17]. Szykmann et al actually provide a relationship object within their schema [17]. Another approach to capturing and understanding the complex relationships in a context independent manner is that of Goossenaerts [2]. His *artifactual wheel-work (AWW)* systematises product life-cycle related concepts in order to derive generic requirements for product life-cycle support services in a ubiquitous information infrastructure. For the life cycle of artifacts and their groupings, a distinction must be made between three kinds of existence: *(i)* the individual or *ens,* e.g., your car; giving rise to the ontogenic wheel *(ii)* the *type* (in the case of things, e.g., the type of your car) or generation (in the case of self-reproducing species), giving rise to the typogenic wheel; and *(iii)* the tribe or family (*phylo*), e.g., the totality of previous and actual types and occurrences of your car, or of all cars, giving rise to the phylogenic wheel.

These wheel works represent true product realization knowledge because they capture the complexity of the relationships in the product realization process and are also context independent. They apply equally well to the realization of aircraft, automobiles, kitchen mixers or software.

Within these relationships, there is another example, namely that relating the decisions made at each stage to the appropriate product values and to the rationale behind those decisions. While it is generally recognized that rationale is important to product re-use, it is not as clear, how it should be incorporated. Wang and Mills have incorporated rationale in their Collaborative Product Representation Model (CPRM), which is essentially a product knowledge schema [18]. This preliminary schema also includes a re-use module which exports the internal methods and data required for appropriate re-use of this component by others, another form of relationships.

Contexts are more difficult to deal with and are typically implicit. Websters Dictionary has several definitions, the most appropriate of which are: "the interrelated conditions in which something exists or occurs"; and "things or conditions that serve to date or characterize an article" [6]. One way we are exploring for representing context in Information Infrastructures is to think of them as a set of environmental parameters that condition how a tool works. Most operating systems, word processors, and CAD software packages provide accommodation for different cultures and preferences which are a type of context. Hale has suggested that documents created during the PRP require a simple type of context such as a date, name of the person involved, the number of the iteration and the tool used to create it [19]. A well known problem in the product realization process which might be possible to solve by considering context is that of multiple Bills of Material: engineering, manufacturing, procurement, etc. Basically, each view is of the same information but in a different context. The proper management of context is an unsolved issue in the management of knowledge.

DISCUSSION AND CONCLUSIONS

Our pragmatic view of knowledge in the product realization process is that it is the relationships in product and process data and information within and across contexts that is as important if not more important than the data itself. This is why CAD systems, PDM's and workflow managers have implicitly started managing them. Each of these separate systems manage only limited relationships and rarely involve contexts. One conclusion, therefore is that we need more comprehensive approaches to managing the manifold relationships in the product realization process. Before that can be achieved however, we need to explicitly identify all the myriad relationships in a meaningful way. The artifactual wheelworks of Goossenaerts might be one approach to addressing this issue.

A second conclusion is that we need to find ways of identifying and taking advantage of contexts. Use of contexts is a natural and powerful mechanism we use implicitly every minute of every day. If we can harness this power in our information infrastructures for manufacturing, we will take the use of information system to new heights.

REFERENCES

[1] K.-D. Thoben, Wognum, P.M., Pels, H. J., Büchner, A.G., Goossenaerts, J.B.M., Ranta, M., Ranke, A.A.M., Gibbons, W.M., Kerssens-van Drongelen, I.C., "From Product Data to Product Data and Knowledge Management - Requirements and Research Perspective".

[2] Goossenaerts, J.B.M., "Industrial Semiosis: Founding the deployment of the Ubiquitous Information Infrastructure," Computers in Industry, to appear, 2000.

[3] Nonaka, I., Takeuchi, H., *"The Knowledge-Creating Company,"* Oxford University Press, New York, Oxford, 1995.
[4] Drucker, Peter F., *"Post-Capitalist Society,"*. New York: Harper Business Press, 1993.
[5] Audi, Robert, "*Epistemology: a contemporary introduction to the theory of knowledge*," published by Routledge, London, 1998.
[6] Webster Ninth New Collegiate Dictionary, Merriam-Webster Inc, 1990.
[7] Ackoff, R, "*Ackoff's Fable,: Irreverent Reflections on Business and Bureaucracy*," John Wiley and Sons, New York, 1991.
[8] Weggeman, M., "Knowledge Management: The Modus Operandi of the Learning Organization" Published in Knowledge Management: Organization, Competence and Methodology, Proc of the 4th International ISMICK Symposium, Rotterdam The Netherlands, Erlon Verlag, October, 1996
[9] Polanyi, M., "*Personal Knowledge: towards a Post-Critical Philosophy*", Harper Torchbooks, New york, 1964.
[10] Wilson, P., "*Consilience*", Random House publishers, 1999.
[11] Feynmann, R., cited in Davis, *God and the new physics*, Simon & Schuster publishers, pp 223-225, 1983.
[12] Bellinger, G., Castro, D., and Mills, A., unpublished article to be found at URL = http://www.outsights.com/systems/dikw/dikw.htm.
[13] Wasserman, A.I., "Tool Integration in Software Engineering Environments:" *Proc. Int'l Workshop on Environments*, F. Long ed., Springer-Verlag, Berlin, pp 137-149, 1990.
[14] Gero, J., "Design Prototypes: A Knowledge Representation Schema for Design", *AI Magazine*, Winter, 1990, pp26-36.
[15] Rosenman, M. A. and Gero, J. S., "Modelling Multiple Views of Design Objects in a Collaborative CAD Environment", *Computer-Aided Design*, Vol.28, No.3, pp.193-205, 1996.
[16] Henderson, M. R., "Representing Functionality and Design Intent in Product Models", in 2nd ACM Solid Modelling '93, Montreal, Canada, 1993.
[17] Szykman, S., Fenves, S. J., Keirouz, W., and Shooter, S. B., "A foundation for Interoperability in Next Generation Product Development Systems," To be published in Computer Aided Design, Special Issue on Product Data Representation and Management, 2000.
[18] Wang, F., and Mills, J.J. "A Product Data Model and Processor Supporting Virtual Product Development," The ASME Design Engineering Technical Conference and Computers and Information Conference, Baltimore, Maryland, September 10-14, 2000.
[19] Hale, M., Georgia Institute of Technology, Presentation to the Gordon Conference on Theoretical Foundations for Product Design and Manufacturing, New Hampshire, June 11-16, 2000.

ACKNOWLEDGEMENTS

The authors would like to thank the National Science Foundation for funding under, the Agile Aerospace Manufacturing Research Institute Grant No DDM 9320949 and the Information Technology Group of the Technology Management Faculty of the Technical University of Eindhoven, for hosting Dr Mills during the writing of this paper.

Requirements on Product Information Management in the Sales and Service Life-cycle Phases of a Plant

K. Jansson, I. Karvonen, I. Salkari, M. Ollus
VTT Automation, Technical Research Centre of Finland
Em: Kim.Jansson@vtt.fi

Keywords Information Management, Life-Cycle, Virtual Enterprise

Abstract The information related to the delivery and operation of a plant is huge. Utilising this information and especially the experience-based knowledge can give business advantages. Operating in a network of organisations means to manage a large variety of processes and products in different locations, working environments and working conditions. This paper discusses some generic requirements on managing information in the sales and service life-cycle phases of Virtual Enterprises. Enterprises should prepare themselves for upstream information management.

1 INTRODUCTION

Selling, engineering, delivering, operating and maintaining a plant or factory are information intensive processes. The demand on efficiency, economy and environmental friendliness is increasing. The competition has moved from local and regional to global markets. To respond to these increasing demands collaboration over time zones and cultural boarders is needed. Operating in a network of organisations means to manage a large variety of processes and products in different locations, working environments and working conditions.

The IMS Globemen project addresses several of these issues. The objective of this paper is to present some findings regarding requirements on information management in the project. Project partners have their individual requirements and special considerations due to differences in products delivered, operation environment, cultural and working environments etc. This paper summarises the individual requirements and discusses some generalised requirements. The generalisation is also based on findings from other ongoing work and past projects.

All discussions concerning present and future IT-tools have intentionally been left out from this paper. It is thus rather obvious that the Internet and Web-based techniques together with mobile and wireless systems will have a dominating role as an enabler for change.

2. SCOPE DEFINITION

The total management of information over the life-cycle of a plant is a huge task. The area is the object for research and development work, which has produced information systems often call PDM and CRM system. The area is too large for a short paper like this and focusing is needed.

The industrial segment of this generalisation is one-of-a-kind products, e.g. plants and large sophisticated equipment. Often the products are so big and complicated that collaboration is needed to market, engineer, deliver operate and maintain the products. Typically enterprise networks and *Virtual Enterprises* (VE) are involved [3].

Another way to focus is on the management of processes and activities to support customers' use of the product irrespective of time and distance, that is to look at the both ends of the life-cycle. Typical activities include sales, services, operation, maintenance and renewal of equipment at customer sites.

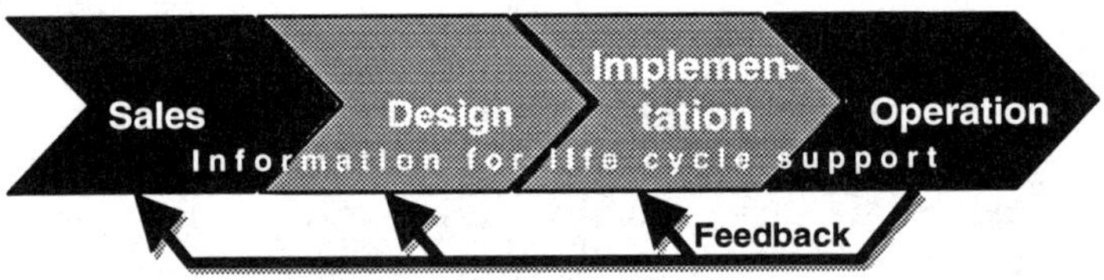

Figure 1. Scope of the paper

3. BUSINESS OBJECTIVES FOR VIRTUAL ENTERPRISE PARTNERS

Depending on life-cycle, the life-cycle phase may contain different actors like project engineers and contractors, equipment supplier, plant operator or local supplier. The business objectives and main business functions can to some degree be different, however many business objectives are coinciding irrespectively of what role the actor represents in the business network. One common and major industrial objective is to deliver products that are leading in their respective field or branch of business. In one-of-a-kind manufacturing there are additional challenges of keeping delivery time short and costs low enough and keeping the quality delivering process at a level that satisfies the customers.

The management of the sales process may support these business objectives by creating for the delivery a good starting point that allows a fast set-up of the delivery project and the virtual enterprise after the contract. The objective of being a supplier of choice also requires supporting the customers to identify their potential in the utilisation and updating of technology and systems. This needs the follow-up of competitors and evolving technologies.

For the equipment provider knowing how a machine or equipment has been used and how the equipment has functioned in customer environments gives valuable information. This information can be used in R&D for improving the equipment or designing the next generation of it.

Project engineers and delivery project contractors have the same business objectives as the equipment provider, that is, the whole plant, it's usage and operational environment. Additionally utilising information related to the project delivery, time and cost is relevant.

Service providers can give value to the customer and thus also increase the competitiveness of the manufacturer by offering advanced service and online operation support for the product

4. DIRECTIONS OF INFORMATION INTEGRATION

Within each life-cycle phase various requirements exist on information access, availability, reliabilities etc. ending up in IT-system requirements. These aspects are of course important and there are still areas within each phase that can be improved.

The integration of information over life-cycle phases is a matter, which holds great potential for improvements. The integration over life-cycle phases has two directions.

4.1 Downstream information integration

This is the "traditional" direction. The integration issues become interesting and more complex when the life-cycle steps involve different organisations in Virtual Enterprises. Figure 2 illustrates two of the dimensions for information integration. First downstream, within one plant life-cycle (Project 1 in the figure) there is risk of loosing valuable information when the product is handed over from the project supplier to the user or operator organisation.

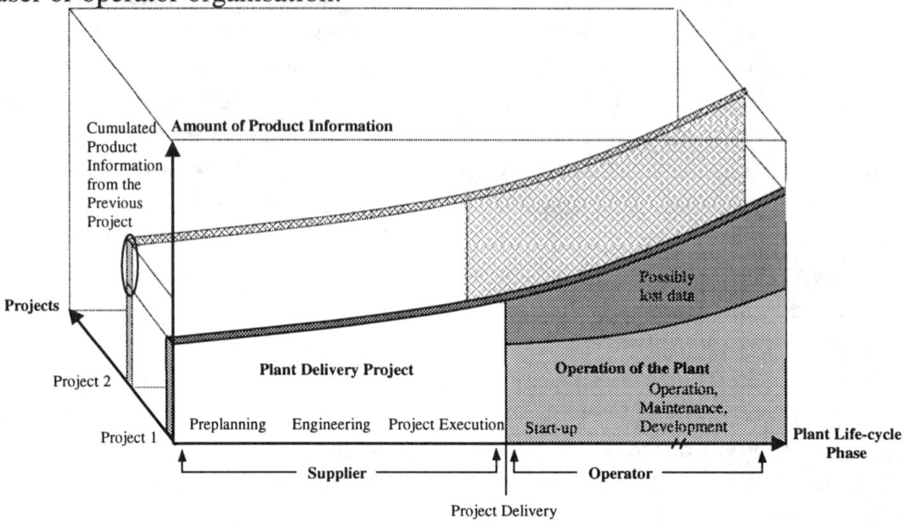

Figure 2. Two dimensions of Information Integration

4.2 Upstream information integration

The upstream integration can be seen in several ways depending on which life-cycle view is taken. In general it is collecting information on how the equipment or plant has been delivered and how the equipment or plant has been used, in which modes of operation and environments. Thus the second dimension shown in figure 2 is how the experience cumulated can be used in subsequent projects (Project 2 in the figure) resulting in a higher degree of knowledge. This knowledge is then used when improving and modernising existing plants

- when preparing contract offers to customers
- when designing new plants
- in research and development
- for operator training
- for supporting delivery projects
- for quick online support and remote diagnostics

5. INTER ENTERPRISE INFORMATION EXCHANGE

The following table gives an introduction to inter enterprise information content in the life-cycle phases under study. Knowledge, information and data are intellectual property and not always free and readily available, nor is it always available in the wanted format. Working according to the VE parading strives to remove some of the still existing obstacle in inter enterprise information exchange. Still there are open issues like:

Table 1. Life-Cycle Phase, Business Functions and Main Information Content Needed

Life-Cycle Phase	Business Function	Main Information Content
Marketing	Market research Sales planning Network / Business partner management	Market segment development Customer needs and future interest Information on competitors Existing competencies at business partners
Sales	Project planning Product configuration Bid preparation	Customer requirements specification Technical solutions specifications Cost information Information on suppliers and their previous performance Experience from previous bids, projects and deliverers
R&D	Technology innovation research Product development and innovation Presales engineering	Technology trends Market development and requirements Operational experience from products in use
Operation support	(Remote) Customer training (Remote) Process optimisation and simulation Service and maintenance (Remote) Diagnostics, troubleshooting On-line information support	Product behaviour, machine performance data Process parameters Operational product Configuration Maintenance needs Manuals and instructions Training data and material
Renewal, Continuous improvement	Feedback to research, development and engineering	Operational data Problems Reasons for design decisions Renewal needs Potential for improvement

- Marketing and sales related information is scattered and no single source exists. The information systems are heterogeneous.
- The information it is subjective and can be interpreted differently.
- In the marketing and sales phases the information is sensitive. Partners are not always willing to give out strategic information before any VE is created or agreements made. Possible business partners may be involved in other/competing networks.
- Contractors do not want to commit themselves to use a certain supplier.

- Supplier my not be willing to revile all capacity and load information. It may affect the price negotiations.
- The operational user is not willing to give performance data to one contractor only. Competition among contractors may be preferred.
- The contractors' organisation is not prepared to take advantage of information collected from different locations.
- The time frame is long from R&D to actual product. It takes time to get operational feedback

6. REQUIREMENTS ON INFORMATION MANAGEMENT

The main principle and idea within the IT user community has for decades been, *"Store information once in one location and use many times"* and it is still valid. Companies are still struggling to meet this requirement and in fact it will never be 100% met. Working in Virtual Enterprises and sharing information over the communication network and organisation boarders have introduced new aspects and levels of difficulty to this.

From the previous section we can conclude that the characteristics of plant related information is manifold. Some characteristics are:

1. large and complex
2. redundant and overlapping
3. loose, imprecise, incomplete and uncertain
4. subjective
5. dynamic with different versions
6. irregular and in different format
7. no direct or logical links exist between information items
8. distributed physically on a global scale
9. reside on different media
10. used by heterogeneous systems
11. outlives the systems that generated and processed it
12. owned and shared by several organisations
13. of different confidentiality and sensitivity level

6.1 Generalised Requirements on Information Management

These information characteristics impose a lot of constraints and requirements on the management. Some generalised requirements on information access and usage are:

Anywhere & Anytime. Today's dynamic working environments should not put any restrictions on this. Companies are working in networks regardless of geographical location and time zones. Information must be available regardless of the local time.

Sensitivity & Security. Trust and confidence between business partners operating in Virtual Enterprises cannot be built over-night. It requires time and successful co-operation to build up. Even business partners co-operating in the same VE have different levels of confidence in each other. In one project companies may work together; on another one they may be competitors. The IT-systems need to have features by which the levels of confidence can be defined "individually" in a dynamic way.

Permanence & Impermanence. No relationship is permanent. To respond to the dynamic characteristics of VEs, features in IT-systems are needed that can give access for a restricted time and then cut the access rights.

Personalise and role based. The amount of information is huge and it is increasing constantly. The user needs services that can personalise what is presented. The environment should be capable to learn and configure itself according to users preferences and habits. The same information will be personalised in different formats to different users through a rich user interface.

Manage Complexity. Future IT-systems should help users in the management of large amount of data of increasing complexity and in integration of information from different sources. Finding relations between data from different sources and possible identification and catching relations between data and rationale for the decision.

Connectivity. There are numerous information systems by different vendors used today in enterprises. A common IT-system platform cannot be agreed even between partners co-operating in one Network or VE. The connectivity between different IT-systems from different vendors is crucial. Reference architectures and standards are needed for inter enterprise information communication.

6.2 General User Requirements for Marketing and Sales

This section generalises the requirements on IT-system supporting the early stages of the life cycle.

Sharing of relevant sales information. For sales related information to be communicated there must be means to share it across non-homogeneous IT domains. There is no single database design or data architecture for the world; neither is there a single data dictionary that would enable disparate databases to be easily combined.

Efficient bid preparation. Usage of tools and knowledge management techniques for faster and more accurate bid process. Better communication between bid perpetration partners and the customer. Improve the reliability of produced bids. The management of product data (previous solutions or product models) helps to prepare a high-quality bid efficiently

Enable downstream information usage. During the sales phase an extensive amount of information is created to define and configure the customer solution. Advanced product design in the sales phase contributes to fast delivery of the product after the contract. The information from the sales phase should be made available for the customer delivery process.

Fast Virtual Enterprise set-up. Large one-of-kind products, like plants, cannot be delivered by one company alone; subcontractors and partners are needed for the delivery. If the set-up of the virtual enterprise for the delivery is started already in the sales phase, the delivery project can be activated faster.

6.3 General User Requirements for After Sales and Service

Below is a summary of generalised requirements on IT-system supporting the after sales and service phases of the life cycle. Like in the marketing and sale phase, also here the advanced management of information is in central position. Upstream information integration in VEs is a novel requirement. The general requirements for IT-support include:

Feedback of product operational information to development, design and manufacturing. Shortening of lead-time at introducing or modifying equipment. Manufacturers of complex one-of-a-kind products, intend to co-operate with their globally spread customers to improve their operation-support and maintenance and to identify the needs for innovation. Thus they have to establish a global service for maintenance and operation support and a proper feed back of information about operation experience to R&D. Therefore they need methods and tools for setting up communication, information exchange and sharing and also for analysing process information.

Fast access to process data for service and maintenance. Shortening of troubleshooting time during maintenance is desired. Manufacturers of complex one-of-a-kind products need appropriate processes and organisational design to shorten reaction time for service and maintenance and to use process data for fault analysis and simulation. This can increase plant availability, product quality and decrease production cost for the customer. The service partners need decreased access time to plant, process and maintenance knowledge, increased quality of provided knowledge and world-wide access to the knowledge.

Up-to-date product document maintenance. The product documentation must be maintained during the operational phase. This includes as well operators interactions, training material, simulation models and change history. Multimedia based documentation is emphasised.

7 CONCLUSION

Downstream information management, even over company boarders, is something that organisations and enterprises have desired for and struggled with for years. The organisational culture is starting to be prepared for this. The upstream inter-enterprise information management is a newer issue. The technology for this is starting to be in place. Delivering the information across enterprise borders is, however, not self-evident. It requires definition of rules and business procedures for creation, sharing, processing and using this inter enterprise information. As the experience based information is cumulated over time, it is also a task that requires longer perspectives and co-operation between the partners.

Are organisations prepared for upstream information integration? What happens if and when experience based information and knowledge suddenly is available? Are organisations prepared to take full advantage of the information?

One of the findings in the analysis work is that upstream information integration is a most interesting topic today. It is also the opinion of the authors that the companies need to prepare themselves to utilise experience based operational information.

8 REFERENCES

[1] Karvonen I., Jansson K. (1999). Project Information Management in Distributed One-of-kind Manufacturing *Proceedings of the NORDNET Managing Business by Projects, Volume 2"* Helsinki 1999

[2] Bernus P., Mertins K., Schmidt G (Eds.) Handbook on Architectures of Information Systems Springer Verlag 1998

[3] Kaas-Pedersen C., Larsen L-B., Vesterager J. Modelling Overview Report No D1.1 Globeman web site http://www.vtt.fi/aut/projects/gm21/demo/

[4] Salkari I, Managing Information over the Life Cycle of a Process Plant, Individual Assignment in Industrial Management, Helsinki University of Technology 2000

[5] Karvonen I., Management of one-of-a-kind manufacturing projects in a distributed environment, VTT Research Notes 2044, Technical Research Centre of Finland 2000

A VR-based CAD System

J.M. Zheng, K.W. Chan, I. Gibson
Department of Mechanical Engineering, The University of Hong Kong, Pokfulam Road, Hong Kong

Keywords (VR) Virtual reality, Geometric modelling, Computer graphics, CAD/CAM

Abstract The performance of current Computer-Aided Design (CAD) systems is still far from the satisfaction of conceptual designers. This paper presents a VR-based CAD system for conceptual design by applying VR interfaces and geometric modelling techniques to improve the human computer interaction. The system uses an electronic data glove as an input device so that conceptual designers are allowed to use hand gestures to conduct various geometric shape operations instead of depending solely on keyboard and 2D mouse. We employ the 3D GUIs for enhancing the gesture interface. The implemented shape modelling techniques offered by the system include: destruction, construction, and techniques for freeform feature creation and modification. The destructive techniques allow the user to arbitrarily sculpt an existing part by means of sculpting tools. The constructive techniques allow the user to assemble an object by using a collection of feature objects, by either adding or subtracting feature objects from the model. The system also offers several novel methods for facilitating intuitive modification of freeform curves and surfaces.

1. INTRODUCTION

Current Computer-Aided Design (CAD) systems are not suitable for conceptual design because they have three significant drawbacks: i) designers are required to specify detailed geometry of the part in a very quantitative manner with tools that are more suitable to a draughtsman than a concept designer; ii) there is a lack of high-level shape operators for designing and modifying model shapes; and iii) the CAD interface is still mainly based on the use of 2D pointing and display devices. This makes the interface unnatural and non-intuitive.

In this paper, we use a CyberGlove, an electronic hand glove developed by Virtual Technology Inc., as an input device to develop a desktop CAD modelling system for facilitating conceptual design. Gestures are used to support intuitive user interface. To develop this gesture interface, we consider that conceptual designers should be given freedom to use different kinds of natural gestures to conduct various geometric shape operations instead of depending solely on keyboard and 2D mouse. For instance, designers can indicate objects or directions simply by pointing with their hand and fingers, and look at an object from different directions through grasping them and turning them into proper orientations and positions. The functional tools can be used for creating, cutting, and editing objects. We also employ 3D GUIs for enhancing the gesture interface. In the virtual environment, 3D menu and "virtual hand" are concurrently used. Various 3D cursors can be used to select menu or manipulate the object. The modelling methods used in the VR-based CAD system include feature based modelling techniques for constructing product model, such as constructive technique and destructive technique.

The VR-based CAD system consists of three parts: input, output and computation engine. The diagram of the system architecture is shown in Figure 1. As shown in Figure 1, the inputs include 2D mouse, keyboard, data glove input control, and position and orientation tracker. The glove input control allows the computer to determine the user's hand gesture and its physical position in 3D space.

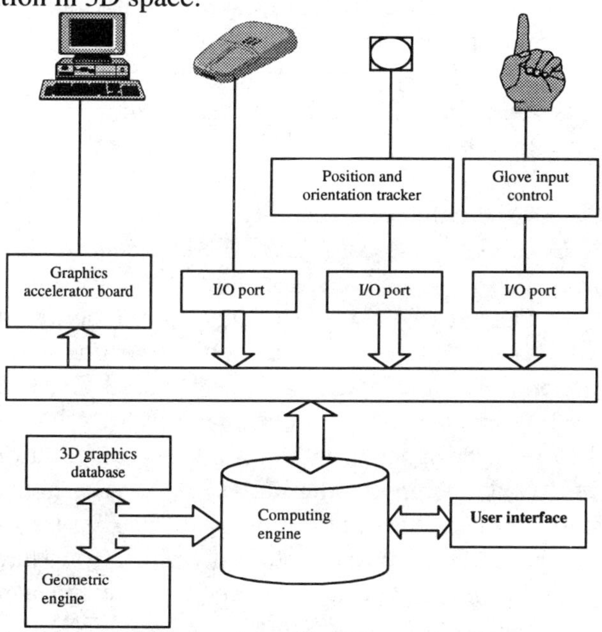

Figure 1 Diagram of the VR-CAD system architecture

2. RELATED WORKS

There are many research activities on VR-based design systems. Based on the interactive ability that the designer can manipulate the product models, current VR-based product design systems can be classified into two main groups: enhanced visualization and VR-based CAD [1].

In the enhanced visualization group, the product part models that are previously created by traditional CAD modelling systems are imported into the VR-based environment through an appropriate transformation. Once the models of the product parts are imported into the virtual environment, 3D interactive devices, such as Dataglove, 3D mouse and 3D display monitor, can be used for the designer to operate the product model and navigate the product model in the virtual environment. Researchers at the University of North Carolina at Chapel Hill [2] presented a system called Immersive Simulation Animation And Construction (ISAAC) for the designer to interactively construct virtual worlds. The system allows the designer to position, orient and scale the pre-generated 3D models that are created by CAD modelling systems or 3D scanning devices in the virtual environment. Fraunhofer Institute for Computer Graphics, Darmstadt, Germany, developed a VR tool set, called Virtual Design II [19], to support the virtual product development applications. The system lets users import data from various sources, preprocess and enhance data, interact with and manipulate data in real time, and present the application using various audio-visual facilities. Other examples include the VENUS project [3] and research at Clemson University [4].

Different from the enhanced visualization group that is limited to just visualizing CAD models, the VR-based CAD systems allow the design activity to occur in the virtual environment. The designer can use 3D devices such as Dataglove, 3D mouse and 3D monitor such as Head Mounted Display (HMD) to create and modify the product models in the virtual environment. The advantage of the VR-based CAD is to provide a more intuitive 3D interface than traditional CAD systems and to offer enhanced tools for the designer to model product configuration. In addition, the VR-based CAD also supports alternative methods of user input, such as voice and gestures. Researchers at the Center for Design Research (CRD) at Stanford University carried out a number of projects on Human Computer Interfaces [5]. Other examples include the VR research at the I-CARVE LAB of the University of Wisconsin in Madison [1,6,7]. Researchers in the Brown University Graphics Group developed human-centered and interactive 3D virtual environments for modelling, scientific visualization, tele-collaboration, and interactive illustrations in a shared visual, spatial, and auditory environment [8,9,10,11,12,13]. The system can offer an intuitive user interface for the user, however, the modelling tools for mechanical design are still limited. The JDCAD system [14] used a pair of input devices and 3D interface menu to allow the designer to define objects. The Conceptual Design Space (CDS) [15], developed at Georgia Technical College, offers a real-time 3D immersive environment for 3D architectural design. The designer can use CDS to create conceptual building designs and modify them, add details, or create new designs, all are immersed in the virtual world. The System Research Laboratory of the Nippon Electronic Company developed a prototype CAD system for the designer to manipulate a CAD model by Dataglove [16]. The system can perform some simple shape manipulation functions but the accuracy achieved is not satisfactory for normal CAD applications.

3. USER INTERFACE DESIGN

3.1 Guideline for user interface design

Consistency : The consideration of consistency means that the interface should be consistent in its requirements for input and have consistent mechanisms for users to make any demands on the system. The consistent interface is one in which the interface elements are uniform and follow a few simple rules. The basic aim of consistency is to allow users to generalize knowledge about one aspect of the system to other aspects.

Feedback: Feedback refers to the process of sending back information to the user about what has been done. The consideration in feedback means that if the user has performed an action that triggered an internal system response or an important message is issued, the computer system should give an indication to show whether the message is accepted by the computer system. Therefore, the feedback can tell users that the requested operation has been completed and the new or modified display can explicitly show the results of the operation. For example, a selected object or menu command is highlighted so that the user can know that her/his actions have been accepted.

The desktop VR-based CAD system attempts to incorporate Windows's user interface elements, such as windows, icons, dialog box, and pull-down menus, so as to provide easy-understood feedback. The feedback involved in the VR-based CAD system can be divided into two aspects: object feedback and position feedback. The object feedback occurs when the user issues a command to select object. For example, when a command for selecting an object is issued, the requested object will be displayed in red color to indicate that the selection is accepted. A button (in which a gesture bitmap is embodied) is down to indicate that the corresponding gesture is recognized and accepted by the system.

Naturalness: Naturalness requires that the interface should not ask for redundant information from the user. It should require the minimum of user input and should provide an intuitive mechanism for the completion of the users' tasks.

Gestures are the main input mechanisms in the VR-based CAD system. It is important for the user interface to eliminate unnecessary memorization so that the users are comfortable to conduct various shape designs via gestures. In the VR-based CAD system, the suggestive gestures are displayed on the prompted dialog box shown on the left side of the screen. The suggestive gestures keep the request for information to the minimum and allow the user to choose different gestures to conduct commands or shape operations.

3.2 Human computer interactive tasks in virtual environment

Navigation: The navigation task provides the user the ability to navigate through the virtual space to view the generated scenes in the virtual space. The hand glove input device can provide a natural method, allowing the users to "physically" hold the object. The users can also grasp the object and dynamically rotate the object so as to allow the object to be viewed from different angles. While the user is navigating the virtual space, the visual output shows the updated 3D images. The user may repeatedly perform navigation operation in this navigation task until he/she satisfies the scene.

User commands: Although there is a different view on the use of user commands in virtual environment, we consider that the need for traditional type of user commands will not be completely eliminated in a virtual reality system because not all tasks are suited to direct manipulation or gestures.

In VR-based CAD system, menu-based commands may be viewed as a supplementary user interface for the virtual environment while the gestures are inadequate for the desired tasks. In order to allow the user to quickly learn the hierarchy tree, the menu items can be structured into groups. For example, the menus in the VR-based modelling system are grouped as the solid feature modelling menu and freeform feature modelling menus. The items in each menu can be grouped by their functions. Figure 2 shows the flow diagram for the user to navigate the hierarchical menu tree. The user first issues a gesture to choose a modelling method, then a 3D menu will be popped up and displayed in front of the user. Finally the user can use his/her hand to select a menu item from the menu. After the menu item selection is finished, the 3D menu will disappear from the screen.

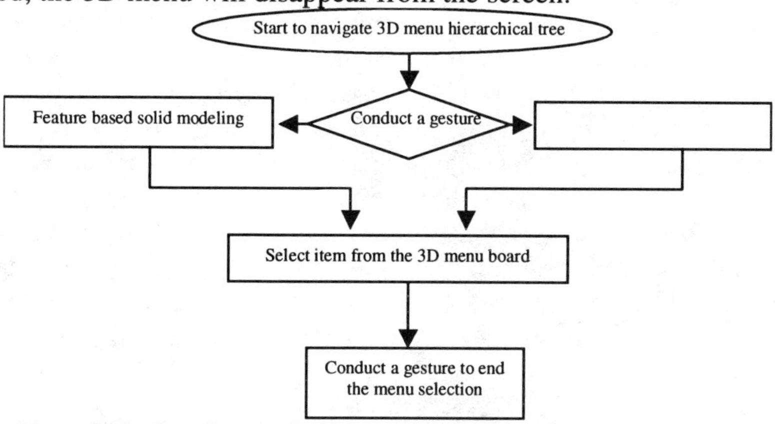

Figure 2 The flow diagram for the user to navigate the 3D hierarchical tree

Object selection: The selection task is that of choosing an element from an object or menu item from the menu hierarchical tree. The intuitive and natural method for object selection in the VR-based CAD system should be to select an object when the user's hand comes into contact with it. The index finger is used to indicate an object to be selected. This method works well in the immersive VR system that can offer natural visual senses. In this situation, it is natural for user to use his/her hand in the same way that he/she would select an object in the physical world. However, this method may not work in current desktop virtual environment. Firstly, the hand motion may not be accurate enough to indicate some small object that needs to be selected. Secondly, the user has to navigate to the distant objects in order to select. Due to the display limitation of a desktop virtual environment, it may be difficult for the user to sense if she/he has touched the object. The ray-casting technique can be used to solve this problem. In the VR-based CAD system, a ray is extended from the user's hand to intercept the desired object, and the first object intersected by the ray is selected. The visual viewer output of the system will check whether the user has indicated the object to be selected by highlighting the selected object.

Object manipulation: The object manipulation tasks in the VR-based CAD system can be cataloged into two groups. The first sub-category of manipulation tasks contains those actions which are symbolic or abstract in nature, such as loading files, selection of colors, and changes of system parameters. Those tasks that do not need 3D spatial input can be implemented by 2D mouse. The second category of manipulation tasks is object creation.

The object creation is the most important subcategory in object manipulation task. Figure 3 shows the steps for creating a cylinder. The user can select the cylinder item from the 3D menu and then a dialog box will pop up to guide the user to issue various gesture-based commands to conduct the object manipulation tasks. The user may use the index finger to indicate the location of the entity that is being created. The visual output feedback can also indicate whether the selected tasks have been accepted and processed. All object manipulation activities in the virtual environment are processed by hand gestures. The system also supports a mechanism for the user to specify accurate shape sizes. When all creation parameters are determined, the user can issue a gesture command to create the object.

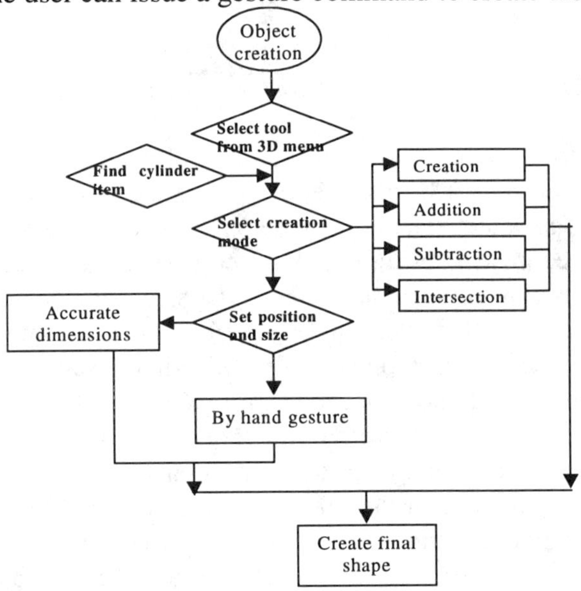

Figure 3 The process of creating a cylinder —an example for manipulation task

Object query: The object query task is to inquire the corresponding information about the selected entity. It can used to specify the dimension, the location and the relationships between the entities. Using object query option, the user can obtain a graphical representation of a model with its corresponding dimensions and expressions. The user can edit these parameters from the query information and update the model to reflect the changes that have been made.

3.3 Gesture interface design

Gesture taxonomy: In the VR-based CAD system, we employ Caroline Hummels' classification [20] to categorize the gestures involved in the system. The gestures used in the system are considered as predefined symbolic commands for a given application in terms of the tasks the user is permitted to perform. We employ this type of gesture interface for the VR-based CAD system because it is easier to recognize a predefined hand posture instead of interpreting the infinite series of daily used gestures. In general, gestures involved in the VR-based CAD system are defined by very simple symbols so that they can be easily recognized. Moreover, the gestures can be divided into indicative gesture and manipulative gesture. The indicative gestures are used for creating and choosing menu item or icon. The manipulative gestures are used to guide the users to define, select, and manipulate the objects. The commands defined by the gestures are interpreted by the icons in the menu item or in the dialog box.

Gesture generation and recognition: In order to capture the meaning of the gesture, gesture generation has to be broken into two phases: analysis and recognition. Analysis is the process of converting the hand movements into a computational representation so as to produce a characteristic (feature) for the gesture. Recognition is a context-dependent process. It is a process to select appropriate data from the whole gesture library, which carries information comprehensive enough about the gesture instances, so that the system can decide if the conducted gesture has been accepted by the system and produce a gesture vocabulary. Generally, the process for gesture creation and recognition consists of three steps: data inputs, gesture feature extraction, and gesture vocabulary output.

Gesture interface: The gesture interface enables a user to give commands by making certain hand gestures. One-handed input gestures are used to provide a command context including menu operations, rigid solid manipulation and freeform surface definition and modification. Each of the gesture recognition systems works sequentially, receiving data sampled from an electronic hand glove device, analyzing the data, and sending them to an interpreter, which responds to user's gestures within an application context.

The main characteristic of the gesture driven interface is that the manipulation commands are pre-defined for a given application in terms of the tasks a user is permitted to perform. It is very easy to define our own set of gestures and to train the system users to explicitly recognize the gestures from this set. Our approach has the following advantages: i) the gestures do not need to be physically performed for the teaching procedure; ii) gestures can be easily defined by the system and used by people.

3D graphics user interface: 3D graphics user interface elements involved in the VR-based CAD system consist of floating menus, 3D widgets, and gesture interface.

Floating menus: Because the menu is well accepted in the 2D community, we still employ the menu as one of the interface elements to provide a large amount of command choices. The floating menu displays two-dimensional menus in the three-dimensional world of VR. These menus are either text-based, describing the available choices with words, or graphics-based, using icons to convey the available choices. The floating menus include: feature based menu and freeform menu.

3D widgets: 3D widgets are objects in the virtual world that present an intuitive, direct manipulation interface to the user. Using 3D widgets as a user interface has many advantages. For example, the action of the "virtual hand", displayed in the virtual environment, directly conveys the action of the "real hand" in the "real world".

4. GEOMETRIC MODELLING TOOLS

Constructive technique: The constructive technique allows the user to assemble a representation of an object as a collection of features, through either adding or subtracting features from the model. It describes a product model into a context dependent feature-based representation. It is a process in which a model is constructed from a library of features, rather than geometric primitives. In this way, features are used as a basis for a number of design activities, including the design of components using standard shape features. The VR-based CAD system offers the user a fixed set of features to choose from. The form of a feature instance can be created on a geometric model by a procedure based on a given set of feature parameters.

According to constructive technique, the user should first extract a set of shape features from the product model, then convert them into a context dependent feature-based model. To convert the model into a context dependent feature-based model, it is necessary to interpret the feature information according to the feature classifications defined by the system. The description of the features in the VR-based CAD system can be divided into two general classes: protrusions and depressions. The protrusion features include solid primitives such as cube, cylinder, sphere, cone, tube, boss, pad, and swept features. The depression features include hole, slot, groove, and pocket.

Destructive techniques: In the destructive technique, feature can be considered as a cutting tool that allows the designer to subtract materials from the target object similar to a machining operation. The VR-based CAD system provides cutting tools for the user to cut a workpiece with virtual hand. The purpose of using destructive technique for solid modelling is to explore the feasibility of using the VR-based CAD system to sculpt an existing part by using a cutting tool. The simulation of sculpting is achieved by subtracting the tool's sweeping volume from the workpiece model. Three essential geometric entities involved in the simulation must be modeled, namely the part, the workpiece and the swept volumes created by the cutting tool. The model of the workpiece is dynamically changing as the swept volumes of the cutting tool are subtracted from it. The graphical display of the tool movement can be shown continuously.

Freeform shape design: The control point based mechanisms are widely used in the current CAD modelling systems to support the design of freeform surfaces. However, their mathematical representation often makes the interfaces unfriendly to the users. Our VR-based CAD system attempts to alleviate those limitations so as to assist the conceptual designer to create and deform freeform surfaces. The VR-based CAD system can allow the use of some non-control point based high-level shape operators to create and deform the freeform surfaces. The interactive sculpting technique allows the user to locally "sculpt" a 3D model. There are two kinds of manipulation tools in the VR-based CAD system: i) parameterization and ii) sculpting. Parameterization refers to the changing of the shape parameter of the surface in relation to the creating mode. Sculpting refers to the arbitrary deformation of the object shape. The system offers three types of sculpting tools: points-based, curve-based, and feature-based. The points-based deformation refers to the movement of points on a surface. Curve-based deformation refers to the movement of a curve that lies on a surface. The feature-based deformation refers to the addition of a feature surface to an existing surface.

5. CONCLUSION

This paper addresses the design of a VR-based CAD system for conceptual design. VR interfaces and geometric modelling techniques are applied to improve the intuitiveness of human computer interaction. From our work, it is considered that VR-based interfaces are more powerful and better matched to human sensory capabilities than those that only depend on traditional keyboard and mouse input methods. The VR-based CAD system is implemented as a desktop based operation system which inherits many user interface styles of Windows operating system, and hence allows the users who are familiar with Windows to use the system efficiently.

6. REFERENCE

[1] Dani, T. H and Gadh, R., Creation of conceptual shape designs via virtual reality interface, CAD, Vol. 29, No. 8, pp. 555-563, 1997.

[2] Mine, M. R., ISAAC: a meta-CAD system for virtual environments, CAD, Vol. 29, No. 8, pp. 547-553, 1997.

[3] VENUS, http://sgvenus.cern.ch/VENUS/vr_project.html, 1999.

[4] Clemson's VR project: http://www.vr.clemson.edu/vr/, 1999.

[5] Chapin, W. L., Lacey, T. A. and Leifer, L., DesignSpace: a manual interaction environment of computer aided design, Human factors in computing systems, the ACM CHI'94 conference proceedings, Boston, Ma, April 1994.

[6] I-CARVE LAB, the University of Wisconsin, http://smartcad.me.wisc.edu/groups/virtual/virtual.html, 1999.

[7] Chu, C. P., Dani, T. H. and Gadh, R., Multi-sensory user interface for a virtual-reality-based computer-aided design system, CAD, Vol. 29, No. 10, pp. 709-725, 1997.

[8] Zeleznik, R. C., Herndon, K. and Hughes, J. F., SKETCH: An interface for sketching 3D scenes, Proceedings of SIGGRAPH '96.

[9] Forsberg, A. S., LaViola, J. J., Markosian, L. E. and Zeleznik, R. C., Seamless interaction in virtual reality, IEEE CG&A, Vol. 17, No. 6, November, 1997.

[10] Zeleznik, R. C., Forsberg, A. S. and Strauss, P. S., Two pointer input for 3D interaction, Proceedings of 1997 Symposium on Interactive 3D Graphics, Providence, Rhode Island, April 27-30, 1997.

[11] Bloomenthal, M., Zeleznik, R. C., Fish, R., Holden, L. S., Forsberg, A. S., Riesenfeld, R., Cutts, M., Drake, S., Fuchs, H., and Cohen, E., Sketch-N-Make: automated machining of CAD sketches, Proceedings of the 1998 ASME 8th Computers In Engineering Conference, Atlanta, Georgia,1998.

[12] Forsberg, A. S., LaViola, J. J. and Zeleznik, R. C., ErgoDesk: A framework for two- and three-dimensional interaction at the ActiveDesk, Proceedings of the Second International Immersive Projection Technology Workshop, Ames, Iowa, May 11-12, 1998.

[13] Miller, T. and Zeleznik, R., The design of 3D haptic widgets, Proceedings of 1999 Symposium on Interactive 3D Graphics, ACM SIGGRAPH, 1999.

[14] Liang, J., JDCAD: a highly interactive 3D modelling system, Computer and Graphics, Vol. 18, No. 4, pp. 499-506, 1994.

[15] GVU Center Virtual Environments Group, http://www.cc.gatech.edu/gvu/virtual/CDS/, 1999.

[16] Kahaner, D., Japanese activities in virtual reality, IEEE CG&A, pp. 75-78, January 1994.

[17] Nobel, R. A. and Clapworthy, G. J., Virtual Sculpting: A new approach to geometric modelling, Proceedings of the 2nd UK VR-SIG Conference, pp. 66-79, December 1994.

[18] Trika, S. N., Banerjeer, P. and Kashyap, R. L., Virtual reality interfaces for feature-based computer-aided design systems, CAD, Vol. 29, No. 8, pp. 565-574, 1997.

[19] Gobel, M., Industrial application of VEs, IEEE CG&A, pp. 10-13, January, 1996.

[20] Hummels, C., Smets, G. and Overbecke, K., An intuitive two-hand interface for computer supported product design, Proceedings of Interna-tional Gesture Workshop, Bielefeld, Germany, pp. 197-208, September 1997.

Dependencies Between Design Product Models and Simulation Models

Reiner Anderl, Sven Kleiner
Darmstadt University of Technology, Department of Computer Integrated Design, Germany
E-mail: kleiner@dik.tu-darmstadt.de

Keywords Concurrent Engineering, design, simulation, product data model

Abstract Multidisciplinary engineering is essential to develop mechatronic products. The key factor is to integrate methodology, systems and models. This paper will discuss the various design representations and the dependencies between product models. An approach for capturing dependencies is presented.

1 INTRODUCTION

According to a survey there is still a big deficiency concerning the application of methods like Simultaneous Engineering, Concurrent Engineering and Systems Engineering in European product development environment [3]. Especially the missing integration of different engineering domains is a handicap of today's product development process. Specialised methods are established in different fields of engineering, but these are unknown or obscure for participants of other engineering disciplines. Hence, a common understanding is of crucial importance for process success.

Besides methodology, information technology affects productivity of engineering work and product quality. Numerous software tools for product design, analysis and simulation support engineering. Tools for design (CAD-systems) respectively computation and simulation (CAE-systems) result in specialised tasks, where skilled experts are required. Engineering tools are divers and mixed, according to requirements of users who work with such tools. This leads to an isolated work concerning design and engineering.

From the point of view of engineering, the mechanical engineering scene has increasingly incorporated electrics and electronics for the control of the mechanical elements. Mechatronics is an integrating discipline that combines the fundamentals of computer science, mechanical and electronic engineering to meet the needs of industries involved in automation/robotics and related areas where there is a need to integrate these technologies. Tools for the design and simulation are needed and play an important role to evaluate designs and to support design decisions. Problems made organisations realise, that they have to co-ordinate the way product developers collaborate and to interconnect engineering tools. Reuse of product models and the mapping of data can reduce the gap between design and engineering.

2 ENGINEERING OF MECHATRONIC PRODUCTS

The methodology of Systems Engineering is part of the systems approach. This means that the system (product) consists of sub-systems, components and parts with borders and relations to the environment [4]. It is important to make a mechatronic design from systems approach. Advanced knowledge of experts from different disciplines is necessary in order to get the best performance. Providing the multidisciplinary design team with valuable methods, tools and information will enhance the productivity.

2.1 Methods and Tools for Design

The feature based and parametric design approach used as modelling techniques in CAD-systems have led to a massive progress in efficient design and engineering of products and represent the state of the art.

Design by features extended solid modelling, by dividing complex geometry in simple elements (features). This features can be classified in body features, form features, operation features and enumerative features, e.g. holes, threads, slots, pads etc. Every feature has its own semantic and could be customised by parameters. Most CAD-systems offer libraries with features, which can be combined and reused in different ways. The totality of pre-defined and user-defined features represents the geometric (static) design of a product. The main aspects of modelling with constraints are structuring a solid model as a history of features, using topology objects and their geometric parameters and applying constraints to these objects. The definition of geometric shapes is strongly supported and can be performed easily. The variation of designed geometry is easy to define [6]. These modelling techniques allow an easy way to modify models, re-use products, create product variants and families besides analysing the model consistency. In future, knowledge-based CAD-Systems will capture, retrieve, modify and create rules to establish designs based on best practices.

2.2 Methods and Tools for Modelling and Simulation

Within the scope of virtual product development, computer simulation allows to evaluate and to compare alternative designs without having an expensive hardware prototype. Modelling of systems behaviour and simulation is important for designing mechatronic systems. In controller design, many different software tools for system analysis and simulation are available [1]. In core these systems are based on very efficient mathematical-numeric libraries for the most important tasks and include functions for modelling, simulation and prototyping as well as for analysis and visualisation. The product is described by an underlying mathematical model, where the coefficients represent the essential product parameters.

The modelling of mechatronic systems is mostly supported by multi-functional, mathematically oriented calculation and simulation packages (e.g. Matlab/Simulink or MatrixX/Systembuild), where the simulation model is presented by a block diagram. A precondition in a block diagram is that different blocks do not constrain each other's properties or parameters. This implies, that parameters of various physical components need to be defined in different (sub)blocks. In most cases, block parameters are defined separately and automatically related to the properties of the block diagram, in order to investigate the effects of parameter variation. As a consequence therefore, system components represented by blocks cannot easily be substituted by others [2].

Object-oriented modelling techniques using iconic diagrams (e.g. electrical network diagrams), energy based modelling approaches (e.g. bond-graph) or more general modelling languages including ready to use model libraries (e.g. Modelica) do not have this problem. Different software packages have been developed, that support modelling and simulation with modelling languages (e.g. Dymola) or bond graphs (e.g. 20-sim, CAMP-G) besides using equations and block diagrams. In future, viewing the mechatronic system in various representations (multiple views) can help to get a common understanding and a good insight of its properties [1].

2.3 Design Product Models and Simulation Models

The design of a mechatronic system can be split into a number of distinct tasks. Different tools to solve these tasks are available, but an integrated design environment is not existing. An integration of simulation causes handling of product data, which have been produced in earlier design phases, e.g. system properties like mass, pressure, load. In other words, the mathematical model for the simulation consists of parameters and constraints, which are predetermined. All these data have to be considered within simulation otherwise there will be a gap between data used in the simulation and data generated during design. Furthermore, new features of CAD-models could represent other product properties as well, which are used for modelling and simulation. In consequence of a physical description of the design that contains all the data that are relevant for a certain class of simulation experiments, designers could interact with this representation to formulate simulation experiments in physical terms [8].

As a result, the need for an interface between design product models and simulation models is one of the key issues for successful integration.

3 CONCEPT OF INTEGRATION

A holistic approach, which is based on integrating design, modelling and simulation, comprises a three-level-architecture. The integration of methods and processes (methodology), systems and models. The integration of methods and processes means supporting methods and operations within departments, companies and co-operations as well as workflow management. The main objective of system integration is a spanning communication in the CAX environment providing (network-)service. Model integration signifies that different tools using proprietary product data models are sharing data using a common data base. The result of model integration is a discipline overlapping product model. To limit the scope of a general product model, a number of special purpose models modularise the common product data model and are tailored to certain design aspects.

3.1 Integration of Systems

Regarding communication between engineering tools and sharing of engineering data the following architectures are popular.

First, tools are connected directly. This architecture implies an one-to-one interface between tools using data access methods of provided application programming interfaces (API). APIs depend on software and software releases. Operation and service of one-to-one interface is very expendable. Second, the interconnection of tools is indirect. The communication between engineering tools is supported by Engineering Data Management Systems (EDM) using data files or a data base, which are defined by neutral data formats.

Besides direct/indirect interconnections, the continuing revolution of standards like CORBA (Common Object Request Broker Architecture) or DCOM (Distributed Common Object Model) will drive architectures to flexible integration concepts based on object-oriented principles, where tools and techniques used can communicate actively their data.

3.2 Integration of Product Models

With the assistance of a common product data model, a data mapping between CAX-tools can be provided. For the integration of different model views a defined data model structure is crucial and entails efficient methods for data mapping between all interesting views. The idea of STEP (Standard for the Exchange of Product Model Data, ISO 10303) is based on a single, standardized product model, that holds all relevant data. However, the missing of parametric, constraint-based and feature-based model characteristics is a handicap of the STEP specification and the described additional relations between design and simulation are not yet part of any STEP product data model (Application Protocol).

The question comes up, how to cover dependencies between different product models and how to establish a common product data model from bottom up. The following approaches require knowledge in both the engineering domains and information modelling to be successful.

3.2.1 Model Definition from the Process View

From the point of view of engineering processes, data exchange between product development tasks need to be analysed first. The model definition task from process view can be participative or expert driven to different degrees. In a participative project the users have a strong role through the entire definition phase. Their knowledge will be utilised to secure a properly working framework. Usually, the engineering designers in the company are the people that have the best knowledge of how the product development process actually works. While most companies have a formally defined product development process it is not always followed in the actual day-to-day work. Data modelling experts lack this knowledge and can thereby only adapt the model framework after the prescribed process (ideal case). However, the expert can provide input on how the product model interacts with the entire CAX environment and supports data modelling.

According to data and information requirements, a first rough application independent framework should be defined after having a model of the product development process. A customised structure of the information content needs to be formed additionally, which should be independent from implementation for further development and modifications. The users should improve the data exchange processes associated with product development before the model definition. It is possible to define a product model from an already existing product development process and data exchange issues, although doing this will not fully utilise the power of new information technology.

3.2.2 Model Definition from the Product View

Data model definition from the product view is based on native data models representing different product views. The main issue is to map the various product descriptions in a common data model. Therefore, native data formats need to be analysed first. The analysis should result in a common structure of the information content, which is the most difficult part. A very important issue is to check model structure and dependencies regarding properties (parameters), structuring data classes and adding important class properties subsequently [7]. Figure 1 illustrates data models from a Simulation System (Matlab/Simulink) and a common reduced data model for design and simulation.

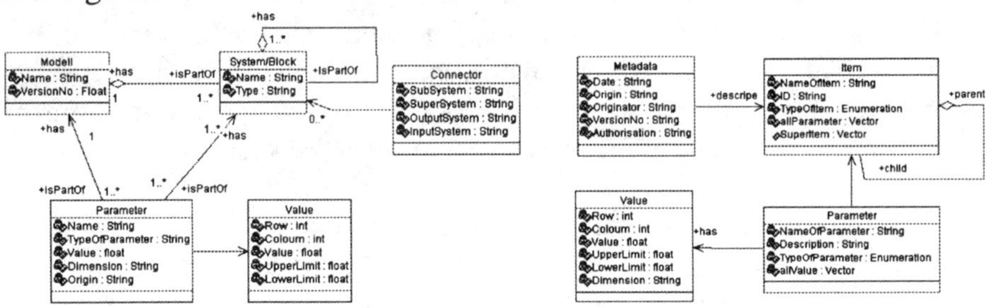

Figure 1: Data model for a block diagram and a merged data model (simplified)

Design product models are used by CAD-systems, whereby different systems are working with different native data models. CAD-models are representing mainly product structure (e.g. assembly hierarchy, bill of material) and product geometry (shape). A model in CAD-systems may be described as a composition of static product properties, which are safeguarded by analysis methods and tasks (e.g. structure analysis, modal analysis, multi-body-analysis). The main purpose of CAD-model is the same as of drawings, they are blueprints for product manufacturing and assembly.

With modelling and simulation programs, the behaviour of dynamic systems, such as electrical, hydraulic systems and any combination of these systems can be simulated. Mostly, systems are modelled using block diagrams. Block diagrams may be described as a composition of lower level submodels. Submodels consist of blocks, which are (pre-) defined icons connecting inputs/outputs and representing equation descriptions internally.

The mapping between different product descriptions must be transparent. The structure of block diagrams and design models are divergent due to different approaches in modelling. Therefore, rules of interpretation (mapping rules) must be obvious for engineers when relating blocks of simulation models to parts of design models (assemblies).

The object-oriented physical-systems modelling approach using model description languages (e.g. Modelica, Sidops+, NMF) supports data mapping from design product models to (object-oriented) simulation models straightforward, especially when using elements from model libraries such as clutches, pumps, motors etc. Object-oriented physical system models do not have an influence because of encapsulation and a non-causal (declarative) model description. Consequently, modelling physical systems in an object-oriented way could easily adapt the structure and properties of CAD-models and hence, share a common product data model (Table 1).

Table 1 Data Mapping between a general Design Model and Simulation Models (excerpt)

	CAD (Design Model)	SIMULINK (Block Diagram)	MODELICA (OOModel)
Product	Assembly	System	Model
Item	Subasssembly Part	Subsystem -	Submodel Component
Element	Feature	Block	Class
Relation	Mating condition Constraint	Line, Branch -	Connection Equation
Parameter	Parameter	Parameter	Parameter, Type

4 IMPLEMENTATION

In accordance with the proposed architectures for the integration of systems and models, neutral or standardised data formats are the key enabler for successful implementation. Within STEP, the translation to and from file formats is supported by off-the-shelf tools (e.g. Caselib from ProSTEP GmbH). In this case, data translation requires the EXPRESS language where both data formats are to be defined by EXPRESS schemes.

In a first prototype implementation of an environment for data management of aircraft actuators, a general conversion tool using indirect connections has been developed. This tools supports data transfer between different applications and has an interface to a PDM-system, which ensures simultaneous work with the same or related information. The neutral data scheme was developed following the process-oriented approach and implemented using Java [6].

5 CONCLUSIONS

Based on the situation described in the previous sections, development of data models drives the integration of basic services for multidisciplinary engineering. Hereby a structure of the information content has been presented according to mentioned product views. The differences between design product models and block diagrams as one description for simulation models has enforced a resulting data model representing only parameters.

The object-oriented simulation methods and models are structured and robust enough to get beyond, integrate multidisciplinary engineering concerning methodology as well as models by enabling automated data mapping with simple algorithms. At this point, the collective approach of design, modelling and simulation of mixed-domain systems as well as the integration of object-oriented modelling techniques require further research.

6 REFERENCES

[1] van Amerongen, Job: *The Role of Control in Mechatronics.* In: Engineering Science and Educational Journal, Vol. 9, Nr. 3, pp. 105-112, ISSN 0963-7346, 2000

[2] Broenink, Jan; Kleijn, Christian: *Computer-Aided Design of mechatronic systems using 20-SIM 3.0.* In: Proceedings of WESIC'99, 2nd Workshop on European Scientific and Industrial Collaboration, Newport, UK, Sept 1st-3rd, 1999

[3] Grabowski, Hans; Geiger, Kerstin: Neue Wege zur Produktentwicklung. Stuttgart: Raabe, 1997

[4] Haberfelner, Reinhard: Systems Engineering, 9. Auflage. Zürich: Verlag industrielle Organisation, 1997

[5] Kleiner, Sven et al: *Data Mangement for Mechatronic Systems.* In: Proceedings of the 6th International Conference on Concurrent Enterprising. Toulouse, France, 28-30 June 2000

[6] Mendgen, Ralf: Methodische Vorgehensweise zur Modellierung in parametrischen und featurebasierten 3D-CAD-Systemen. Aachen: Shaker Verlag, 1999.

[7] Rumbaugh, James et al: Object-Oriented Modelling and Design. Englewood Cliffs, New Jersey: Prentice-Hall, Inc. 1991

[8] Sahlin, Per: *A tool for data mapping from design product models to object-oriented simulation models.* In: Proceedings of the 11th European Simulation Symposium: Simulation in Industry, October 26-28, 1999, Erlangen, Germany

A Manufacturing Information Infrastructure to Link Team Based Design to Global Manufacture

R.I.M.Young, J.M. Dorador, J.Zhao, W.M.Cheung
Loughborough University, Loughborough, Leicestershire, LE11 3TU, U.K
Email: R.I.Young@lboro.ac.uk

Key words Information Infrastructures, team based design, global manufacture

Abstract: The moves towards team based design and global manufacture have led to the need for substantial improvements in the information support systems to aid the decision making of product designers and manufacturing engineers. This paper describes the findings of a research programme which has investigated the manufacturing information infrastructures required in both product and manufacturing models. A data model to represent global manufacturing facilities is proposed. Further a product data framework has been defined which can support multiple manufacturing views of a product. The machining and assembly processes have been used to illustrate these.

INTRODUCTION

While product model technology has advanced significantly in recent years, the current status of information model structures is limited when the multiple viewpoints of members of design teams need to be supported [1]. The limitation of current information infrastructures is extended further when the need to provide manufacturing information from multiple, globally located, facilities is considered. This paper discusses a research programme entitled 'Manufacturing Information Models' which explores the information infrastructures necessary to support the related activities of design for manufacture within a concurrent engineering environment and post design manufacturing planning [2].

While web based technology offers radical new approaches to accessing information [3], this information can only be of value if it can be interpreted and used by the appropriate decision-makers. This is particularly relevant where software applications utilise the information, as a clear definition of the information structure is essential. Hence there is a clear need for well-defined information infrastructures which can be used to aid the sharing of information across team based design and global manufacture.

The extensive work of the STEP community is progressing slowly towards standards for data exchange between CAD systems. They recognise that this is an initial stage in the progress towards information sharing in product design and manufacture systems [4]. The work reported in this paper started by assuming that information structures were the key to providing support to team based design. This has now developed to the belief that while product information structures can provide a basis for integrating team based activities, there is a clear need to link knowledge models into such a system.

Whilst the range of information and knowledge involved in supporting the range of decisions to be made in new product development is substantial, this paper focuses on manufacturing information and knowledge infrastructures. In particular it uses machining and assembly as example processes against which to explore information infrastructure requirements.

THE MANUFACTURING INFORMATION MODELS CONCEPT

Informing Decision Makers

The premise behind this paper is that computation systems in integrated design and manufacture should provide support to engineers by offering them quality information on which to base their decisions. This lies between the two extremes of automation, which typically provides inadequate results, and simple information retrieval routines, which are likely to result in information overload.

A further basis behind this paper is that if broad based, integrated solutions are to be provided, then there is a need to separate the range of information and knowledge from the software applications, communication mechanisms and people which will utilise it. The basic concept of the research is illustrated in figure 1.

Supporting Design for Manufacture and Manufacturing Planning

Both design for manufacture and manufacturing planning require an understanding of the manufacturing resources which are available to manufacture a product and how to use them. This is the case for single factory operations and for global supply chain systems. Within information systems there is therefore a need to define infrastructures which can capture information and knowledge about manufacturing. This can then be used by both design for manufacture and during manufacturing planning.

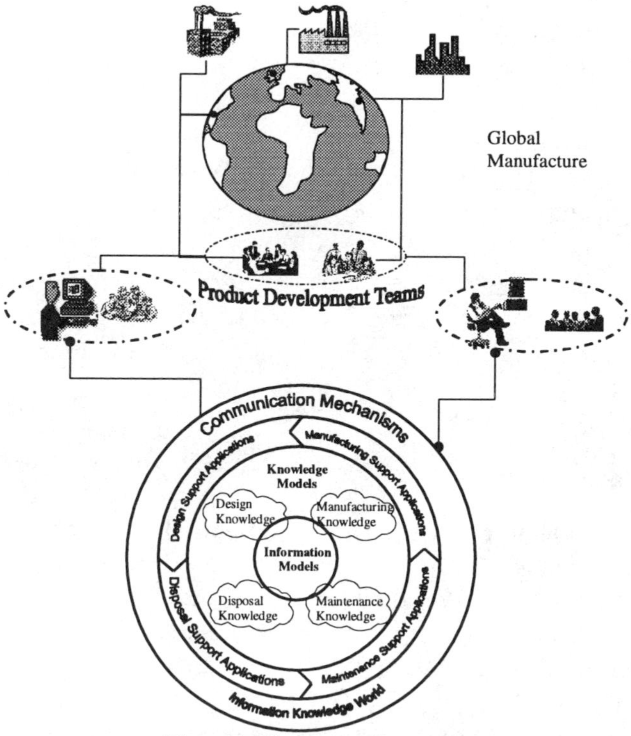

Figure 1. The research concept

Following the concurrent engineering philosophy leads us to perform design for manufacture as early as possible in the design process. It also raises an issue in terms of the relationship between design for manufacture and manufacturing planning. If manufacturing input is captured during design then how much remains to be done during manufacturing planning?

This paper proposes that ranges of manufacturing functions can be defined. The extent to which they are used during design is dependent on the extent to which particular manufacturing resources can be specified during the design process. Critically, the decisions made at any stage must be stored and managed in relation to the product under development.

Manufacturing functions to support design can provide general feedback on likely manufacturing processes or at a more detailed level can provide suggestions for design changes which improve the ease of manufacture of the product. Decisions at this detailed level require a clear understanding of manufacturing processes and are closely related to process planning.

Typical functions relating to detailed machining decisions can be listed as:

- Identify machining processes required
- Identify possible machine tools to be used
- Identify cutting tool types required
- Identify fixturing requirements
- Select machine tools
- Determine setup plans
- Determine cutting tools
- Determine cutting parameters

The order in which these are performed is largely sequential, with the earlier functions being open to suggest design changes. Performing the later functions in the list is dependent on having chosen the facilities to be used for manufacture.

Manufacturing Information and Knowledge Requirements

Manufacturing information can be classified in a number of different ways. It can be considered in terms of how to produce a particular product or in terms of how a particular enterprise can use its facilities to manufacture things. The former being part of a product model and the latter being a manufacturing model. Within the enterprise, information can be considered at a total enterprise level, or at specific factory, shop, cell or station levels. At each of these levels manufacturing information can be considered in terms of the resources which are available and the processes which can be performed. In addition knowledge of how these resources and processes can be used is important. For example, at the global level, knowledge about global manufacturing strategy can be captured. At the individual station or machine level, knowledge of how the machine can be used to perform particular processes can be held.

The information and knowledge about different manufacturing processes, e.g. machining and assembly, will be specific to the process. However, the structures used to hold this information and knowledge are a combination of generic and process specific relationships. The following section describes the product and manufacturing information structures which have been developed in the research to support both machining and assembly.

THE DESIGN OF A MANUFACTURING INFORMATION INFRASTRUCTURE

Product Model Structures

A UML class diagram showing the general structure of a product model resulting from the work of this research is illustrated in figure 2. Typically product models are used as a source and repository for information concerning product geometry, dimensions and tolerances, and product material descriptions. These are captured under the Characteristics class in this work. A significant aspect of this figure is the Views class. This offers the facility of viewing and using the product model from different perspectives, which is critical to success if team based product development is to be successfully supported. In this example, the structure offers design views and manufacturing views. Views of the product which relate to particular manufacturing process views of the design can be held within the design views i.e. design for machining and assembly views. It is in these views that information relating to feature technology descriptions would be stored. The results of manufacturing designs made during the design process or during manufacturing planning are stored within the manufacturing views of the product.

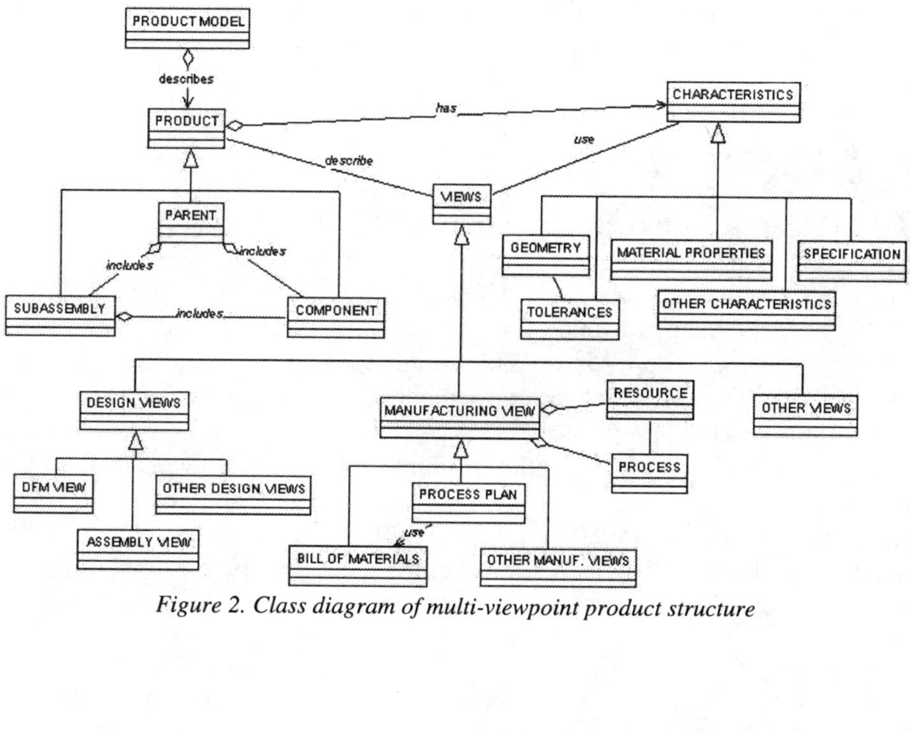

Figure 2. Class diagram of multi-viewpoint product structure

Manufacturing Model Structures

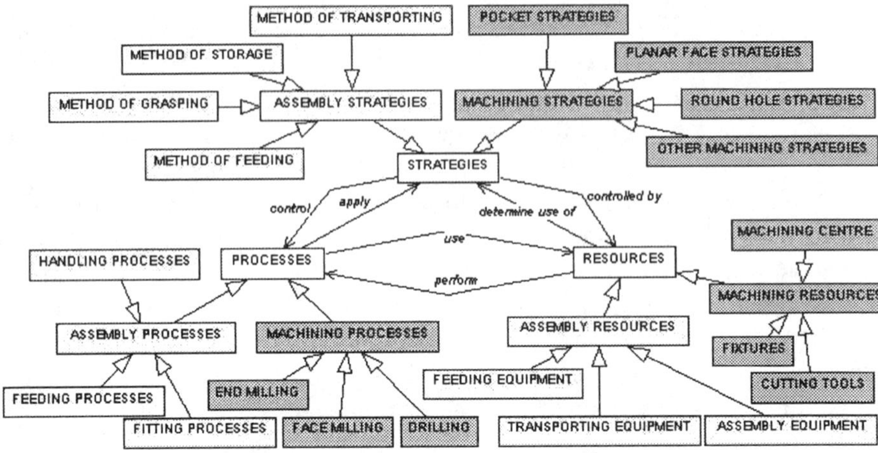

Figure 3. A class diagram for machining and assembly facilities

A general class structure in which to model a global manufacturing facility in terms of its resources, processes and strategies has been defined and can be found in [5]. A more detailed structure which can be used to capture machining and assembly capability is shown in figure 3. This illustrates examples of particular resources, processes and strategies for the two processes in question. It also highlights that the relationships between the resources and processes can be defined and that the strategies provide definitions of how the resources and processes can be used. The strategies provide the knowledge element of the manufacturing model.

Relationships between the model structures

While the product model and manufacturing model can be considered to be two independent models, the relationships between the two model structures are interdependent. This interdependence is defined by the common information which the two models use, and by the relationships between information views.

Figure 4 illustrates how the resource and process structure in the manufacturing data model can be used in the product data model. The difference in the information content is that the manufacturing model captures all the processes and resources available, while the product model only captures those to be used in the manufacture of the product under development.

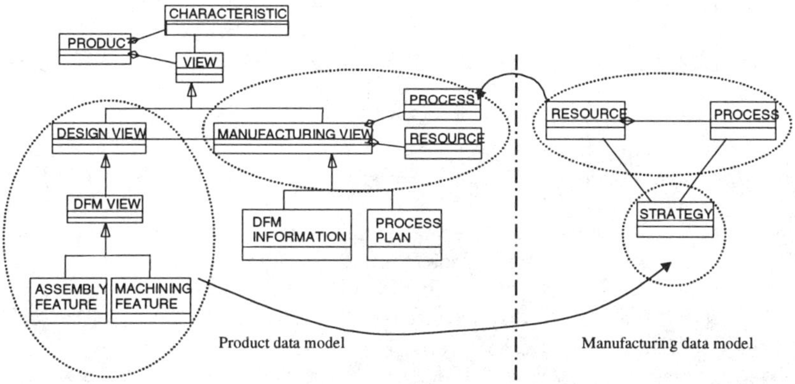

Figure 4. Relationships between product and manufacturing model structures

Figure 4 also illustrates that the relationship between design for manufacture views and manufacturing strategies is significant. The design for manufacture view captures the features which are significant on the product while the strategies class provides a representation of the ways in which a process can be used. This therefore provides a flexible means of relating feature based design descriptions to methods of manufacture.

CONCLUSIONS

Clearly defined information infrastructures are critical to the communication of useful information between product development team members, especially where integrated software solutions are required. This paper has provided a contribution to the definition of product model structures to support the multiple views of information required by product development teams. It has also shown how models of global manufacturing capability can be generated to cope with multiple, globally dispersed, manufacturing processes. Significant relationships between information model structures have been defined to support information sharing across team based design and global manufacture.

ACKNOWLEDGEMENTS

The authors wish to thank the EPSRC and their industrial collaborators. This work has been performed under grant number GR/L41493 entitled "Manufacturing Information Models". The research undertaken by J.M.Dorador is funded by UNAM and CONACYT, Mexico.

REFERENCES

[1] Rosenman M. A., Gero J.S., (1999) Purpose and function in a collaborative CAD environment *Reliability Engineering and system safety* **64**, 167-179.

[2] Young R.I.M, Canciglieri-Jnr O., Costa C.A., Dorador J.M., Zhao J., Cheung W.M. (2000) Information Support in an Integrated Product Development System *Integrated Design and manufacture in Mechanical Engineering* (IDMME'2000), CD-Rom paper MA3.1

[3] Tegel O. (2000) Information and communication technologies to support cooperation in the product development process. *New Product Development and Production Networks*, Ed. U.Jurgens, Pub. Springer, pp389-406. ISBN 3-540-64172-6.

[4] Fowler J, (1995). STEP for data management, exchange and sharing. (UK, Technology Appraisals), ISBN 1871802369.

[5] Zhao J, Cheung W.M, Young R.I.M, (1999) A manufacturing data model to support virtual enterprises, *International Journal of Agile Management Systems*, **1**(3)150-158.

Non-destructive Tracing of a Product Life Cycle Through Geometry Extraction from Radiographs

J. Hefele, R. D. Bolton
Los Alamos National Laboratory, Los Alamos, NM, USA
Em: jill@lanl.gov and rbolton@lanl.gov

Keywords Photogrammetry, Geometry Rconstruction, Non-destructive Testing

Abstract Tracing the internal changes seen by a product assembly during its useful life cycle without disassembling or destructively testing the product provides many challenges. When an assembly is disassembled for evaluation, the effect of component interaction is lost, particularly if operations such as press-fitting have been used in the assembly process. It is undesirable to destructively evaluate a product when it is still in its useful life cycle. Los Alamos National Laboratory is developing a process in which radiographic information is used to characterise an entire assembly for life cycle evaluation and the geometric information is extracted for input into a Computer Aided Engineering (CAE) software package. Digitized radiographs have been reconstructed into solid models using FotoG™ geometry extraction techniques and ProENGINEER™ solid modelling capabilities. Executed simultaneously, FotoG™ translates the extracted geometry into a format useable by the ProENGINEER™ software. Once the solid model is constructed, it is used for creation of finite element analysis input, virtual environment model creation, and tooling design. The creation of the finite element model allows us to forestall future problems using predictive analyses. Virtual environment creation provides us with the capability of creating a learning environment for technicians performing maintenance or disassembly of problem units. And tooling design provides us the capability to safely transport severely damaged assemblies in the event of an accident scenario.

1 INTRODUCTION

As a product matures through its useful life cycle, characterisation of the product's current state becomes increasingly essential in order to forestall function and servicing problems. "As-Built" and "As-Is" characterisation methodologies have been developed recently [1] [2] [3] using inspection data as the raw input in modelling a product. This type of characterisation works well for portrayal of a component's current state. However, once components are assembled into a final product, the geometry can easily be modified due to the processes it has been subjected to during assembly or because of interactions with other components owing to use.

Characterisation of a product in its assembled state provides it's own challenges. Capturing the "As-Is" state of components in the assembled configuration requires non-destructive methods of evaluation. Radiography and Computed Tomography [4] can provide the raw data necessary for this type of characterisation. We have concentrated our initial development efforts on the use of radiographic data because much of our historical data exists in this state. Changes can occur to component geometry as a product moves through its useful life cycle and these variations from the "As-Built" state can also be captured using non-destructive evaluation techniques, thus allowing life-cycle data points to be collected as an assembly ages.

This paper will examine the data collection and manipulation processes for characterisation of a product as it moves through its useful life cycle using photogrammetric, radiographic, and computer-aided engineering modelling techniques. We will also inspect issues related to data collection methods employed to obtain the best possible data for use in product model reconstruction.

2 DATA COLLECTION PROCESS

The data collection process for model reconstruction of a product consists of three steps; initial survey of data collection environment, documentation of environment through digital photographs, and radiography of the product being examined.

2.1 Initial Survey

The initial survey of the data collection environment establishes a network of control points to be used as reference between the digital photographs taken in the next step and the real world. A theodolite is used to measure the location and orientation of the control points placed in the data collection environment prior to photography. A minimum of four control points is required. Stick-on targets are also set and located using the theodolite or by measuring the distance between control points/targets and establishing an arbitrary but correctly scaled co-ordinate system during this phase. The targets are used as references in the documentary photographs.

Theodolite measurements can be made at this point to locate the radiographic film holder and the x-ray source.

2.2 Environment Documentation Using Digital Photographs

Once the control points and targets have been established, the environment can be documented using digital photography. The assembly to be evaluated must be visible in at least two of the photographs taken with different perspectives. It is essential that the camera/lens combination used to take the photographs in this step have been calibrated using test photographs of a specially designed calibration field. Analysis of the digital photographs is benefited through inclusion of a fair amount of overlap in the photographs taken. Section 4 discusses this in more detail.

2.3 Radiography

Radiographic data provides the true basis for reconstruction of the assembly of interest. The same principles apply to radiographic documentation for assembly reconstruction as do to environment documentation for site reconstruction; i.e. more overlap in radiographic images, the better.

3 DATA MANIPULATION

Once the raw data is collected in the form of survey data, digital photographs, and digital (or digitised) radiographs, the reconstruction can begin. Analysis begins with identification of the camera/lens combination used to collect the digital photographs. This provides the FotoG™ software with the necessary information for resection of the images. Images are mathematically resected as a set with ties to control points. The analytical process is well-documented on Vexcel's web site [5]. Radiographic images can be over-laid into the digital photograph site reconstruction data or can be reconstructed as a separate data set if survey data has been collected documenting the location of the x-ray source and radiographic film frame.

4 DATA COLLECTION ISSUES

Data collection techniques effect the results obtained during the reconstruction process. The type of targets used, for example, determine the ease with which points are identified in the digital photographs while processing the data. Additionally, the precision of the transit system used, completeness of the survey techniques, choice of camera locations, number of control points identified, number of images used in data processing, and type of image geometry used all effect the final results obtained in the data set analysis..

4.1 Targets

Figure 1 illustrates examples of types of targets used. Of the three types of targets shown, we have found the easiest to work with in the photogrammetry environment is on the extreme right.

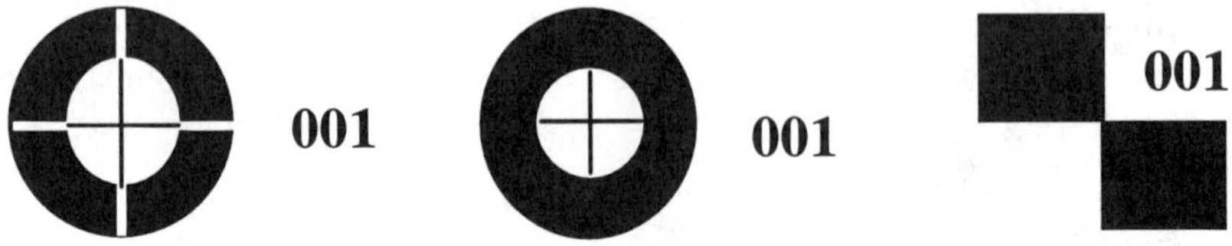

Figure 1 Examples of Target Geometry.

4.2 Transit System Measurement Precision

The precision inherent in the theodolite used to take measurements during the initial survey stage will effect the precision in measurements taken digitally in the reconstructed data set. It is, therefore, essential to chose a transit system with as high a precision as possible. Figure 2 compares the same measurements taken in two different data sets by the same theodolite. The error, 2-3 mm, is within the stated precision of stated by the manufacturer.

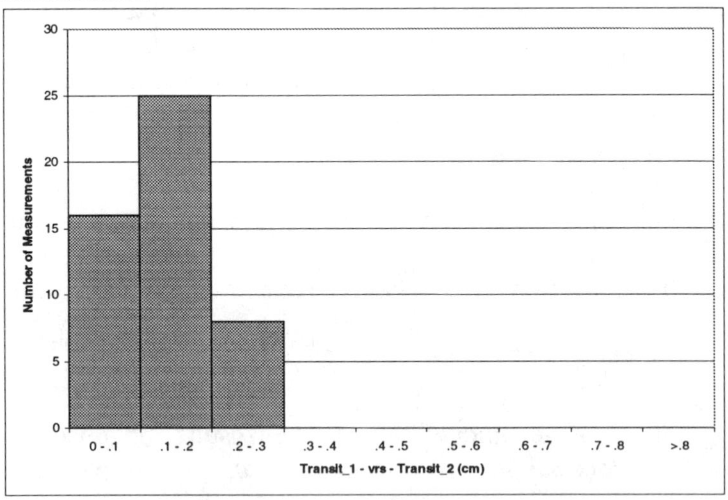

Figure 2 Comparison of Transit Measurements.

4.3 Photogrammetry Measurement Precision

The precision of the photogrammetric software reconstruction is also an important issue when choosing tools to be used for reconstruction efforts. Figure 3 compares the differences observed in measurements reconstructed from digital images using the FotoG ™ software with the same measurements made with the theodolite.

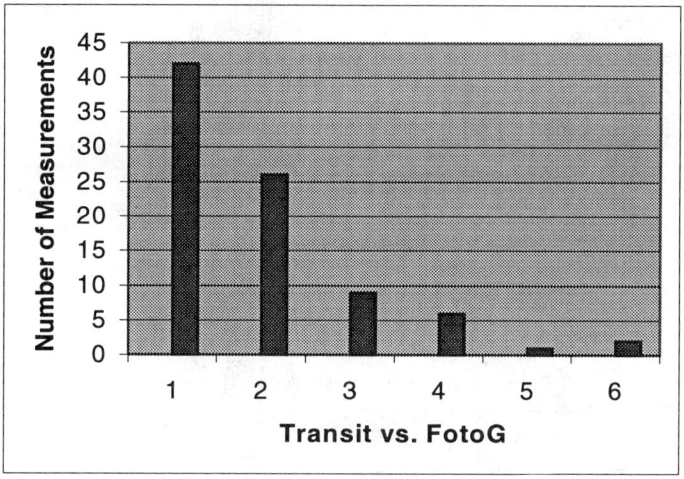

Figure 3 Comparison of Transit Measurements with FotoG Software.

4.4 Number of Control Points vs. Precision

The system of control points established in the initial survey during a reconstruction effort establishes the reference between the digitally documented environment and the "As-Is" condition that exists in the real world. The number of control points used has an effect on the precision attained during reconstruction. A minimum of four control points is required. Figure 4 shows the effect of increasing the number of control points used during a reconstruction from four to eight to sixteen to thirty-two. In general, the more control points used, the better. However, there is a break even point where increasing the number of control points has only a small benefit to increasing precision.

4.5 Number of Images vs. Precision

The data collection process can be the most time-consuming element of a reconstruction effort. Documentation of the site and assembly environments in a complete manner is extremely important to the resulting analysis. The number of images, however, is not nearly as important as the quality of the images taken. A few images with good overlap, say greater than 60% coverage, and strong geometry, approximately 90% convergence angles, will give the best results in a reconstruction.

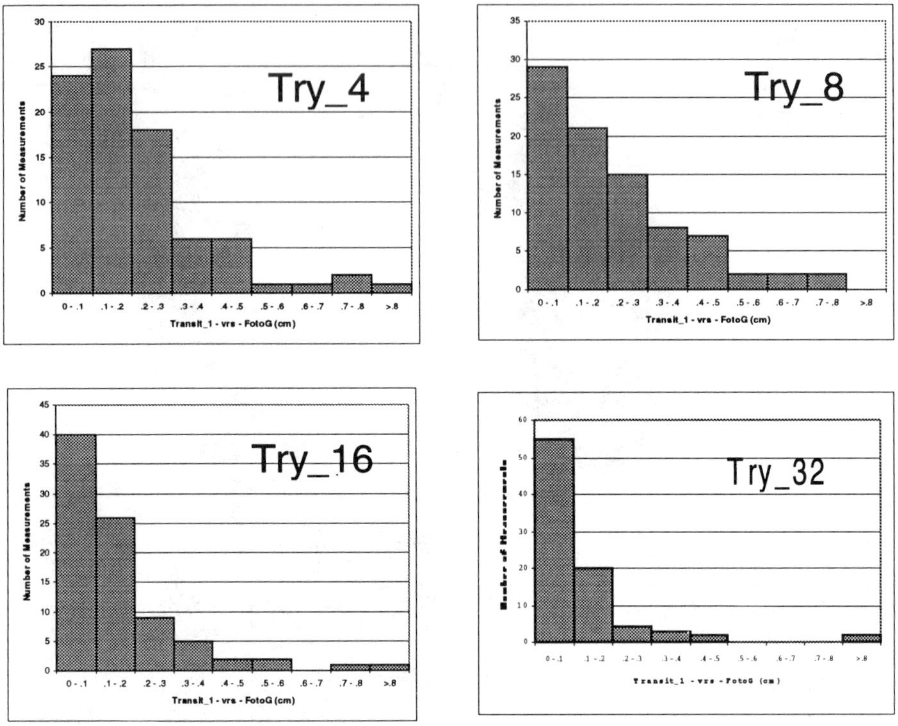

Figure 4 Effect of Number of Control Points on Precision.

The upper two graphs in Figure 5 illustrate the increase in precision from using three images with good geometry as compared to two. However, as illustrated by the lower left graph, adding a fourth image with not quite as strong geometry had a negative effect on the precision. The analyses pictured in the two lower graphs in Figure 5 both employed the use of four images, but the analysis pictured on the left used images with stronger geometry than that on the right.

5 CONCLUSION

Photogrammetry, radiography, and reverse engineering reconstruction techniques are effective tools in the non-destructive evaluation of the product life cycle of an assembly. These techniques can be used to forestall problems before they become catastrophic. Research and development continue to improve processes and software development in this area.

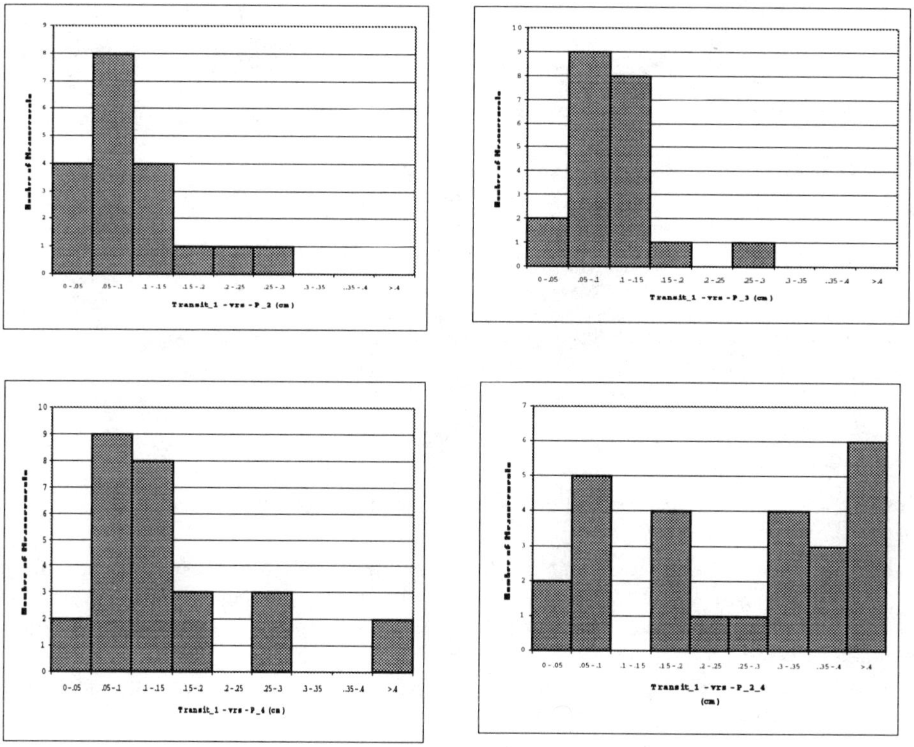

Figure 5 Effect of Number of Images on Precision.

6 REFERENCES

[1] Dolin R. M., Hefele J. (1997). As-Built Engineering: A Philosophy for Representing Product Definition, *LA-UR-97-3631 Proceedings Nuclear Explosives Design Physics Conference*, Livermore, CA.

[2] Hefele, et. al., (1997). Product Characterisation Using Inspection Data in an As-Built Engineering Environment, *Proceedings Second Biennial Tri-Laboratory Engineering Conference on Modelling and Simulation,* Los Alamos, NM.

[3] Hefele, J. Dolin R. M. (1998). Density Effect of Inspection Data in As-Built Modelling of Parts, *Proceedings 2nd International Conference on Engineering Design and Automation*, Maui, HI.

[4] Kelley T. A., et. al. (1998). Reverse Engineering Using Computed Tomography, *Proceedings 25th Annual Review of Progress in Quantitative Nondestructive Evaluation*, Snowbird, UT

[5] Close Range Photogrammetry – Learn About Photogrammetry, *http://www.vexcel.com/fotog/learn.html*

Planning for Manufacturing – Managing Connective Designs and Asymmetric Designer Knowledge in Product Consortia

Martti Meri
Department of Computer Science and Engineering, Helsinki University of Technology
Em: martti.meri@hut.fi

Keywords Planning, Co-ordination, Organisation theory, Concurrent Engineering

Abstract This paper presents production management as a three dimensional concurrent engineering (CE) problem, where the third dimension, besides product and process, is the organisational structure. The need of the organisation theoretic dimension is inspected from a point of view of ideal organisational types and an extended concept of equifinality analysis taking the special considerations, like flexibility, of network organisations into consideration. Methodologically we stress the need to define the ideal organisational roles using measurable micro-level quantities. We exemplify the need to model the organisational dimension by presenting a program that constructs the set of possible networks and uses it to illustrate how approaches using global organisational measures or using no measures at all are both insufficient approaches.

1 INTRODUCTION

Knowledge has been recognised as one of the key forms of wealth of the new organisation. In the information era, organisation transforms into a specialised agency that needs to cumulate, manage, develop and protect its core competence. In economic science literature, theories of intellectual capital promulgate the need to make this knowledge explicit. This does not solve the management problem in network organisation because information asymmetry limits possibilities to generate global explicitly presented knowledge. In manufacturing industry the knowledge concerns intimately the practical ways of how product components are really produced and delivered. In the meanwhile the increasing global competition sets high requirements for the products, which need to be advanced and highly integrated. How to overcome this paradox of simultaneous organisational dispersion and increase in the integration degree of products?

The main result of this paper is a novel reformulation of the concurrent engineering problem as a planning problem. While the classical requirement is one of making the manufacturing process and the product match, in our approach a match between organisational action and components of product structure is established in a level of finer granularity and enhanced semantics. In the networked organisation the process dimension becomes scattered and requires improved representation formalisms. The new formulation for the concurrent engineering problem calls for tools that support concurrent design of the product and the organisational network responsible for the production activities.

Project as a model of business behaviour is increasingly penetrating to new business domains. Therefore we believe that the results of the study can be generalised to many areas. One of the motivations for the research was that information flow diagrams, typically used as the formalism in the initial phase of the behaviour modelling are inadequate for network organisation. It was soon discovered that an enriched representation formalism that can refer to product and design scopes was needed.

The paper is organised in the following way. In section 2 we discuss the problems of asymmetry in network organisation. The following section, section 3, lays out the framework. Illustrative experimental case is presented in section 4. Section 5 contains the conclusions.

2 ASYMMETRIC KNOWLEDGE

Classical organisational structure categorisation distinguishes hierarchies and markets. Network organisations aim at combining the desirable features of the two classical ideal extremes. However, network organisations can not be considered as an intermediate form of the two, because it requires elements that are not as such encountered in the other forms. Trust of a different type is one of the distinctive features. Network organisations balance stability against flexibility, specialisation against generalisation, and centralisation against decentralisation (Alstyne 1997). Specialisation leads to information asymmetry, more restricted scope, economies of scale and cost efficiency, whereas generalisation leads to information symmetry, economies of scope, and efficiency of greater resource utilisation (Alstyne 97).

Asymmetric knowledge means that agents have different knowledge structures. Typically a component supplier as a customer to sub-component supplier knows less about the sub-component than the sub-component supplier. Still the component must be an integral part of the whole. Let us model this situation by listing product component features that agents must know roughly (and call them desires) and product component features that agents must know in detail (and call them capabilities).

Different measures of organisational design (Doty et al. 1993, So and Durfee 1996) have been developed both at the macro-level and the micro-level. We define the following micro-level measures.

Table 1 Local measures of organisational entities.

Measures	Description
Span of desire (sod)	Agents are characterised by a set of goals. This measure describes the demand of the agent for final products and sub-component deliveries
Span of capability (soa)	Agents are also characterised by a set of goals they can reach by them selves
Ideal type	Agents can set a strategy concerning their policy of doing themselves the activities needed for producing the products and components, or getting them from the network environment. Value 0 means that the agent's desires and capabilities do not intersect, whereas value 100 means that they coincide.

At the macro-level *sod* and *soa* are used to state the worst case sizes of span of desire and span of control. In the experiment the macro-level measures are used when generating experimental random cases that are within the limits of these global measures.

The typology by Miles and Snow (1978) identifying four ideal types of organisation: the *prospector*, the *analyser*, the *defender*, and the *reactor*, has empirical backing (Doty et al. 1999). Wood (2000) identifies eight organisational types that fall into three categories: *shapers* are capable of changing significantly the aspects of their environment, *adapters* reflect the environment, and *reactors* are generally resistant to change. Of these shaper clearly has both knowledge and control over facts of the environment. The theory does not however support modelling the knowledge structures and aid in pinpointing any specific control mechanisms. Common to these categorisations is that they are based on the emphasis placed on the external domain as opposed to the internal domain. Actual definitions of these types are too lengthy to be presented in this paper, but we redefine them in terms of a set of measures of the knowledge structures.

In our constructive approach we redefine two of the original types of Miles and Snow in a way that allows us to study the internal and external world of agents and the interaction of these worlds in network formation:

- Defensive organisation is focused on its capability scope and lets it direct the external scope. This is a resource-based strategy. The ideal type value is close to 100.
- Prospective organisation is focused on its external environment and does not let its internal capability scope limit it. This is a market-based strategy. The ideal type value is close to 0.

The main part of the environmental structure, that organisations share, is in the product domain. It might be that the structure is abstracted in order for it to be understandable by the set of organisations forming the organisational structure. Specialisation leads to asymmetry and asymmetry together with the reciprocity requirement leads to the criteria that the system at some level and in some time horizon is to be expected to have a balance of dependence.

At the strategic level, organisations express their desires and capabilities in terms of facts concerning the environment. Combined desires of a set of organisations form the goal-state of the potential consortium formed by the organisations. Assume an initial state differing from the goal-state, and action schemas describing ways to bridge the gap. Ways of doing this, plans, are located at an intermediate layer in between the strategic layer and the operative layer.

Equifinality means that a system can reach the same final state from different initial conditions and by a variety of paths. In economics literature this usually means that different organisational structures have the same organisational effectiveness. The model presented here allows a more general interpretation of equifinality, because the planning problem representation of the strategic decision making task allows a more explicit setting of the goals. This forms an integral framework for studying both efficiency and flexibility of organisational designs.

3 THE FRAMEWORK

Planning is seen as an intermediate level of co-ordination between the strategic level and the operative level (E. Durfee 93). The structure of our CE- framework conforms to this basic division of levels. So and Durfee (1996) mention subsequent research where organisational structures were experimented and the results supported the contingency theories. Problems with this research included the fact that the research was embedded in complex task-environments, with agents whose abilities and behaviours were difficult to clearly characterise, and with performance measures that were not clearly articulated (So and Durfee 1996).

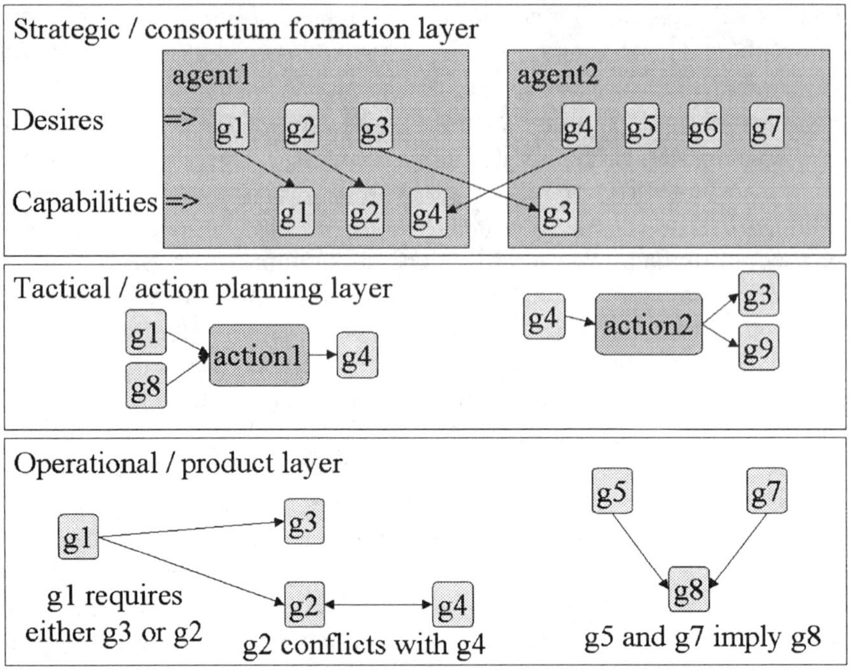

Figure 1 The framework

Prospector and defender organisational types are characterised in terms of the measures *sod* and *soa*, which can be calculated as the number of desires and number of capabilities an agent has. The type of the organisation defines the strategy for deciding what abilities to develop within the organisation. If the strategy is that of a defender, the tendency is to develop capabilities that produce things that are among the set of goals needed by the organisation itself. In figure (Figure 1) agent1 is a defender because it maintains capability to manufacture components marked by goals g1 and g2 and depends on others for g3. Agent2 is pure prospector because it relies on other on all its desires. There is a reciprocal need for collaboration between agent1 and agent2 because they exchange services (g3 and g4). Consortium is a network where the members jointly can reach all the goals they mutually desire. Changes in desires and abilities change the intersection (Figure 1) of these two sets, size of which is interpreted as indicator of the ideal type of the organisation. Because changes also effect the desire-capability arrows and their direction, we try to avoid, as a policy here, drawing means-ends hierarchies, because they are scenario dependent.

At the tactical level, special solvers can be built to solve the planning problem, which consists of three elements, namely the initial state, the goal-state and a set of operations schemata. Operations schemata describe production activities and their external service interfaces including the preconditions and effects. In the recent years advances in the planning algorithms and increases in the computation power have made it possible to solve large enough planning problems to make the approach practically feasible to real world problems of moderate sizes.

In a network environment characterised by asymmetric desires and capabilities, states presented in product terms and activities defined in behavioural terms call in addition for modelling of the organisational structures, in order for the model to represent the components that are used as elements in formation of the manufacturing process. These elements can be used for constructing the scenarios (Wood 2000) of strategic business planning, as well as the information infrastructure that is capable of supporting all such relevant scenarios.

Process-paradigm is good for many things but its deconstruction is necessary for the flexible network organisation. Activities are situational and it is necessary to model preconditions of activities in terms of enriched domain model structures and varying scopes. Also parallelism needs to be based on clearly defined rules for mutual interaction and interference. Using process-paradigm it is not well understood what effects temporal squeezing has. Process diagrams can be used as normative models of the plans generated at the tactical layer.

The use of systematic reference to facts of domain state, i.e. the product facts (denoted by the g-starting variable names in Figure 1.), makes it possible indirectly to integrate the strategic planning and operative systems. This is possible because the tactical layer can act as an intermediary.

4 AN EXPERIMENT

An experiment was conducted by generating random case environments and counting cumulative occurrences of prospectors and defenders in the scenarios of those environments. Agent count was fixed to 7. Ideal type of the agents was taken from uniform distribution [Min,Max]. Figure 2 presents 3 data-sets with 80 cases each. The values in the data sets for [Min ,Max] were [0,40], [0,100] and [60,100]. Desires were allocated randomly to agents so that their number has taken from a uniform distribution [0, sod]. Then capabilities where allocated randomly so that their number is from [0, soa] and they are selected from among the desires, if a random number at selection time belongs to the interval [0,ITA], where ITA is the ideal type of the agent.

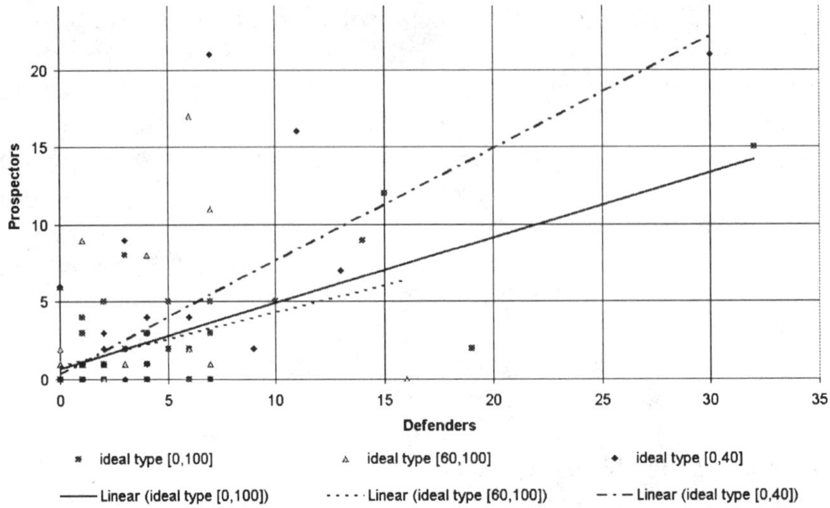

Figure 2 Prospector-Defender ratios in 3 data sets with characteristic ideal type each having 80 cases (show on diagram). Counts cumulated over all scenarios within each case.

For each case there can be many scenarios, each corresponding to an operational alternative consortium, but often the case presents none. A program in C-language was written for the scenario generation. A Prolog-program was written to form the consortia and analyse their organisational constitution.

The experiment indicates that the macro-level measures are not good indicators of the situations. There is a large average-variation within the data sets and the regression lines do not differ that much. The explicit micro-level models of the specific knowledge structures are needed. Further research is needed, but the preliminary results indicate that strategic non-mainstream positioning and analysis of the explicit structures are worthwhile efforts for the agents. Strategy selection should take into consideration the relative overall structure of the desires and capabilities - if *soa* > *sod* be a prospector and if *sod* > *soa* be a defender. The actual structure of the desires and capabilities model effects the number of scenarios in a case and the participating potential of single agent strongly. From the individual agent's perspective besides the abundance of networking possibilities, the membership in those and the lack of solutions that exclude the agents are of interest.

The foreseen advantages of the planning approach presented in the paper for the manufacturing industry include better activity based life cycle planning, capability based product consortium formation, and deeper insight into inter-organisational co-ordination requirements.

5 CONCLUSIONS

The primary contribution of this paper is a framework for 3-dimensional concurrent engineering. The third dimension besides product and process is the organisation. Organisational dimension makes it possible to combine strategic planning and manufacturing planning. Grounding of the goals and actions to the state has the effect that congruence of goals, interferences of actions, and semantics of the both are better managed. Also the goals become easier to update and action representations become generic, because abstracted action schemas are used.

The role of the organisational dimension in the framework is exemplified using an experimental case study of ideal organisation types and network organisation version of equifinality. The experiment shows that the type of things provided by such a framework are required when reengineering the networked manufacturing business.

6 REFERENCES

[1] Van Alstyne M. (1999). The State of Network Organization: A Survey in Three Frameworks, *Journal of Organizational Computing*, **7**(3).

[2] Conner K. R., Prahalad C. K. (1996). A Resource-based Theory of the Firm: Knowledge Versus Opportunism, *Organization Science*, **7** (5), pp. 477-501.

[3] Durfee. E. H. (1993). Organisations, plans, and Schedules: An Interdisciplinary Perspective on Coordinating AI Systems. *Journal of Intelligent Systems*, Special Issue on the Social Context of Intelligent Systems, **3** (2-4).

[4] Doty, D. H., Glick W. H., Huber G. P., (1993). Fit, equifinality, and organisational effectiveness: A test of two configurational theories. *Academy of Management Journal*, **36**(6). pp.1196-1250.

[5] Levitt R. E., Thomsen J., Christiansen T. R., Kunz J. C., Jin Y., Nass C., (1996). Simulating Project Work Processes and Organizations: Toward a Micro-Contingency Theory of Organizational Design. *Management Science* **45** (11) pp. 1479-1495.

[6] Miles R. E., and Snow C. (1978). Organizational Strategy, Structure, and Process, *Academy of Management Review*, **3** (3), 1978 July, pp. 546-562.

[7] So Y., H. Durfee E. H. (1996). Designing Tree-Structured Organizations for Computational agents. *Computational and Mathematical Organization Theory*, **2** (3) pp. 219-246, Fall 1996.

[8] Wood R. (2000). Managing Complexity, Profile Books Ltd. London.

Product Modelling and Rationale Capture in Design Process

J.P.T. Mo
CSIRO Manufacturing Science and Technology, Australia
Em: John.Mo@cmst.csiro.au

Keywords Product Modelling, Design Rationale, Knowledge Capture, Global Design

Abstract Product design is an intuitive activity and generates a lot of information supporting the product in the rest of its life cycle. These information consist of both foreground (explicitly recorded) and background (normally unconsciously referred in practice) information. This paper describes a system developed based on the hierarchical structure of product modelling methodology for capturing an important part of the un-written information, design rationale. Developed in a Java environment, the system is platform independent and is designed to be extensible into a global design support system via the internet.

1 INTRODUCTION

In recent years, the philosophy of global manufacturing has been developed among aggressive manufacturers who are prepared to take on the challenge of managing their activities in the most effective fashions in different parts of the world [1]. More specifically, product development is an essential operation involving business partners and customers geographically distributed [2]. During the design phase, designers need to cooperate in all aspects of the design. With the help of the latest CAD technologies, design teams are using various kinds of product models to represent their products. STEP and other international standards have helped to foster information transfer among different players [3].

However, designers in different parts of the world have the problem of communicating their design idea to their colleagues elsewhere when they distribute their CAD models. Information represented by CAD models are primarily explicit information of the product. The reasoning behind the design are often forgotten when the product model is distributed to the other members of the design team. In other words, the intent (or rationale) of the feature or details of the design is not recorded due to lack of such facility.

The system described in this paper is a Java based design intent capture and management system which captures the design intent (what the design thinks) along with normal design information. The system uses the hierarchical structure of product modelling methodology and provides an active tool during the design process to help recording the rationale information from the designer.

2 GLOBAL INFORMATION MANAGEMENT SYSTEM

The purpose for the Global Information Management System (GIM) is to cater for the need of non-interactive sessions between 2 or more geographically separated design groups in different time zones. While the use of tele-presence tools such as video conferencing is expected to provide an interactive environment for several design groups to work together in a specific time window to communicate, this only occurs in a short period within the 24 hour clock product development cycle [4]. In the other time periods, only people in one time zone will be on-line and all the others will take their rest.

The research to generate the Global Information Management System is focussed on the issue of asynchronous access to information. In a global concurrent engineering situation, design teams will be spread across different time zones, and will thus need to access information from other sites when they are off-line, or will need to know what happened at others site when they themselves were off-line.

The system is aimed to provide support to the on-line design group to search, retrieve, view, edit, be alerted, be managed and more importantly, capture design rationale. To do these, an understanding of the system architecture and elements required to build such a system is required [5]. It is based on an object oriented object structure of the product and fits the implicit information of the product into relevant locations of the data structure. By definition, explicit information are those related to the design of the product such as user specification, material, overall dimensions and functions. These are captured by CAD systems. Implicit information are those information not normally expressed explicitly by the design. Typical examples are the reasons for certain material selection or the size of some mating features. These information are captured through a series of windows linked to the product data model so that the rationale behind the designed features can be recorded accordingly.

3 METHODOLOGY OF SYSTEM DESIGN

The methodology applied to develop the system is to treat all design activities as activities manipulating design objects. The definition of design objects originates from the concept of data abstraction for compute aided process planning [6]. There are three levels of data abstraction in the framework of objects which is required to develop a unified system linking CAD model data to manufacturing. It makes use of a universally acceptable set of objects which can interact with process planning modules (which are themselves objects) to achieve the required output to manufacturing. With this approach, a part is composed of a series of objects which possess certain uniquely identifiable characteristics. As the design process progresses, the objects are moved to higher levels where more attributes are added or inherited from other objects.

In the design of GIM, the relationships of the objects are taken into account. In most cases, the cardinality of the object relationships are multiples (n-to-m). When the relationships are defined in the product model, object methods can be developed to pop-up windows (e.g. to alert designer of conflicts) through which the designer can enter information relevant to that relation. This situation can be illustrated in Figure 1.

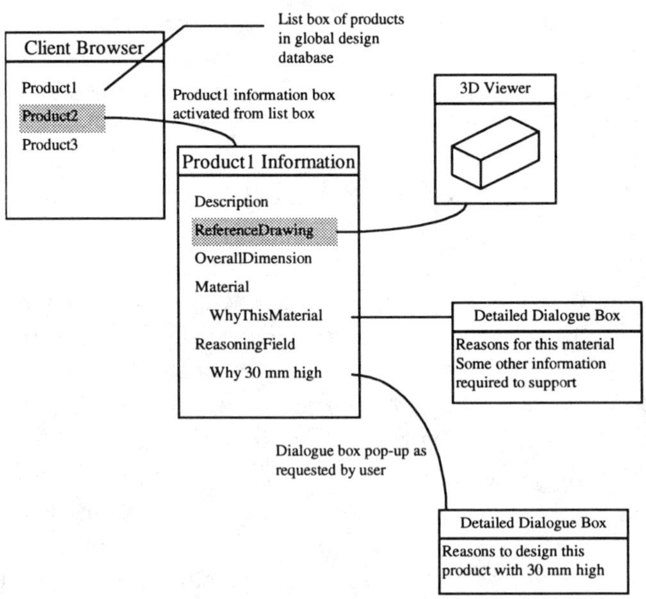

Figure 1 System User Interaction Scenario

4 AN EXAMPLE PRODUCT – DINING TABLE

The design rationale information has to be recorded in a form which is easily retrieved and displayed. One important characteristics of this database is that it must be object oriented. Figure 2 shows the outcome of a product model illustrating some of the basic ideas in OOM of product. It describes the relationships and attributes of the design of a "Dining Table" which is formed by packing a table and a table cloth as a product.

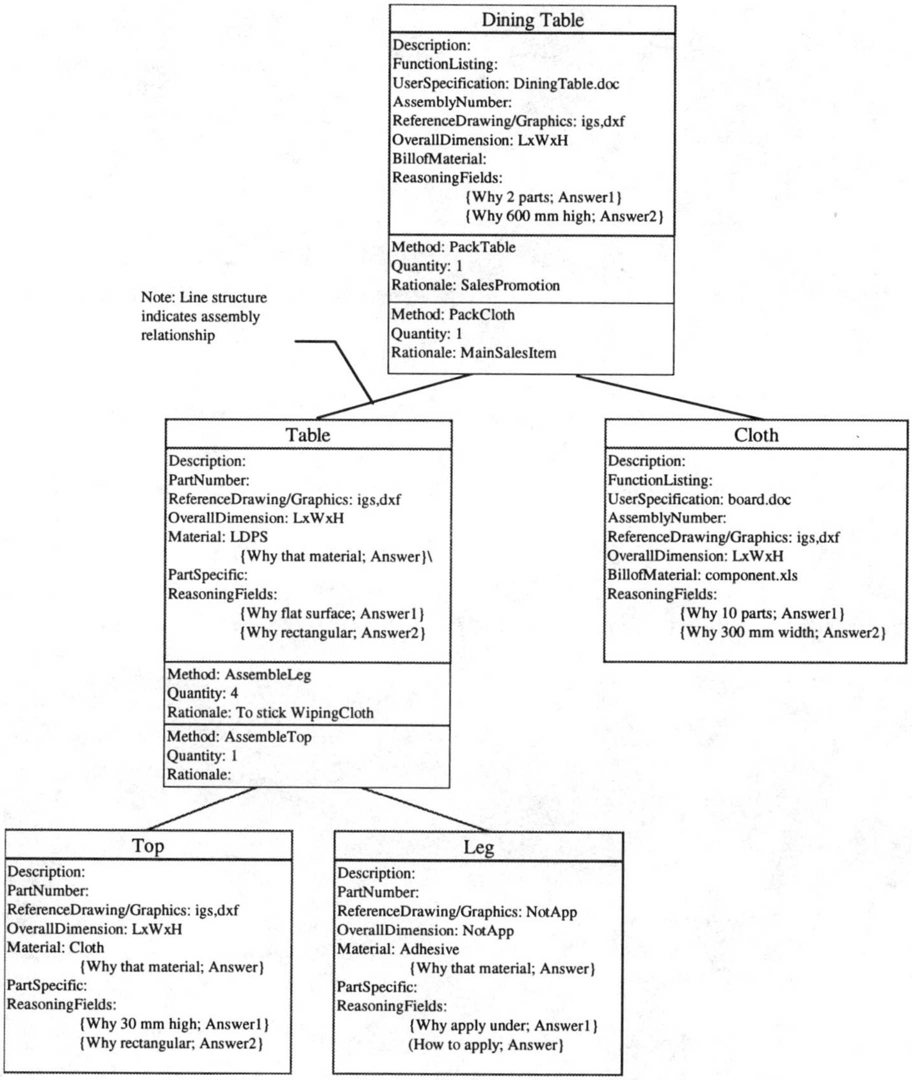

Figure 2 The Dining Table Product Model

5 IMPLEMENTATION

Using the tree objects in Java, the product model structure can be implemented as shown in Figure 3. Assembly objects are given the same set of attributes. At the top level, the dining table is the root of the tree. Description information relevant to that assembly is displayed when the level is highlighted.

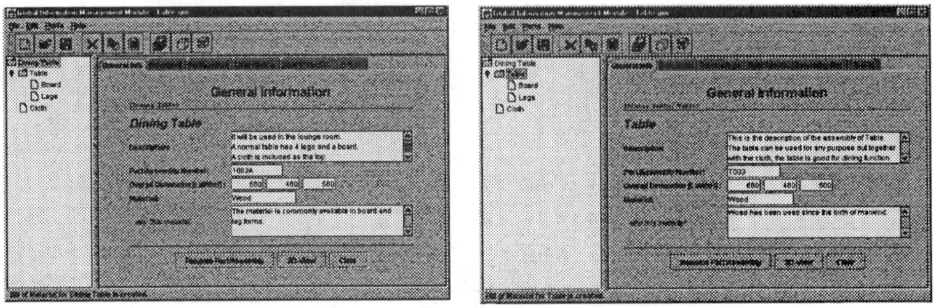

(a) *(b)*

Figure 3 Top level model information screens

A product has functions to perform. The important implicit information of a product are not usually recorded in a CAD environment. These are captured in the other tabbed-panes. The "Function Listing" screen (Figure 4) captures the function of the object or assembly. Reasons for various aspects of the design (can be totally unrelated to CAD) are captured in a freely entered screen. By highlighting the reasoning question in the list, the answers are captured and displayed at the bottom window (Figure 5).

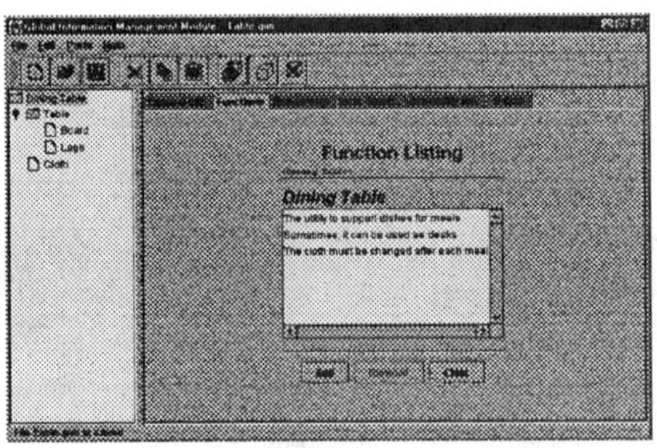

Figure 4 Function Listing

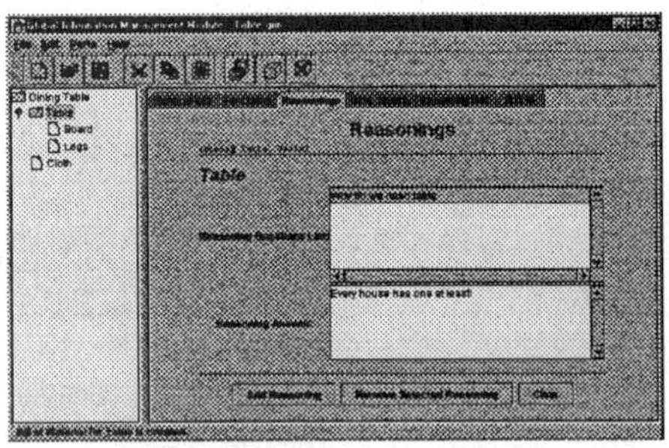

Figure 5 Reasoning capture

To help the designer, a simple product viewing capability is also included (Figure 6). It is planed that the viewer will be integrated with a global CAD system developed in a complementary project later [7]. Customer information are usually supplied in a wide variety of format. At this stage of development, a HTML browser is included (Figure 7). Browsers of other formats can be incorporated into the system easily.

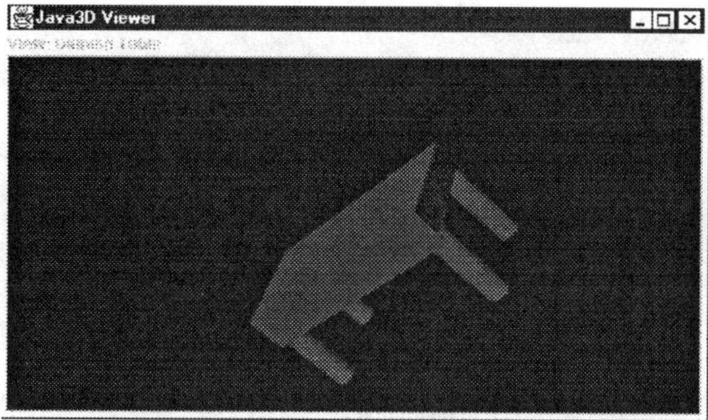

Figure 6 3D product viewing window

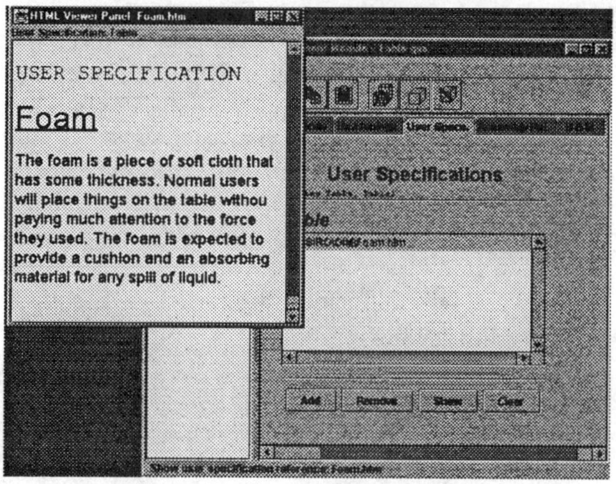

Figure 7 Customer specification

6 CONCLUSION

A system for capturing design rationale has been created using an object oriented product modelling structure and the internet based programming language, Java. Developed in an object oriented environment, the system captures explicit information of the design as well as implicit information of the product according to the level of the object in the model. Further work is planned to expand the capability to interface to CAD systems in a more dynamic way. It includes the design and implementation of an enhanced system architecture which interacts with the user while he/she is working on the design. For example, an automatic pop up query based on the product model structure will appear to capture the intent of the designer at the time when the designer is making the drawing. These features can be done readily with the product modelling background of GIM.

7 REFERENCES

[1] Redman J., Mo J.P.T. (1999). "Process Modelling for Global Work Team Creation and Management", 2nd International Conference, Managing Enterprises '99, 18-20 November, Newcastle, Australia, pp.275-280

[2] Shinonome M., Hashimoto H., Fuse A., Mo J.P.T. (1998). "Development of an information technology infrastructure for extended enterprise", ", IFIP TC5 WG5.3/5.7 Third International Conference on the Design of Information Infrastructure Systems for Manufacturing (DIISM '98), Paper 23, May 18-20, Fort Worth, Texas, U.S.A., pp.353-364

[3] Chen Y.M., Hsiao Y.T. (1997). "A collaborative data management framework for concurrent product and process development", *International Journal of Computer Integrated Manufacturing*, Vol.10, No.6, pp.446-469

[4] Mo J.P.T., Kovacek M., Cirocoo L (1998). "A Framework for Round the clock Design and Support", Proceedings of the 10th International IFIP WG5.2/5.3 International Conference PROLAMAT 98,.9-12 September, Trento, Italy, pp.439-450

[5] Ooi S.T., Bertok P., Zhao L.P., Jiang H.C., Mo J.P.T. (1999). "Patterns as Reengineering Front-End in Distributed Environments", Proceedings of 2nd Australian Workshop on Software Architectures, 23 November, Melbourne, Australia, pp.75-82

[6] Mo J.P.T., Ranganathan D. (1996). "From Design to Manufacture by Objects", *International Journal of Flexible Automation and Integrated Manufacturing*, Vol.4, No.2, 1996, pp.95-110

[7] Jiang H.C., Mo. J.P.T. (1999). "Internet Based Design System for Global Concurrent Engineering", 2nd International Conference, Managing Enterprises '99, 18-20 November, Newcastle, Australia, pp.150-156

Agent Design for LCC Information Gathering

T. I. Zhang
Computer Systems Engineering, Royal Melbourne Institute of Technology, Australia
Em: t.zhang@rmit.edu.au

H. C. Jiang
Object-Oriented PTY. LTD., Australia
Em: harveyj@oopl.com.au

E. A. Kendall
The School of Network Computing, Monash University, Australia
Em: kendall@infotech.monash.edu.au

Keywords Multi-agent system, Information gathering, Life cycle costing (LCC)

Abstract To respond to the challenge of global economic competition, manufacturers are searching for ways to bring high-quality products, system, and structures into being in response to established needs. Simultaneously, as cost is a key factor among many physical and social factors to determine the success of a product, they attempt to reduce costs during every phase of the product's life cycle. The major obstacle for LCC models is data gathering from a highly distributed heterogeneous environment with a huge number of information sources. This paper presents a system analysis approach to design agents for information gathering for CASA model that is one of popular LCC models.

1 INTRODUCTION

Cost is a key factor to determining the success of a product [10]. The life cycle costs of the product can be the total costs at phases of research and development (R&D), production and construction, and operation and support (O&S) [3]. To respond to the challenge of global competition, manufacturers need to reduce costs during every phase of a product's life cycle [18].

However, there are obstacles to use LCC models [2]. One major barrier is data gathering from a highly distributed heterogeneous environment with a huge number of information sources in an organization. When data-processing systems are distributed in various formats, manufacturers have to search them separately and manually integrate information from flat files, relational databases, and remote supplier parts catalogs. CASA model is a typical example [19]. Due to the explosion in the amount of information, it is more useful for collectors to understand customer needs, develop a product to meet these needs, and bring that product to market quickly and at fair value [11]. In recent years, a new wave of changes to the business environment has emerged [13]. The exponential growth of the Internet throughout the last decade has led manufacturing companies to move into a globalized business environment. They can interact with business partners and customers around the world over the Internet. As information from the Internet is diverse, this barrier becomes outstanding.

To overcome this barrier, agent technology is more advanced to deal with information gathering [17, 21]. That is because an agent is able to carry out activities in a flexible and intelligent manner that is responsive to changes in the environment without requiring constant human guidance or intervention [4]. In this paper, we present a system analysis approach for an agent-based system. We then introduce a layered conceptual model for information gathering based on the architecture of InfoSleuth [20].

2 A SYSTEM ANALYSIS APPROACH

Object-oriented (OO) methodology with use cases has been widely used in software development and use case analysis has proved to be useful and successful for requirement specification of OO systems. But as agent technology is getting more popular today, we need an agent-oriented methodology for multi-agent system development. However, as an agent is autonomous, social, reactive and proactive [22], we can not directly apply the OO methodology to agent-oriented software engineering. Current research in role models shows promising results for agent analysis and design [16]. We have combined these two methodologies to specify and develop an agent-based information gathering system for product life cycle cost estimation.

Figure 1 depicts the process of building models for specifying agents. In this figure, each activity represented by a solid box uses several different models, transforming and refining them from activity to activity. The box is given an ICOM (Input, control, output and mechanism) representation adopted from the functional model of IDEF [5,6]. Note that IDEF is a standard modelling tool widely used in manufacturing industry. The thick lines with arrows to connect activities represent interaction collaboration between these two activities.

All activities as shown in Figure 1 can be classified into two categories: agent-oriented analysis (bounded by a dashed polygon) and OO analysis including activities (not bounded by the dashed polygon). OO analysis consists of four activities: "Identify Actors", "Identify Use Cases", "Identify Objects" and "Determine Business Objects". These four activities use the traditional methods developed by Jacobson et al [12].

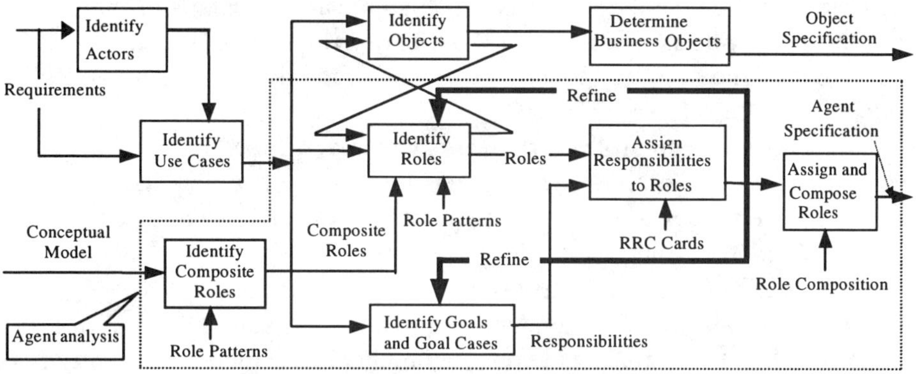

Figure 1 *Agent-oriented analysis process*

The identified use cases are also fed to the activity "Identify Goals and Goal Cases" in agent-oriented analysis processes. A goal is an objective or a desired state that can be achieved or ascertained by an agent. We use $G = \{g_i \mid i \in N, g_i = goal\}$ to define a set of goals. A goal case is a goal-based use case and it belongs to a set of goal cases defined by $U = \{u_i \mid i \in N, u_i = goal\ case\}$. A goal case is a collection of scenarios about the agent's interaction with a system. An agent can start a goal case when the corresponding goal is triggered. We use the existing method [15] to identify goals, goal cases and present them in the hierarchical diagram.

Role models that commonly occur can then be documented as role patterns [14]. One important method is to use role responsibility and collaboration (RRC) cards as shown in Table 1 to specify responsibilities and collaborations of agents, especially in the early phase of agent-oriented software development.

Table 1 Template for a RRC card

Role type		Collaborator
Names of Roles	List all responsibilities	List all collaborators

As a result, all the roles represented by R can be defined as $R = 2^S \times 2^C$ where $S = \{s_i \mid i \in N, s_i = responsibility\}$, 2^S is a power set of S,

$C = \{c_i \mid i \in N, c_i = collaborator\}$ and 2^C is a power set of C. Furthermore, responsibilities can be refined as $S = G' \times U'$ by assigning all potential goals (G') and goal cases (U') for a system.

A conceptual model is used to organize a system in a structure with the functionality to be best supported. We can apply role patterns to this model for identifying composite roles that is a classification of roles for the system. This classification can be instanced in an application. Details of the activity for our application is given in §3. To identify roles from use cases that are the output of the activity "Identify Use Case", the activity "Identify Roles" follows such steps [16]:

- Instance composite roles if they are available;
- Examine role patterns from the existing role patterns. The determined role patterns can specify types of interaction and collaboration of the role with other roles.
- If there are no relevant patterns, partition goals to form roles.
- Determine all roles for the identified interactions and collaborations.

After identifying goals and constructing goal case models, the activity of "Assign Responsibilities to Roles" starts at the bottom of the hierarchy goal diagram, of which the goals are very detailed and would not have any sub goal. This activity determines goal cases that each role can achieve and assigns them as responsibilities of that role.

Once responsibilities are assigned to roles in a multi-agent system analysis, we can partition either identified roles (R) or identified goals (G) and goal cases (U) for agent design. Based on G and U, we can partition $B = 2^{(GxU)} \times 2^C$ for designing a multi-agent system represented by $\bigcup_\beta B_\beta = B$. To partition R, we can design another multi-agent system that is expressed by $\bigcup_\alpha A_a = R = 2^S \times 2^C = 2^{(G'xU')} \times 2^C$. Note that A_α is a partition in R and stands for an agent. This role partition implies that agent design is to compose elements (roles) that belong to R into different categories (agents) which are subsets of $2^{(GxU)} \times 2^C$. We call this partition procedure as role model composition that is the integration of the relevant role models in an application [1]. The role composition is more than just the sum of the constituent patterns. It captures the synergy arising from the different roles an agent plays in the overall composition structure. $\because R \supseteq B$ $\therefore \bigcup_\alpha A_a \supseteq \bigcup_\beta B_\beta$. Therefore, role composition method for agents is better and more systematic than directly combination of goals, goal cases and collaborators that are identified from use cases. To assign and compose roles to agents, we have to have the view of the whole agent organization.

- Design agent in appropriated size i.e. to assign appropriated number of roles to each agent.
- Compose the roles for the agents with differentiation emphasizing the specialization that is goal oriented. Split an agent if it has too many responsibilities for the different sub-goals. Merge agents if they got similar responsibilities.
- Compose the roles to agents with good quality of collaboration, for which some aspects are essential such as cohesion, lower coupling, and minimum need for communication.

3. LAYERED CONCEPTUAL MODEL

We develop a conceptual model to organize our system in a structure with the functionality to be best supported. The Layered Architecture pattern [7] has been widely accepted as a standard in network design and software engineering. Here, for our conceptual model, the information-gathering domain is classified into different layers as shown in Figure 2.

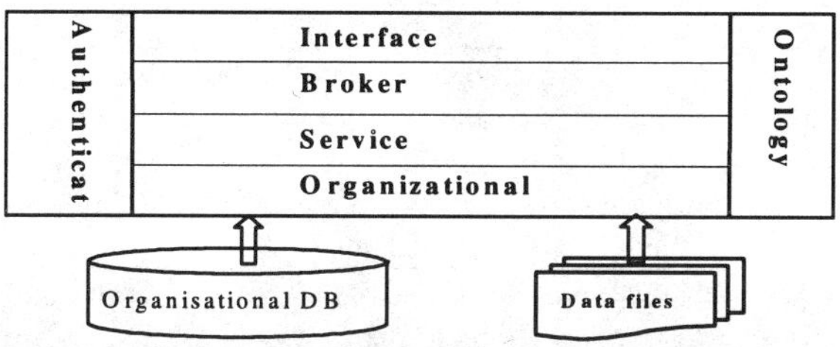

Figure 2. Conceptual model of information gathering system

The ontology layer collectively maintains a knowledge base of the different terminology and concepts that are employed over the whole organization. This layer thus describes language that would be used for specifying and translating requests for information. The authentication layer performs the task of checking and validating users. The interface layer is used to predict the user's intentions and to request services provided by the remaining modules. This layer acts on behalf of users to relay specifications and obtain results. The broker layer predicts or models the intentions of the overall organization and then provides services to users via the interface layer. The service layer is used to provide services, which differs from the organizational layer that controls resources. The service layer represents and provides the high level services that can be formed by encoding expertise and by utilizing the organizational layer. The organizational layer can be used to manage the organizational resources. Its main task is to gather data from the various sources.

We have identified composite roles by using role patterns such as Observer, Broker, Master/Slave, Manager, Bodyguard, and Adapter to this model [24]. These composite roles that can instanced in our application are tabulated in Table 2.

Table 2. Summary of the roles for the layered model

Composite Roles	Role Types	Descriptions
Interface Role	• Observer (Observer) • Client-proxy (Broker) • Client (Bodyguard) • Client and target (Adapter)	• Observe user request • Request to provide services. • Obtain result
Broker Role	• Broker (Broker) • Client (Bodyguard) • Subject (Bodyguard) • Client and Target (Adapter)	• Match user's request to service and send the request to service provider • Reply result
Service Role	• Master (Master/Slave) • Server-proxy (Broker) • Client and Subject (Bodyguard) • Client and Target (Adapter)	• Accept the request • Provide service • Distribute work if need • Get final result Send result
Organizational Role	• Manager (Manager) • Slave (Master/Slave) • Client and Subject (Bodyguard) • Client and Target (Adapter)	• Manage information resources • Query information • Reply result
Authentication Role	• Bodyguard (Bodyguard) • Client and Target (Adapter)	• Verify and validate user • Send out permission
Ontology Role	• Adapter (Adapter) • Client (Bodyguard) • Target (Bodyguard)	• Get help request • Find alternatives • Translate Terms

4 AGENT DESIGN

Our application can be modeled by the simplified use case:

(1) The user observes a request about operating and maintaining a product and then waiting for results.

(2) When receiving a request, the maintainer sends requests to a planner for a maintenance plan and to a cost estimator for operating and maintaining (O&S) cost.

(3) The planner distributes work to a relevant information keeper who manages the database for product breakdown structures and requirements.
(4) The cost estimator distributes work to information keepers who manage data for labor, equipment, and material and then calculates the O&S cost. If a material is not available, the estimator distributes work to the purchaser who manages to get it from suppliers.

This use case model describes the system requirement briefly and simply. By using information in Table 2 and interrogating this user case, we have identified roles as shown in Column 2 of Table 3 that are instances of composite roles. By applying rules demonstrated in §2, we partition these instances into different categories separated by dashed lines in Table 3. These categories shown in Column 3 of Table 3 are agents that play the roles that have been instanced.

Table 3 Agent identification

Composite Roles	Instanced Roles	Potential agents used
Interface Role	User/Customer	User agent
Broker Role	Maintainer	Maintenance agent
Service Role	Estimator	Estimation agent
	Planner	Planner agent
Organizational Role	Projects Infokeeper,	Project manger agent
	Labors and Equipment Infokeepers	Resource agent
	Materials and Supplier Infokeepers	Inventory agent
	Support & Management Infokeeper Miscellaneous Infokeeper	Support agent
Authentication Role	Security	Security guard agent
Ontology Role	Term and Location Helper	Information agent

All the agents in our research are implemented by using Jack Intelligent Agents [8, 9]. The organizational databases are accessed using Java Database Connectivity JDBC [23].

5 CONCLUSION AND FUTURE WORK

This paper has proposed a system analysis approach, which provides systematic processes for agent-oriented software development. By applying this approach, we have designed agents used to gathering information for O&S cost by using CASA model. The further research is to develop such a system for the costs of other phases of a product life cycle.

6 ACKNOWLEDGEMENTS

This research is supported by Cooperative Research Centre for Advanced Composite Structures Ltd. (CRC-ACS).

7 REFERENCES

[1] Andersen, E. (1997). Conceptual Modelling of Objects: A Role Modelling Approach, *PhD Thesis, University of Oslo*

[2] Benson, S. (1998). Life Cycle Costs and Dic Pump, preprint, *http://www.discflo.com/lccart.html*

[3] Blanchard, B.S., Fabrycky W.J. (1990). *System Engineering and Analysis*, Prentice-Hall, Inc., New Jersey, USA.

[4] Bradshaw, J M. (1997). *Software Agents*, Menlo Park, Calif.: AAAI/The MIT Press.

[5] Bravoco, R.R., Yadav, S.B. (1985) Requirements Definition Architecture –An Overview, *Computers in Industry*, **6**, 237-251.

[6] Bravoco, R.R., Yadav, S.B. (1985). A Methodology to Model the Functional Structure of an Organisation, *Computers in Industry*, **6**, 345-361.

[7] Buschmann, F., Meunier, R., Rohnert, H., Sommerlad, P., and Stal, M. (1996) *Pattern-Oriented Software Architecture: A System of Patterns*, Wiley, USA.

[8] Busetta, P., Ronnquist, R., Hodgson, A., and Lucas, A. (1999). JACK Intelligent Agents-Components for Intelligent Agents in Java, *Agent Link News* **2,** January.

[9] Cross, M., Ronnquist, R. (1999). A Java Agent Environment for Simulation and Modelling, *SimTech 99*, Melbourne, Australia.

[10] Fabrycky, W.J., Blanchard, B.S. (1991). *Life-Cycle Cost and Economic Analysis*, Prentice-Hall, Inc. New Jersey, USA.

[11] Hennecke, F. (1999). Life Cycle Costs of pumps in chemical industry, *Chemical Engineering and Processing,* **38,** 511-516

[12] Jacobson, I., Christerson, M., Jonsson, P., and Overgaard, J. (1992). *Object-Oriented Software Engineering – A Use Case Driven Approach*, Addison-Wesley.

[13] Jiang, H. C., Mo, J. (1999). Internet Based Design System for Global CE, *Proc. of the 2nd International Conference on Managing Enterprises*, Newcastle, Australia, 150-156.

[14] Kendall, E. A. (1998). Agent Roles and Role Models: New Abstractions for Multiagent System Analysis and Design, *International Workshop on Intelligent Agents in Information and Process Management*, Germany, September.

[15] Kendall, E. A., Palanivelan, U., Kalikivayi, S. (1998). Capturing and Structuring Goals: Analysis Patterns, *European Pattern Languages of Programming*, Germany, July.

[16] Kendall, E. A. (1999). Role Modelling for Agent System Analysis, Design, and Implementation, *First International Symposium on Agent Systems and Applications (ASA'99), Third International Symposium on Mobile Agents (MA'99)*, Palm Springs, Oct.

[17] Knoblock, C. A., Ambite, J. L. (1997). Agents for Information Gathering, *Software Agents*, Edited by Jeffrey M. Bradshaw, AAAI Press/The MIT Press, 347-373.

[18] Li, Y., Huang, B., Wu, C. (1999). Virtual Enterprise Information System, *Proceedings of the 1st Asia-Pacific Conference on Intelligent Agent Technology,* 493-497.

[19] Manary, J. M. (1996). DSMC's CASA model Still Going Strong, *Article in PM:* Jan.-Feb.

[20] Nodine, M., Perry, B., Unruh, A. (1998). Experience with the InfoSleuth Agent Architecture, *Proc. of AAAI-98 Workshop on Software Tools for Developing Agents.*
[21] Sycara, K., Zeng, D. (1996). Multi-Agent Integration of Information Gathering and Decision Support, *ECAI96: 12th European Conference on AI,* Edited by W. Wahlster.
[22] Wooldridge, M., Jennings, N. R. (1995). Intelligent agents: theory and practice, *The Knowledge Engineering Review,* **10**(2), 115-152.
[23] Zhang, T., Kendall, E. A. (1999). Agent-based Information Gathering System for LCC, *Proc. of the 1st Asia-Pacific Conference on IAT (Intelligent Agent Technology),* 483-487.
[24] Zhang, T., Kendall, E.A. (2000). System Analysis of Agent-Based LCC Information Gathering, *Proceedings of the 1st Pacific Rim International Workshop on Intelligent Information Agents (PRIIA),* Melbourne, September.

Information Technology and Telecommunication Infrastructure: Network Applications for Hong Kong Business and Service

A. A. Shabayek, Kin-man Wan
University of Hong Kong, Hong Kong, China
Em: aaeshaba@hkucc.hku.hk, Em: kmwan@graduate.hku.hk

Keywords Broadband Network, Information Technology, Multimedia Infrastructure

Abstract Hong Kong is an unique city in being very well served by robust communications networks; in particular the telephone and cable television networks can provide high capacity optical fibre links to virtually all premises at low cost. Hong Kong needs to make the best use of the ever-advancing information and communication network to improve products as well as services for public and private sectors. With the network in place, the missing link is integrating different elements in the system. Integrated multimedia system is an example of a niche market that is yet untapped. This paper focuses at the elements that encompass the integration of multimedia infrastructure for Hong Kong industry and service and also outlines applications demonstrating the linkage of different networks.

1 INTRODUCTION

Information technology have been dominated by the advanced industrialised countries, and large companies, but recent technological advances are opening up large market sectors and permitting new players to participate [1]. Our very up-to-date communication network, the compact size of the territory and its position as an important commercial and financial centre combine to give Hong Kong the unique opportunity to develop and implement new communication services to become a leading digital city in the world. This opportunity is presented as an example to illustrate our thesis that, even with limited R&D investment, very significant new business prospects could be generated. In the information and telecommunication sector, many businesses are already flourishing in Hong

Kong. They have a synergistic effect on the potential of developing new businesses. Each existing business already has its invested capital and growth potential and in some cases, also significant R&D know-how. The key issue is to evaluate, against the current background, the additional investment needed in R&D in order to reach various selected business opportunities. Then the specific capital investment requirements for a chosen new business can be determined.

In early 1990s the telecommunication network was primarily telephony based. In recent years, technological advances have extended the information-carrying capabilities of the network first through the introduction of digital techniques and then through all-electronic switching. Many new switched data services have appeared [2].

With the graduate widespread installation of optical fibre system in the trunk and distribution areas of the public and private networks, the stage is set for the networks to handle a vast range of new services. These services may differ significantly in nature from the switched telephony service. The network would be capable of handling broadband signals such as video, thereby introducing a degree of conflict between TV distribution and telephony [3]; the conflict could also be viewed positively as opportunity to merge certain services.

2 NETWORKING CONSTITUENCIES

Mainly equipment makers, network operators, and Service vendors are the key players supporting the networking system. Traditionally, equipment makers in the telecommunication industry work closely with the carrier network operators and in anticipation of the service demands of the customers; they design and manufacture customer premises, systems and network equipment [4]. Vast investments are made in technology R&D with the aim of extending the capabilities of the network for service delivery at better quality and lower cost. This healthy and productive marriage between the equipment makers and the network operators has benefited the customers well.

Common carriers operate in a regulated environment; they are obliged to connect all customers, however remote, to the telecommunications network. In the current environment, the customers may connect approved equipment such as telephones, computers and fax machines to the network. Carrier operators charge the calls by time, distance and bandwidth at a rate approved by the regulatory commission. The approval rate on justifiable capital and operation costs and acceptable profit margin. The common carrier operators strive to satisfy all customer demands at a low charge per call, with little or no waiting time, and with many convenient call features. This is achieved by working with the equipment suppliers to improve call features and to reduce cost per call through technological improvement [5].

3 TELECOMMUNICATION-RELATED BUSINESS IN HONG KONG

The First Pacific group is entering the telecommunication business along several avenues. Pacific Link Communication is the newest of the three cellular phone operators. First Pacific Mobile Communication offers trunked radio service and has obtained a license to provide paging service. In partnership with three other companies, it is bidding for a license to offer CT2 service in Hong Kong.

Hutchinson Telecommunications is expanding forcefully and rapidly in several fields of telecommunications. It owns 50% of the communication satellite AsiaSat 1 and will be broadcasting TV programs to a large part of Asia. Hutchinson Telephone operates two cellular networks with the dominant market share among the three cellular operators. Hutchinson Paging offers paging service to 50% of the market and is bidding with partners for a CT2 license. Hutchinson iNET offers a value-added electronic mail service and Hutchinson Mobile Data offers financial information through a radio broadcast to subscribers.

In 1990, Hutchinson offered the first trunked radio communication (TRC on 800 MHz frequency to provide more efficient radio communication to taxis and commercial fleets. Hutchinson is also expanding overseas. For example, it has established a foothold in the UK by buying the paging company Millicom Information Services and by seeking a controlling share in BYPS, a CT2 service provider [6].

There are four cellular networks in service in Hong Kong: the UNITACS system of HKT CSL; the ETACS system of Pacific Link; and the AMPS and TACS systems of Hutchinson. As of year-end 1999, there were 670,000 subscribers, or more than 10% of Hong Kong's population, yielding the highest penetration rate of hand-held cellular sets in the world. With an annual growth rate of 47%, spurred by up to 50% annual price drops on handsets, it is expected that the four systems will be saturated by year-end of 2000. To provide a long-term solution, The Hong Kong Government is providing the digital cellular system via the American USDC and the European GSM standard.

4 IT APPLICATIONS VIA BROADBAND TELECOMMUNICATIONS NETWORKS

Information Technology (IT) emerges to make Hong Kong becoming one of the leading digital cities in the world. The Technology applies in many different areas, which support the enhancement of the industry as well as services and maintain Hong Kong at the competitive edge in the information arena. In order to enhance work's accuracy and efficiency on preparing and retrieving client records, centralised databases system and client/server computing environment enable manufacturers, hospitals, banks, and government department in Hong Kong to check and update their records efficiently at no time.

4.1 Medical Service

Due to the development of advanced information technology in medical services, the Hospital Authority in Hong Kong is continuously up-dating the medical service by installing sophisticated information systems and devices through the medical broadband network, e.g. medical imaging and scanning machines. The medical network in Hong Kong has a number of Computerised Tomography scanners (CT) and Magnetic Resonance Imager (MRI). Other medical imaging tools include ultrasound and Nuclear Medicine (NM).

Given Hong Kong 's scarcity of CT's and MRI's, its need to maximise the utility of limited medical expertise, and the needs of the hospitals to upgrade and/or computerise their management and technical capabilities, a medical broadband network would be useful. The main purpose of such a medical broadband network are to share resources, provide timely access to data and pictures, and allow easier access of expertise to support medical decision. The network allows major hospitals to pool their radiology/pathology facilities and share expensive and scarce medical expertise. Broadband communication can enhance the utilisation of such equipment and such specialists' time and lower the pre-patient cost by spreading the overheads a lager user community.

Due to the large number of films and operational necessity to physically transfer these films among departments and doctors, an estimated 30% of these records are lost. A computerised picture archive and communication system (PACS) would process and store digital records of X-ray, ultrasound, CT, MRI and NM images in optical disk storage [7]. Figure 1 depicts the general concept of a broadband bus that links hospitals to centres of shared facilities.

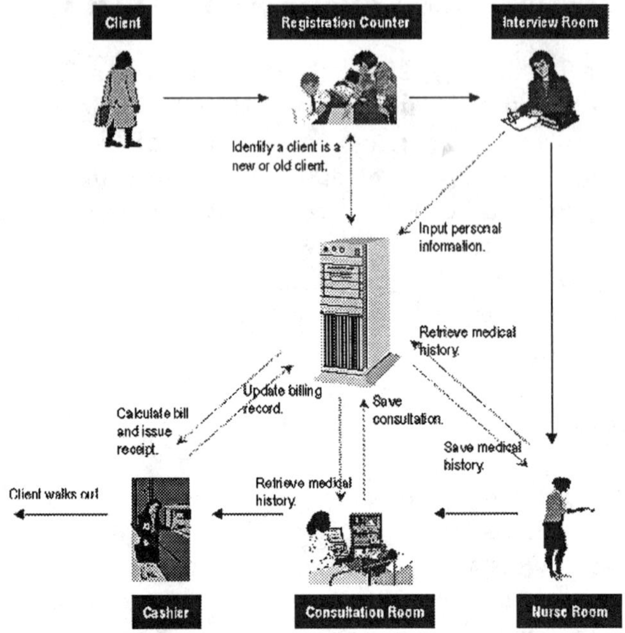

Figure 1 Medical Broadband Network Linking Hospitals, Clinics and Doctor Offices to Centres Shared Facilities, e.g. Radiology & Pathology Labs.

4.2 Transportation Infrastructure Network Via IT & Smart Card System

Hong Kong has very reliable transportation infrastructure via development of information technology into their system. The Transport Department is now expanding the transport system with the introduction of an up-to-date intelligent support system to reduce traffic congestion and travelling time. All of these facilities can bring a great convenience to the citizens of Hong Kong. Traditionally, the contact magnetic card is the only ticketing system of public transportation. After the introduction of the automated fare collection system, the process became much easier and user friendly. By using the Octopus transit card and the imaging processing technology, the passengers can travel conveniently on different modes of public transportation systems. The system design is based on the integration of ticketing network on different modes of public transportation. As a result, the operating and maintenance costs can be reduced.

The octopus transit card has an embedded microprocessor that stores, updates, and transfers information, in real time, to the transient networks.

For the passengers' convenience, information transmission to the reader of the transportation module occurs instantaneously even with no direct contact with the smart-card, e.g. scanning the passenger's wallet which contains the octopus card is enough for communication. Once communication between imbedded antenna (smart card) and external antenna (reader) started, the card's chip becomes energised due to the magnetic induction and the magnetic waves and radio signals transmit the data.

Each time the smart card is used, its credit status is updated and transaction data is processed immediately to computerized station of the main system [8]. The system then sends the data to credit analyzers, add-value machines, auditor gates and integrated processors. There is a central clearinghouse that connects all the computers using wide area network (WAN). All transaction data is transmitted to the clearinghouse for checking, processing and distributing. If passenger wants to add value to her/his smart card, she/he can use the add-value machine, to reload extra credit into the octopus's microprocessor. At the same time, the updated data will be transmitted to the whole system, i.e. credit analyzers, processors, etc. A flow chart summarizes this process is shown below in Figure 2.

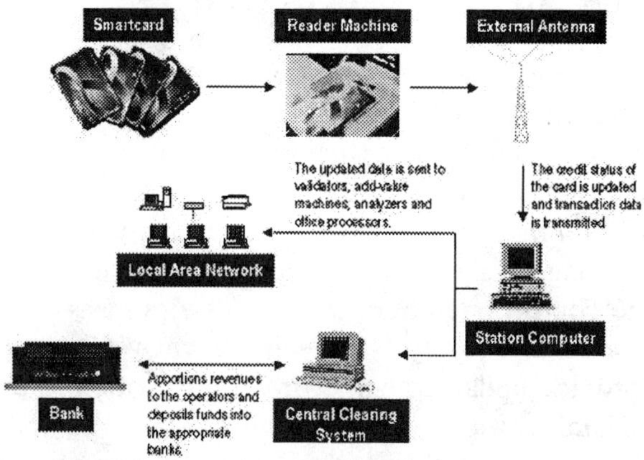

Figure 2 Flowchart for Octopus Transit Network.

4.3 Intelligent Traffic System

Hong Kong, as a whole, comprises group of islands that are interconnected by bridges and underwater tunnels. In order to optimise traffic on highways more effectively, Hong Kong is developing more intelligent transport system such as traffic control system, automatic toll collection (auto-pass) system, and satellite speeding control system. The traffic control system acts as a safety valves in the traffic networks; it is equipped with infrared sensors that measure traffic congestion and control traffic lights accordingly. While the main function of the automatic toll or auto-pass system is to effectively and automatically identify each passing vehicle through toll roads, bridges or tunnels in order to process the charge electronically. Electronic road pricing system is a new technology that offers an optimised way to use roads and highways. The system uses microwave signals via local area network (LAN) as well as wide area network (WAN) and microwave towers to detect the vehicles' identification number (ID) that cross the pricing location for automatic billing. Usually, the signal block sends a signal to the passing vehicle, which is equipped with a smart card placed at the vehicle's front. Once the signal scanner scans the smart card, the signal reflects back the information to the system with the required fee debited from the smart card's account. Similarly, the satellite speeding system uses a radar detector for speed measurements and also identifies any over speeding vehicle; consequently the system transmits the information electronically via WAN and microwave towers to the traffic bureau's main computer for processing [9].

4.4 IT for The Airport Flight System

The Hong Kong new International Airport, at Chek Lap Kok, implemented the most sophisticated IT in its ground and flight operation systems. The operation system includes check-in ticketing system, gate arrangement system, flight information monitoring system, luggage checking and handling system, Cargo transportation system, etc.

The Hong Kong Civil Aviation Department applies different information technologies to control operations within the airport as well as air traffic in order to enhance flight safety. Advanced radar system and communication network can provide updated and accurate information to the pilots' cockpit and hence maintain orderly flows of air traffic. The radar flight system provides important information such as accurate aircraft position, road map for takeoff and landing, and the movement of vehicle on the runways. Such information is also transmitted to traffic controllers for flight routing arrangement. With this sophisticated system, pilots and controllers would be able to maintain a smooth and safe traffic even under low visibility conditions. The Hong Kong Civil Aviation Department is continuously improving the air traffic control by using satellite system for monitoring the flight routing and air traffic [10].

Radar and satellite systems enable the pilots and controllers to exchange information in real-time basis. This information includes weather condition, altitude, wind velocity, relative position of aircraft, etc. A geosynchronous satellite, moves with the earth at the same speed and direction, calculates the position of the aircraft and disseminates the information to the whole system as air traffic signals, e.g. giving map for another route in order to avoids collision with close-by aeroplanes. Figure 3 outlines the flow of information for flight monitoring system.

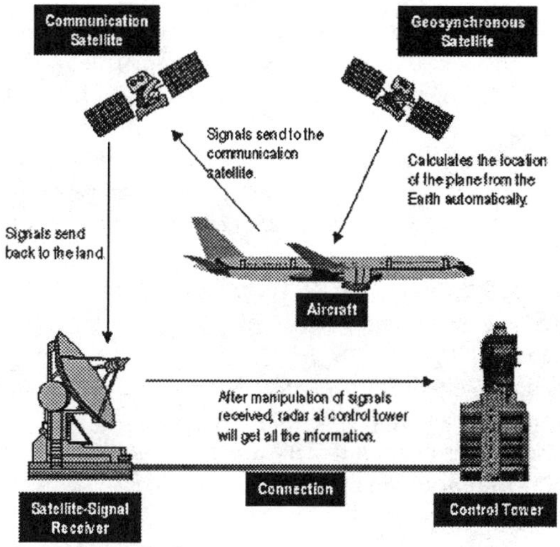

Figure 3 Flight Routing & Monitoring Systems.

5 CONCLUSION

This paper identifies integrated multimedia information system for small as well as large business to be one niche market. Hong Kong s ideally placed to enter the market because the system is designed/application intensive and there is ready access to low-cost manufacturing and software development in China.

In real economic terms, the proposed development will create business opportunities for a core of networks and communication equipment manufacturers. The results of R&D will spread to related industry and downstream manufacturers. In the long term, a road map and well thought out plan would minimise the development risk and maximise the product evolution flexibility. For example, hardware development should make use of application-specific integrated circuit (ASIC) and software for ease of upgrade.

A graphic database service could be the next service for development. The distributed vector databases and graphic CAD tools would give rise to many unfamiliar operational issues, here in Hong Kong, and possibly a range of new customer premises equipment. This would be the basis for a range of service that could improve design and production efficiencies for various industries and would open up the path towards developing a host of multimedia services.

6 REFERENCES

[1] National Science and Technology Council. (1996), Technology in the National Interest, Office of Technology Policy, U.S. Department of Commerce, Washington, DC

[2] Industrialist, (Jan. 1999), Monthly Industry Magazine Published By The Federation of Hong Kong Industries, 37-39

[3] Telecommunication Internetworking (1999). Louis P. J, McGraw-Hill Publishing Group

[4] Chen Y.B., Li X., Orady E. (1996). Integrated Diagnosis Using Information-Gain-Weighted radial Basis Function Neural Networks, Computers and Industrial Engineering, 30(2), 243-255

[5] Sollenberg N. R. (1997). Business & Technology Trends , Business & Technology Information Quarterly, Hong Kong, 3(1), 16-20

[6] IEEE Spectrum, Sept. (1996)

[7] Hong Kong Health and Welfare Bureau, *http://www.info.gov.hk/hwb/index.htm*

[8] Hong Kong Transport Bureau, *http://www.info.govhk/tb/*

[9] Hong Kong Highway Department, *http://www.hyd.gov.hk/*

[10] Hong Kong Airport Authority, *http://www.hkairport.com/*

PART SEVEN

Knowledge Management

Background and Foreground Knowledge in knowledge management

J. Zheng, M. Zhou, J. Mo, A. Tharumarajah
CSIRO Manufacturing Science and Technology, Australia
Email: Jeffrey.Zheng@cmst.csiro.au

Keywords Implicit, Explicit, Tacit, Knowledge in Knowledge Management, Foreground, background, Executable, Non-executable, Dynamic Model of knowledge

Abstract In this paper we propose a new dynamic model for knowledge in knowledge management. To analyze problems and relevant restrictions, two well-accepted knowledge representation models are discussed. The new model, Executable Knowledge Model (EKM), can distinguish foreground and background knowledge and it defines a determinable boundary for executable and non-executable knowledge in knowledge management. This model provides strong executable and testable properties of knowledge in knowledge management to guide further knowledge management development.

1 INTRODUCTION

What is knowledge? This fundamental question has raised such deep debate since the commencement of human history, inspiring the greatest philosophers such as Confucian, Plato, Aristotle, Russell, who have dedicated their lives to answer this question. Knowledge itself is a very slippery concept with many different variations and definitions. The nature of knowledge or what it means to know something is an epistemological question. Do we simply acknowledge that we are in an ambiguous area and do the best we can? No, we must each make these choices in an informed rational mindset for manipulation. In this field, there is no unique correct answer, only theories and opinions. With the rapid technology revolution, many researchers and commercial companies are starting to explore new management systems to handle and manipulate knowledge in order to support business activities. Knowledge management system seeks to efficiently handle practical knowledge. Since many knowledge models are mainly based on human beings and their experiences, there are intrinsic problems to directly use human oriented knowledge representation models to guide knowledge engineering practices in design and implementation of Knowledge Management Systems (KMS).

Knowledge models are applied to both nature and human beings. It is necessary to have executable principles to distinguish feasible functions in knowledge management with clear theoretical boundaries between human and machine. In this paper, a new model is proposed to satisfy these requirements. Two models (Nonaka 1991 and Nickols 2000) in knowledge management are evaluated. A new model (Executable Knowledge Model, EKM) is then proposed and applied for identifying foreground and background knowledge among management systems.

2 TWO MODELS OF KNOWLEDGE MANAGEMENT

Knowledge is the internal state of an agent following the acquisition and processing of information. An agent can be a human being, storing and processing information in his mind, or an abstract machine including devices to store and process information. To categorize human knowledge, Polanyi distinguishes that human knowledge has two major components: tacit and explicit knowledge [1,2].

2.1 The Tacit and Explicit Loop

Following Polanyi's concepts, one of the well-accepted theories about knowledge is proposed by Nonaka [3,4]. According to Nonaka, tacit knowledge consists of personal relationships, practical experience, shared values and explicit knowledge consists of formal policies and procedures. In Nonaka's knowledge model, the creation and classification is part of the externalization process. After having created or acquired tacit knowledge, human will put their ideas on paper. Retrieval is part of the internalization process. To show the dynamics for knowledge in knowledge creation, Nonaka's model can be illustrated in Figure 2.1.

In this model, the socialization has been approached in the context of organization culture. Combination is largely studied by the data processing (processing of formalized information, or data), internalization has been covered in the framework of the organization learning, but the last dimension, externalization (tacit formalization), has not been well developed yet. In order to do that, we can use techniques implemented inside the group-ware systems to capture information to identify elements. We can also use mechanisms implemented in tools for meeting that allow capturing part of formulization of these elements (externalization). For a system, the aim of which is to help to capitalize enterprise knowledge and to create new knowledge (strategic resource), we must take into account not only the two static knowledge dimensions, but also the four dynamic knowledge interactions.

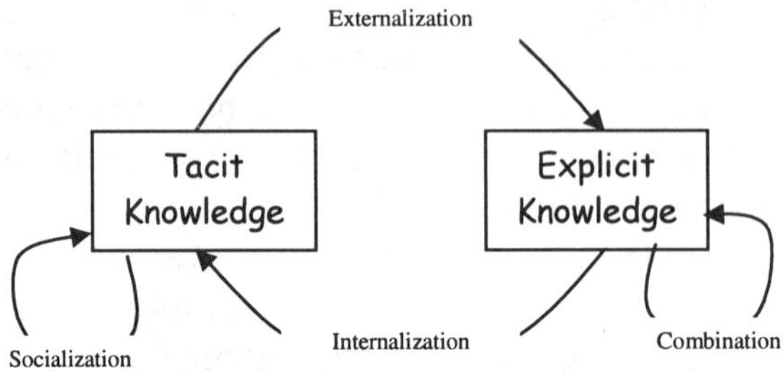

Figure 2.1. The Tacit-Explicit Knowledge Cycle

Anderson [5,6,7], Davenport [8], Harris[9], Lechner [10], Mahe [11], Stanoevska [12] and Thompson [13], have shown that the Nonaka models provide great benefits to design and implementation of real KMS in cooperative and social effects in complicated practical applications.

2.2 Nickols Model of Knowledge in the Knowledge Management

An alternative theory was proposed by Nickols[14]. A testable model for knowledge include the following:

- Explicit Knowledge
- Tacit Knowledge
- Implicit Knowledge
- Declaration Knowledge
- Procedural Knowledge

An important property of his model is to provide a testable procedure to distinguish different terms. There is an integration procedure to illustrate intrinsic meanings of terms in Figure 2.3.

The procedure can be explained as follows: When knowledge has been articulated, then it is explicit knowledge. Otherwise, another question is raised: Can it be articulated? If the answer is yes, then it is implicit knowledge. If the answer is no, then it is tacit knowledge. Facts and things (tasks and methods) have describing properties belong to declarative knowledge. All declarative knowledge is explicit. However, motor skills (mental skills) are doing things that correspond to procedural knowledge and all procedural knowledge is tacit.

Nickols provides this testable procedure as the foundation of knowledge. Since other terms (declarative, procedural and strategy knowledge) can be merged into three essential terms (Explicit, Tacit and Implicit) [14], only these terms need to be handled for relationships among the components. If we view Nonaka's model as an automaton, there are four interactions among explicit and tacit knowledge (socialization from tacit to tacit, externalization from tacit to explicit, combination from explicit to explicit and internalization from explicit to tacit).

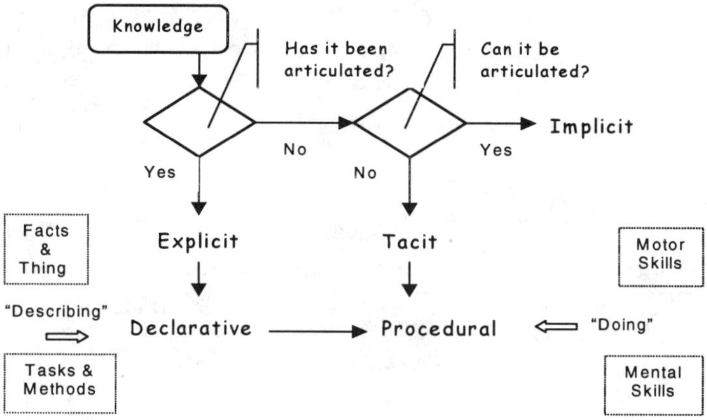

Figure 2.3. A Framework of Knowledge in Knowledge Management

Similar to Polanyi's distinctions, Nickols model is a static procedure to identify knowledge. To investigate more complex system behaviors, a dynamic model is required to integrate all three essential components of knowledge.

3 EXECUTABLE KNOWLEDGE MODEL

3.1 Basic Model

Considering three knowledge components: explicit, implicit and tacit knowledge. Explicit knowledge describes basic facts and storable document sets; tacit knowledge corresponds to personal experience, instinctive ability, spiritual knowledge and any other skills which cannot be expressed on paper or articulated. Different from above two extreme components, implicit knowledge has a bridge property that links together the explicit and tacit components. When a knowledge expert has implicit knowledge, such knowledge can be represented as explicit knowledge through knowledge discovery and description procedures. For example, a scientist developed a new idea after many years of research, the procedure is the same as to transfer tacit knowledge into implicit through externalization processes.

When an inventor has articulated and recorded relevant knowledge as papers and description documents. Relevant implicit knowledge is transformed into explicit knowledge during procedures. Combining these considerations, we propose a new model, ***Executable Knowledge Model*** (**EKM**), for knowledge in knowledge management shown in Figure 3.1.

Executable Knowledge Model (*EKM*)

The EKM is composed of three components corresponding to three basic knowledge collections. There are four interactions among the three components under certain conditions. In addition to distinguish three basic collections, the most interesting properties of this model are its dynamic behaviors among various transformations.

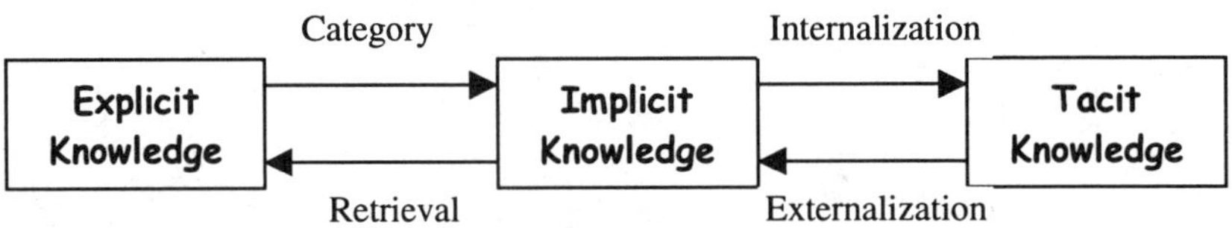

Figure 3.1. Executable Knowledge Model

Category: From Explicit to Implicit

Explicit knowledge can be manipulated and organized by various concepts and methodologies to construct higher level relationships through conceptual organization, category or classification operations. These complicated organization structures make explicit knowledge implicit. A typical example is to categorize books such as "Alice in Wonderland". Children think it is a storybook or a fairytale while mathematicians believe that it is an awesome book for training logic thought.

Retrieval: From Implicit to Explicit

From implicit knowledge, we may use content-based retrieval or creative articulation to make implicit knowledge explicit. In normal situations, there are various conditions and assumptions to add on our efforts to make such a transformation. We can imagine that a knowledge expert articulate knowledge. Complicated implicit knowledge may take much longer to be retrieved.

Internalization: From Implicit to Tacit

Internalization is a process of embodying implicit knowledge into tacit knowledge. A human expert with implicit knowledge may use internalization methodology to transform implicit knowledge to tacit. For example, a show pilot, amazed at the skill and poise of an Olympic diver, he tried to incorporate the diver's repertoire into his own performance. After months of studying video footage of platform divers, the pilot was inspired to perform his own twists, tumbles and pikes – all using his airplane.

Externalization: From Tacit to Implicit

Externalization is a process of presenting tacit knowledge into implicit knowledge. One example is the discovery of Benzene (C_6H_6, the simplest aromatic hydrocarbon) structure. In 1865, German chemist August Kekule had a strange dream that six snakes were swallowing each other each head biting into another's tail. Inspired by this tacit model, he proposed the hexagonal formula with alternate single and double bonds to make this new discovery.

3.2 Executable Model

Using implicit knowledge as an intermediate state of knowledge, this makes possible for us to distinguish executable and non-executable collections from KMS shown in Figure 3.2.

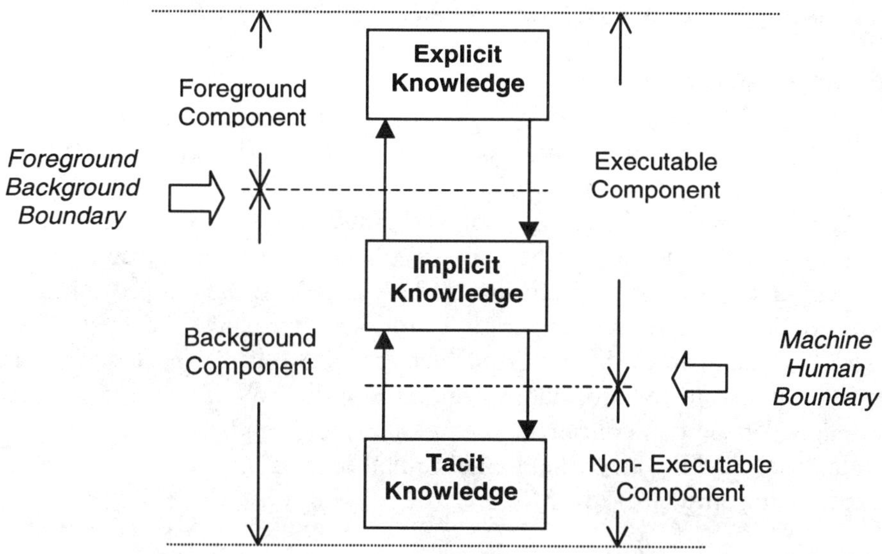

Figure 3.2 Foreground and Background, Machine and Human Knowledge Boundaries

Foreground Component

Selecting any knowledge subject belong to explicit knowledge as an object, excluding the object the rest parts of knowledge in the system correspond to its background. In general, a foreground object depends on the user's viewpoint. From a systematic consideration, foreground and background is corresponding to two complementary partitions. They are relative and complement each other. The relationship of these pair relations can change in different consideration under various circumstances.

However, foreground subjects can only be selected from explicit knowledge for any KMS. We cannot make a knowledge subject as foreground, if there is no way to explicitly express it. Because the reason, it is more convenient to describe the entirely explicit knowledge as foreground in KMS. Foreground component is the collection of all explicit knowledge in the KMS.

Background Component

Implicit and tacit subjects are unable to function as foreground objects in KMS. Under this consideration, we can define two collections of implicit and tacit knowledge as background component. Any knowledge subject in the implicit and tacit will be intrinsically corresponding to background. The interface between explicit and implicit knowledge is the boundary distinguishing background and foreground components in the system. This provides a natural organization to apply intrinsic knowledge properties distinguishing foreground and background components in KMS.

Executable Component

The EKM can be partitioned by another criteria. Explicit knowledge collection is composed of real materials that can be manipulated by computer or other mechanical tools. From this viewpoint, any explicit knowledge object is an executable object. However, there are hidden links among explicit knowledge objects and implicit knowledge objects. As the meanings of implicit knowledge, it will be possible to have a procedure to make implicit knowledge explicit. Therefore, when we have a super powerful mechanism to do such transformations, we can expect that it can complete almost all transformations among implicit to explicit and explicit to implicit. Under this condition, both explicit and implicit objects will be executable. Collecting all explicit and implicit objects, they are composed of an executable component in KMS.

Non-Executable Component

There is no guarantee to have a general mechanism structure to transform tacit knowledge into implicit and vice versa. From this viewpoint, the tacit component is the non-executable component of KMS. From an engineering viewpoint, it is more important for us to focus our attentions on the executable part and let human beings to use their intelligence in order for all non-executable parts to be handled correctly.

5 CONCLUSION

This paper provides basic descriptions of EKM for knowledge in knowledge management. The EKM is constructed for KMS from an engineering viewpoint. Since many existing models of knowledge in knowledge management mix the processes among human beings and machine executable processes, the compounding system properties create severe difficulties for knowledge engineers to design support environments for knowledge management. The major issue in EKM is to separate executable parts from non-executable parts in KMS. This separation makes a clear boundary to handle extremely abstracted problems in various conditions.

6 REFERENCES

[1] Polanyi, M. (1969). *Knowing and being*, The University of Chicago Press.

[2] Polanyi, M. (1997). "Tacit Knowledge," Chapter 7 in *Knowledge in Organizations*, Laurence Prusak, Editor. Butterworth-Heinemann: Boston.

[3] Nonaka, I. (1991). "The Knowledge Creating Company." *Harvard Business Review*, November-December: 96-104.

[4] Nonaka, I. (1995). *The Knowledge Creating Company*. Oxford University Press.

[5] Anderson, J.R. (1976). *Language, Memory and Thought*. Erlbaum: Hillsdale.

[6] Anderson, J.R. (1993). *Rules of the Mind*. Erlbaum: Hillsdale.

[7] Anderson, J.R. (1995). *Cognitive Psychology and Its Implications* (4th Edition). W.H. Freeman and Company: New York.

[8] Davenport, T. and Prusak, L. (1998). *Working Knowledge*. Harvard Business School Press: Boston.

[9] Harris, D. (1996) *Creating a Knowledge Centric Information Technology Environment*, http://www.htcs.com/ckc.html

[10] Lechner, U. and Stanoevska, K. (1998) *The NetAcademy – A New Concept for Online Publishing and Knowledge Management*. ACOS 98 Workshop.

[11] Mahe, H. Rieu, C., Beauchene, D. "An Original Model to Organize Know-How in a Benchmarking Context." http://spuds.cpsc.ucalgary.ca/KAW/KAW96/mahe/maherieu.html

[12] Stanoevska, K. Hombrecher, A., Handschuh, S., Schmid, B. "Efficient Information Retrieval: Tools for Knowledge Management." http://www.netacademy.org

[13] Thompson (1997) "Corporate Memories." *Byte*, September 1997.

[14] Nickols, F. (2000). "Knowledge in Knowledge Management." *Knowledge Management Yearbook*, Butterworth-Heinemann: Boston. http://home.att.net/~nickols/Knowledge_in_KM.htm

[15] Belkin, N. and Croft, W. (1992) "Information Filtering and Information Retrieval: Two Sides of the Same Coin?" CACM, Vol. 35, No. 1

Standardised Model Data Exchange for Dispersed Systems Engineering Design Teams

David Harris
Systems Engineering and Evaluation Centre, University of South Australia
Email: david.harris@unisa.edu.au

Keywords Systems Engineering; Data Exchange; Teams.

Abstract The engineering of systems has always been a multi-disciplinary activity carried out by teams of engineers operating within different domains of expertise. The paper discusses work in developing standardised approaches to systems engineering tool data exchange and developments which will support globally dispersed teams. The paper begins with a brief overview of the domain of systems engineering and the issues involved in exchanging Systems Engineering (SE) models. It discusses the SEDRES Project, a European project which developed an appropriate SE data model and transitioned it into the ISO, where it has become Application Protocol AP-233, within the ISO 10303 STEP environment. Other issues were raised within the SEDRES Project, and the results of experiments into the effectiveness of the system are discussed.

1 SYSTEMS ENGINEERING

As Alice in Wonderland's hookah-smoking caterpillar may have said "When I use the words 'Systems Engineering' they mean exactly what I want them to mean, no more, no less". Almost every text on the topic has its own definition, and there is so little agreement that the final issue of one of the principal Systems Engineering standards, ANSI/EIA 632 [1] uses the title "The Engineering of Systems" and abandons attempts to define Systems Engineering. However, to give some bounds to this paper, the following definition, from the Australian Dept of Defence, will be used:-

> *Systems Engineering is the design, realisation and implementation of Engineering Systems, where such a system is an integrated set of hardware and software elements which, when operated together, perform a useful process of which the resulting output is greater than the output of the individual parts*

The SE Process is described in several standards. One of the most used is ANSI/EIA 632 but there are others, most deriving from early military specifications, in particular MIL STD 499 [9]. An overview of the SE process is shown in Figure 1.

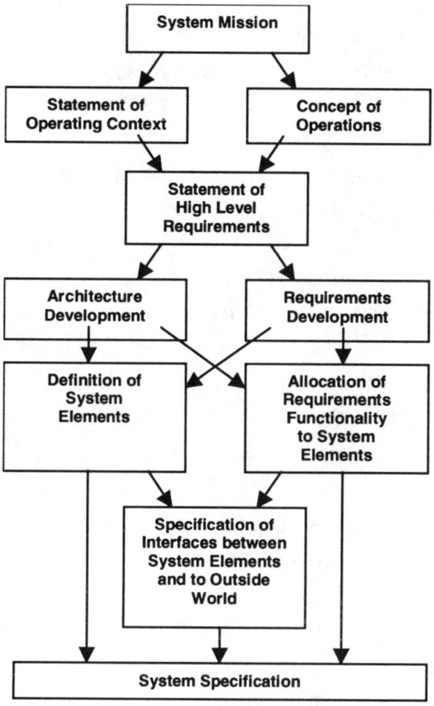

Figure 1 The Systems Engineering Process

Systems Engineering addresses complexity. It takes a holistic view of the system, viewing it from the top down and from the bottom up, leading to an integration of elements to achieve an intended, overall result. Taking a holistic view tends to fly in the face of the Cartesian method of decomposition of large problems into smaller, more easily solvable ones. The problem is that decomposition can lose sight of the connectivity and mutual influence of the elements.

On the other hand, problems must be brought down in size so that, at the lowest level, they can be within the scope of what can be held within one human brain. The methods of systems engineering, then, attempt to reconcile the need for decomposition while retaining the decomposed elements within a structure where their mutual effects remain at the forefront. In fact, integration is one of the major functions of the systems engineering activity.

2 SE DATA EXCHANGE NEEDS

The world of engineering is being dramatically affected by the growth of information technology, and SE is no exception. In fact, SE is affected in two ways - in the systems themselves and in system creation. It is now very difficult to think of a system of any complexity which does not include large amounts of software which control every phase of its operation. There is a tendency in the software community to see large software suites as "systems" in their own right. The view of the more engineering oriented Systems Engineer is that, at the very least, software programs require a (hardware) computer to run them. There is considerable tension over these issues between the two communities.

On the other hand, the creation of all systems, including software-only systems - now requires the use of ever more sophisticated software tools. In many cases, the same tools are used in both fields. The kinds of tools in use reflect the activities involved in SE. They include:-

- System Modelling tools (System Dynamics Modelling, State Transition Tools)
- Requirements Management Tools (Text, Structured text)
- Functional modelling tools (Data flow generation, causal chain modelling)
- Behavioural modelling tools (event driven modelling, time-stepped modelling)
- Simulation tools and environments

There are many others. Interested readers are referred to the web-site of the International Council on Systems Engineering (INCOSE) [9,10]. As SE becomes more tool-centred, the potential for using network-based teams for the engineering of systems is increased. Teams are now frequently distributed over companies, over countries and even over the world. This is not, of course, unique to SE, but SE is entering this environment later than other engineering disciplines, and the rules of the game are still being worked out.

To indicate the data exchange situation faced by the SE team, Figure 1 illustrates the activities involved in a typical prime-contractor to sub-contractor exchange. In this illustration, the Prime works in environment A and releases a part of the total system design to be implemented by the Sub-contractor in environment B. The activities run concurrently, so that this is an implementation of concurrent SE.

1. A piece of the system model (created in Domain A) is identified for transfer to Domain B for development. The broad, outlined arrow indicates that this is a model transfer from Tool A to Tool B. This is contractual information crossing the boundary.
2. Note that the model changes form to reflect the fact that it is now in a different tool domain, with different model representation.
3. Work proceeds concurrently on the system model and the sub-system model. During this process there is a series of information exchanges. The broad arrows, crossing from tool domain to tool domain, carry tool information, but this is not model data (it could be, for example, screen dumps) and is not contractual but is used for explanation between the engineers. The narrow arrows do not penetrate the tool domains but use other media for human-to-human communication, again for explanation, clarification etc.

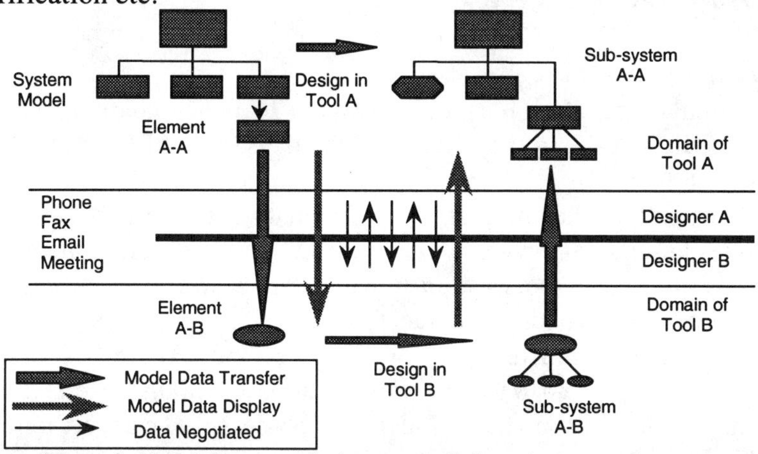

Figure 2 Model data transfer between collaborating network agents

4. At the end of the sub-system design it is shown as being transferred back from Tool B to Tool A, using the broad, outlined "contractual information" arrow. This is a model transfer. Note that the model elements change "shape" from the representation of Tool B (shown here as ellipses) to that of Tool A (shown as rectangles).

There is a lot which can be read into Figure 1, but for the purposes of this paper we will only make a few points:-

- The "thing" transferred is a piece of a system model. This is done on the assumption that this is the best way to transfer Designer A's intentions to Designer B. Before network based computing, this would be done by written documents.

- The model is represented differently in the two tools. It is almost certain that some information existing in the sending domain will not be representable in the receiving domain and will be distorted or lost.
- The boundary between Tool A and Tool B is a contractual boundary. Too free a flow of communication across this boundary can upset contractual arrangements and configuration management.

In a typical network environment this scenario will be repeated many times, and there will be interactions between them. The transfer between tools requires a translation between the representational models used by the tools. This transfer is performed by a data exchange interface and it can be seen that there will be clear benefits if there are standards for this model data transfer which can be used by interface developers.

3 STATE OF THE ART

3.1 What to transfer

Communicating SE information between team members appears to be more difficult than in other fields such as CAD/CAM. Why is this? Because:

- Communication techniques for SE data are immature. Model representations vary widely, there are no standard communication data models and many tools use proprietary databases. The environment is not conducive to robust data transfer.
- The use of data transfer is not well understood. There is often a view that transfer of data between tools is a "good thing" in itself without looking at its role in the complete SE process. The result tends to be technological focus on transfer of data rather than a focus on the functionality of the team members using the data.
- What the recipient needs is not product data but interface data. Viewed in the light of the whole process it can be seen that the reason for transferring a model is in order to provide the recipient with a clearer statement of the interface between the sender's model and what the recipient is expected to provide. What the recipient needs is a clear set of interface requirements which may be provided by a model, or by other means.
- The transfer must achieve a sufficient level of shared meaning for the two parties to contribute successfully to the overall task. This requires the transfer of data, and the subsequent connection of this data into the world view of the recipient in such a way that a correct meaning is generated.

3.2 Usage Scenarios

The INCOSE working group on Tool Integration and Interoperability has produced a document detailing the way it sees the Systems Engineer of the future interacting with others in a fully Integrated Systems Engineering Environment (ISEE). This document - INCOSE Report on Scenarios for an ISEE [8] - sets out seven scenarios in which SE data exchange will be used, and details the kind of transactions which will be required. The scenarios identified are:-

Scenario 1: Tool Migration

One-off transfer of a tool database from one (legacy) tool to another (new) tool.

Scenario 2: Parent-Child Integration

This is the situation of prime to sub contractor relationship, which is described in Figure 2.

Scenario 3: Peer to Peer Integration

This scenario describes communication between people at the same level in the hierarchy of the system design network.

Scenario 4: Tool to Tool Traceability

Maintaining traceability between activities carried out in different tools by different people will become important in support of concurrent engineering.

Scenario 5: Data Transformation/Views

The objective here is to provide access to the developing system database in a form which is familiar to individual users. It is intended to provide "windows" onto the database, presenting a range of tailored views.

Scenario 6: Integrated CM Process across tools

In network-based design, configuration management is a major issue. This scenario addresses the activities required to achieve this.

Scenario 7: Navigation

To use the full capability of network-based design the individual engineer should be able to find needed information anywhere on the network. Network navigation activities are covered in this scenario.

3.3 User Requirements

At the highest level of requirements, any SE data exchange standard must:-

- Benefit the SE process.
- Support Integrated Product Teams and Configuration Management.

- Support the concept of a Logical Design Data Repository for network-based design.
- Provide a usable and efficient data exchange capability.
- Lead to International standard.
- Be supported across a number of systems engineering design tools.
- Obtain tool supplier involvement.

4. THE SEDRES PROJECT

4.1 Overview

The System Engineering Data Representation and Exchange Standardisation (SEDRES) Project [13] was an initiative of five major European aerospace companies and three universities, supported by a grant from the Commission of the European Community under the ESPRIT Scheme. Those involved are shown in Figure 3.

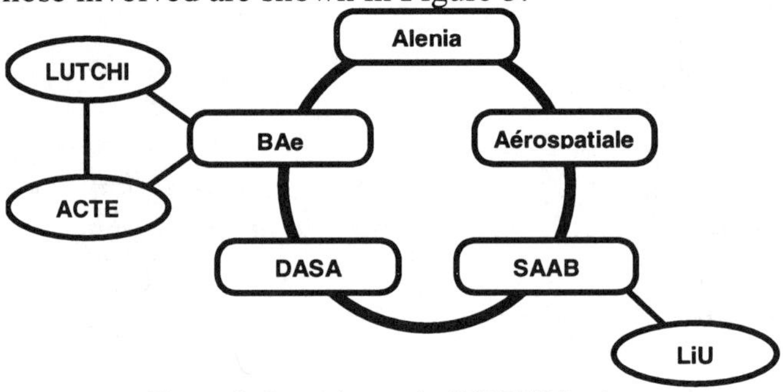

Figure 3 Participants in SEDRES Project

SEDRES addresses the issue of the representation of data held in System Engineering tools with the aim of standardising the exchange of this data [5,6]. SEDRES Project developed and demonstrated a data exchange capability, based on the STEP, ISO 10303, the Standard for the Exchange of Product Data [2,11].

The AP233 data model follows the STEP standard, providing a structure and a set of object classes onto which the elements typically found within a systems engineering tool are to be mapped. A major activity within the SEDRES project was to define a set of appropriate objects to cover the domain of SE tools. The STEP standard defines the methods by which data models will be structured and the way in which data transfer files will be constructed from these objects. The consequence is that an AP233 interface can accept a file (ie, it will not simply generate an error) from any STEP Application Protocol although of course, in most cases, it will not be able to construct a model in the receiving SE tool (Figure 4).

The representation of the model in the Source Tool is mapped, by the SEDRES export interface, onto the SEDRES Data Model, and thence into a standard flat ASCII file for transmission. In the STEP terminology, this file is known as a "Part 21" file. The Part 21 file is then transmitted to the receiving, or 'sink' tool, and is received by its SEDRES Import Interface. This maps the Part 21 file onto the SEDRES Data Model, and thence onto the internal representation of the receiving Tool.

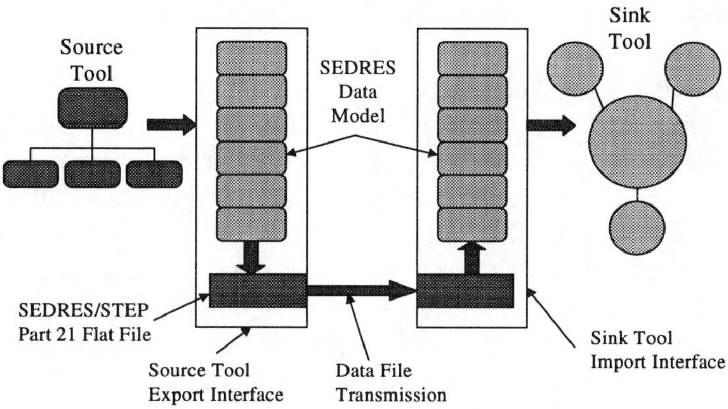

Figure 4 Model data exchange between dissimilar SE tools

Figure 4 illustrates the fact that the model representation in the two tools may be quite different, both in its internal, structural representation and in the way it is presented to the user, via its user interface.

4.2 Implementation

During the SEDRES Project a number of research-level interfaces were developed. Not all covered the same range of capabilities (Table 1).

Table 1. SEDRES exchange interface capabilities

Tool	Vendor	Import	Export	Reqt	Func	Behav	Confg	Gph
LABSYS	AS		□	□			□	□
CoRE	BAE	□	□		□	□	□	□
Matrix-X	ISI		□		□			
SCADE	Verilog	□	□		□	□	□	□
StateMate	i-Logix		□		□	□	□	□
StP	Aonix	□	□		□	□	□	□
TeamWork	Cayenne	□	□		□	□	□	□
Workbench	SES	□			□		□	□

As can be seen, in some cases only an import or an export interface was developed. Others covered only some of the capabilities of Requirements management, Function, Behaviour and Configuration Management. The "Graphics" column indicates that at least an attempt was made to transfer the screen placement of objects. In general, the interfaces demonstrated that the SEDRES data model could support the required level of data transfer, but only some of them were developed to the point where their effectiveness could be measured. For a fuller treatment of the accomplishments of this work readers are referred to [6]. Further information on the effectiveness of this data transfer is provided in [4].

4.3 SEDRES to AP-233

The SEDRES Data Model was primarily the work of Linköping University and was used as the basis of the 11 interfaces written for 8 tools (export and import interfaces count separately) during the SEDRES Project. During the currency of the Project a submission was made to the ISO for development of a new standard, based on the SEDRES data model, and falling within the family of STEP standards covered within ISO 10303. This was accepted by ISO as a New Work Item in 1996, with the proposed new standard given the title of Application Protocol AP-233.

4.4 SEDRES-2 and SEDM

The research work of the SEDRES Project is continued in two further projects, one in Europe and another in Australia.

SEDRES-2 [14] carries forward with most of the original partners, with additional new partners and with European grant support. Its aim is to consolidate the work of SEDRES, to carry out further evaluation of its contribution to the SE process, and to extend its uptake beyond the aerospace industries.

SEDM (Systems Engineering Design Methodologies) is an Australian project which uses the work of SEDRES and AP-233 to develop tool data exchange interfaces targeting the needs of its Australian partners. It is linked into SEDRES-2 through the provision of interfaces useable within that project as well as providing another evaluation facility based on database exchanges.

5. EFFECTIVENESS

In evaluating the "effectiveness" of tool-to-tool model data transfer between network-connected team members, we need to address, not the transfer itself, but the use to which it is put in furthering the design task. It is also useful to compare it with the common current methodology of transfer of text-based requirements documents.

5.1 Requirements Transfer

Although there is currently some use of model data transfer, it is comparatively rare at present. Far more common is the transfer, from one domain to another, of a document setting a list of requirements for the "deliverable" to be supplied. The document may be transferred electronically and may include more than simple text (it may be structured). This does not affect its method of use, which is:-

- The sender builds a model of the system in the tool appropriate to this domain.
- The sender identifies an aspect of this model which is to be passed to another party for supply of an appropriate deliverable item.
- The sender prepares a document listing the requirements for that deliverable. The preparation of this document may be partially automated (it may be a report generated from the tool) but it is important that the requirements are well structured and provide a valid basis for the recipient to carry out the necessary work.
- The recipient will read the requirements document and gain an understanding of what is required of the deliverable. The recipient must re-generate the intent of the sender, or the "meaning" which the sender had in generating the requirements document.
- The recipient will use the meaning so generated to construct a model (by hand) in the tool in use and appropriate to the domain in which the deliverable is to be created.

We can see that the recipient must now derive the "meaning" of the sender by analysing the function of the model which has been sent, and now resides in the receiving tool.

5.2Measures

In evaluating the effectiveness with respect to the process, a convenient start point is to take a high level view of any process innovation and see if can do the job "better, faster, cheaper". We will change the order slightly.

Better implies that the received model forms a better basis for work on developing the deliverable than a model input by the engineer on the basis of an understanding generated from the text requirements. This subject has not, to the author's knowledge, been investigated.

Cheaper seems to imply "faster", with possibly some contribution from the quality of the model in the receiving tool (if it is, indeed, "better").

Faster is measurable, and measurements were made in the SEDRES Project and reported in Britton et al [3]. This paper reports time savings in generation of the model in the receiving tool of up to 8-fold. However, it does not take into account other effects (model usability, time taken to comprehend the model etc). It appears that considerable gains are possible, but more work is needed to determine their magnitude.

6. SHARED MEANING

The objective in transferring information (requirements, tool models etc) between participants is to transfer meaning. The "engineer part" of the sending agent (an engineer using a tool) has a meaning - or an intent - within his mind which has been built into a construct within the tool.

Take away the engineer (with his brain) and the meaning disappears and what is left is structured data or text. In a tool-to-tool transfer, it is just this data which is transferred, and usually distorted on the way (data more than text, but text structures can suffer too). It becomes meaning again when the receiving engineer hooks it into his world-view within his brain and creates a meaning from it. The big issue (and the cause of so much heart-ache) is that these two meanings rarely coincide. They don't need to coincide exactly (in fact, they can't) but they need to coincide well enough to get the job done satisfactorily. The important issue is not the receipt of data between tools but the development of a sufficiently coincident meaning. This is further discussed in [7].

It is, of course, fundamentally important to the sharing of meaning that the model data is transferred correctly and that it builds a valid model within the receiving tool. To the extent that this does not occur, the two engineers will spend time unravelling the errors (ie, wasting time) rather than constructing a shared meaning. As a consequence, the work outlined above in obtaining agreement on an international standard for data transfer is critical. However, the next step in generating increased effectiveness is to address the shared meaning issue directly.

7. CONCLUSIONS

Data exchange standardisation for Systems Engineering is taking shape and will most likely be built on the STEP suite of standards, ISO 10303. At present, an appropriate standard has been under development within the ISO framework and will emerge as a Committee Draft in September, 2000. With this standard in place, it can be expected that commercially viable data exchange interfaces can be built, providing the basis on which a body of users can grow.

With this technology in place, it will be necessary to address the creation of new system designs as a system in itself. With this view it can be seen that many of the issues are not technological but social, with an overlay of business issues. This is the next major area for research.

8. REFERENCES

[1] ANSI/EIA Standard 632 1999, "Processes for Engineering a System" Electronic Industries Alliance, January 1999

[2] AP233, 2000 http://www.sedres.com/ap233/sedres_iso_home.html, 2000.

[3] Britton, J.H, Candy, L and Edmonds E.A, 1998, Software Support for Usability Measurement: An Application to Systems Engineering Data Exchange Development. Proceedings of HCI'98, People and Computers XIII. H. Johnson, L. Nigay, C. Roast.(eds). Springer-Verlag Berlin, 1998, pp 37-52, 1998. ISBN 3-540-76261-2

[4] Harris, David D and Candy, Linda, 1999. " EVALUATION IN THE SEDRES PROJECT: Measuring the Effectiveness of Model Data Exchange between System Engineering Tools" Proceedings of the Annual Symposium of the International Council on Systems Engineering, INCOSE 99, Brighton, UK, July 1999.

[5] Johnson, JFE, 1998, "The SEDRES Project (Systems Engineering Data Representation and Exchange Standardisation); extending STEP from structural definition to product functionality. Proceedings of the Annual Symposium of the International Council on Systems Engineering, INCOSE 98, Los Angeles, USA, July 1998.

[6] Johnson, JFE, 1999, "The Systems Engineering Data Exchange Standard SEDRES / AP-233: Relevance to Requirements Engineering & Capability Maturity". Proceedings of the Annual Symposium of the International Council on Systems Engineering, INCOSE 99, Brighton, UK, July 1999.

[7] Harris, D D, 2000. "Supporting Human Communication in Network-based Systems Engineering", Proc European Conference on Systems Engineering (EuSEC), Munich, September 2000.

[8] INCOSE 1997. TIIG Report "Tools Integration Interest Group Report: Scenarios leading towards a concept of operations for an integrated systems engineering environment", Proceedings, INCOSE '97 conference, Los Angeles, July 1997

[9] INCOSE 2000. http://www.incose.org, 2000

[10] INCOSE TIIWG, http://www.incose.org/tiiwg/, 2000

[11] ISO, 1994, ISO 10303–1, Industrial automation systems and integration - Product Data Representation and Exchange - Part 1: Overview and fundamental principles, ISO, 1994

[12] MIL STD 499, United States Department of Defense standard.

[13] SEDRES, 2000a, http://www.ida.liu.se/projects/sedres/, 2000.

[14] SEDRES, 2000b, http://www.sedres.org, 2000.

Knowledge Creation at FORTUM Engineering

P. Valikangas, J. Puttonen, M. Sulkusalmi
Fortum Engineering Ltd, Finland
Em: pekka.valikangas@fortum.com

A.S. Kazi, M. Hannus
VTT Building Technology, Finland
Em: Sami.Kazi@vtt.fi

Keywords Knowledge Mining and Creation, Information Infrastructure, Virtual Enterprise, Fortum Engineering

Abstract The creation and management of knowledge is considered a key enabler of value added business process delivery. This paper presents the findings of the main efforts towards knowledge creation at Fortum Engineering (FE), an engineering, procurement, and construction provider of power plants. Dealing with different levels of knowledge, it is necessary for FE to provide an environment that is able to translate and codify its tacit knowledge into explicit knowledge to support the definition of power plant configurations using expert applications. Where data or information is not available in a tangible form for translation, some mapping methodologies may be used to render the tacit knowledge meaningful and usable. FE's suggested, knowledge creation and management environment involves; mining of data structures to capture and transfer information into the appropriate knowledge level, manual and automated inputs and rule definitions to capture and manage knowledge on different knowledge levels, and the correlation of knowledge and its data or information source using mind mapping tools. The focus of this paper is to first justify the need for knowledge creation at FE and then to illustrate the envisioned knowledge creation environment. Key phases of the knowledge creation process are furthermore explained.

1 INTRODUCTION

Knowledge is an organisational capital that needs to be exploited to its full potential for value added business process delivery. The aim being to both increase and enable an individual to participate in decision-making based on value-added information in addition to being in a position to exercise control over person's work domain.

One of the key success factors for an engineering company is the ability to share and, where possible, formalise its tacit knowledge into a meaningful form. This is then to be used by both its knowledge workers and seekers alike, in addition to being stored and used within the company's own expert applications for power plant design. The core underlying purpose being to "first sieve the wheat from the chaff and then to sieve out the real drops of wisdom relevant for the topic under concern" [1].

2 NEED FOR KNOWLEDGE CREATION AT FE

Fortum Engineering (FE) is part of the Fortum Group, a diversified group of companies in the energy industry. Three Fortum companies are co-operating in the main business area of FE: FE, Fortum Power and Heat Ltd and Fortum Services Ltd. FE is selling power plants as a constructor by managing engineering, procurement and construction. Fortum Power and Heat is selling energy as the owner of power plants and is a client of power plant delivery. Fortum Service is an operator of power plants and selling more coverage to energy production and at the same time taking part in the bid from a total life cycle aspect.

Figure 1 below describes the total life cycle of a power plant from the viewpoint of information management in the Fortum Group. This is the forum, where the knowledge is supposed to be available especially for tender design purposes.

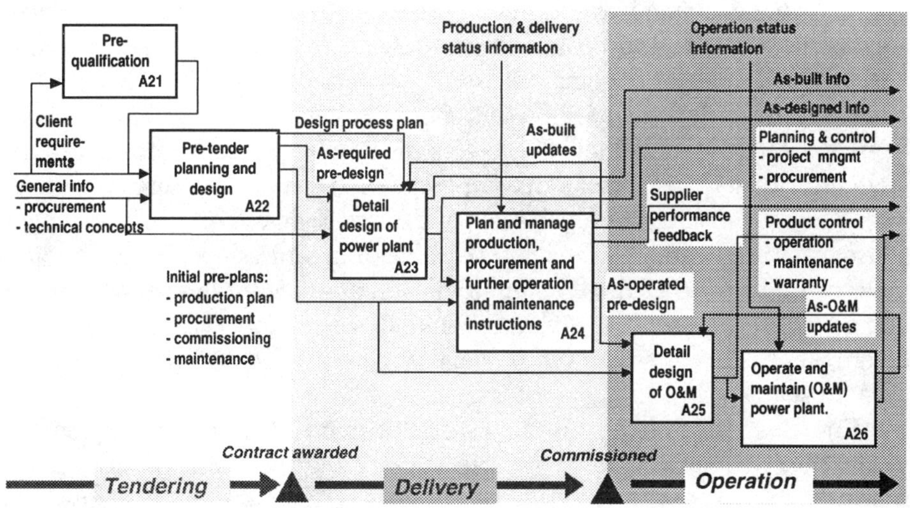

Figure 1 Manage power plant information (IDEF0 level A2 diagram)

The need for efficient co-operation with affiliated companies and clients under high demands on total delivery of power plants and low margin of profit drive the company to develop constantly its engineering technology and associated applications. FE is developing and implementing an expert system for tender phase design in order to define suitable power plant products with cost estimates for different clients. The expert system is using formalised knowledge based product models and design rules.

The collection and development of formalised knowledge is both laborious and tedious among the collectors, developers of the expert system and sources of information, the technical experts in the company. It is also difficult for people from different technical backgrounds to understand each other. The need therefore is evident for a knowledge creation environment where the tacit knowledge of key persons could be formulated to be in the form of structured rules for the expert system. The known explicit knowledge (in addition to some remaining tacit knowledge) may be linked to the relevant product descriptions through the expert system, thereby providing pointers to key sources of information and knowledge.

The creation of knowledge is a joint venture within the Fortum Group. The road ahead is through the development and exploitation of knowledge technology.

3 DEVELOPMENT OF KNOWLEDGE TECHNOLOGY

The development of knowledge technology is seen as the next major challenge for engineering businesses. A brief look at recent history makes it possible to understand a common tendency towards the development of engineering science and design tools. Figure 2 shows three basic trends of uprising new kinds of applications for design.

The Finite Element Method and Computer Aided Design have already demonstrated the potential of new computer based technology in their straightforward development and implementation in engineering companies. This work has resulted in high levels of efficiency in engineering design and consequent implications to the performance of organisations. FE has been on the forefront in taking these technologies on board from the very beginning and has been successful in its endeavours. Consequently the willingness and determination to continue with such engineering development work is omnipresent.

Knowledge technology is the emergent trend towards bringing in new kinds of possibilities and tools to engineering business. This trend is probably even stronger and larger than the previous ones, because it is developing in accordance with the evolution of Internet technology. These trends have similar development features in the form of strong theoretical and educational development, clear growing market needs and substantial amount of developers and vendors.

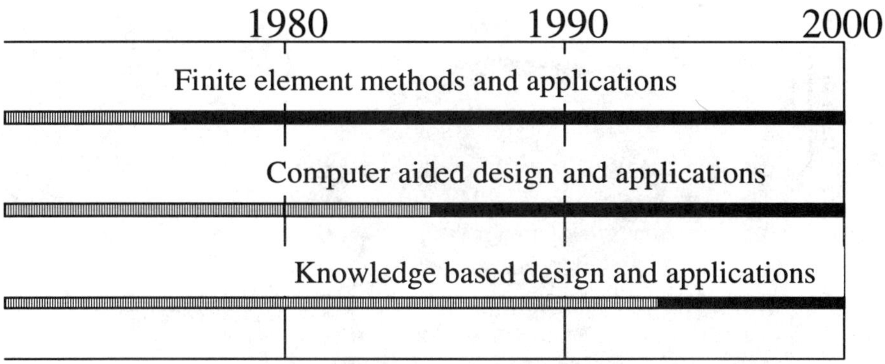

Figure 2 Trends of engineering science and design tools.

Good examples and encouragement have been found from the experiences of the Japanese industry as studied and presented by Nonaka and Takeuchi [2]. Davenport and Prusak [3] have provided valuable insights to knowledge management whereas von Krogh et al. [4] have done so for knowledge creation.

The problem of FE is that even if there seems to be a common vision of knowledge creation in accordance with knowledge based expert systems no suitable tools are available for quick implementation into FE's process. Therefore the focus is to develop methods and implement tools to get an effective environment for knowledge creation and management by which a better support can be offered during the tender phase of power plants.

4 KNOWLEDGE CREATION ENVIRONMENT

The development of a knowledge creation environment is a continuous development effort at FE. Figure 3 presents the role of the knowledge creation environment, where the existing product data, construction and operational experiences are used to enhance existing and create new value added company knowledge. This will further support tender phase application development so that the client's requirements can be taken into account more efficiently.

It is suggested that knowledge evolution goes through a sequential process from tacit knowledge to rules in the expert system for the tender design of different kind of power plants. The goal is to get an environment, which also enables the creation of new knowledge - see Figure 4.

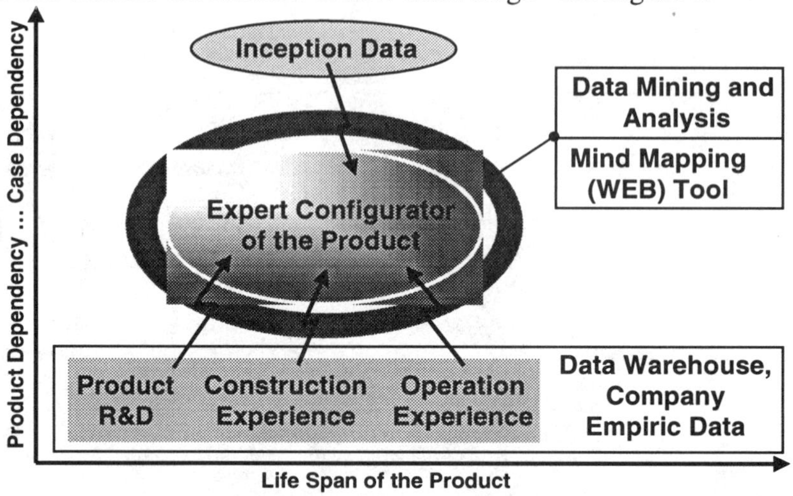

Figure 3 Knowledge creation environment in Fortum Engineering.

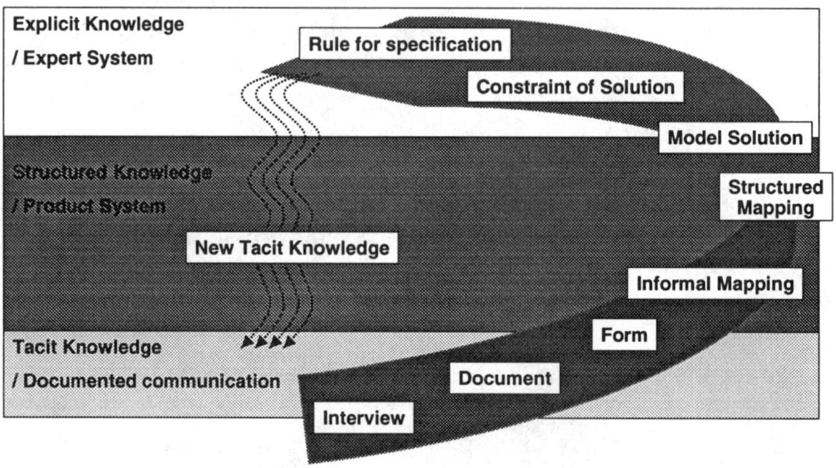

Figure 4 Evolution of knowledge

Table 1 lists the different forms of knowledge and describes their correspondent formats. Table 2 lists the users of the knowledge creation environment and describes their roles.

5 KEY PHASES IN KNOWLEDGE CREATION

5.1 Informal Mapping

Mind mapping is a method for collecting and structuring information into informal knowledge. Figure 5 shows an example of how the design knowledge of equipment may be mapped.

Table 1 Evolution phases of knowledge.

Phase of knowledge	Format / Description
Rule for specification	Code of Lisp and JAVA in expert systems for rule specification
Constraint of solution	Code of Lisp and JAVA and formulas of checkpoints in expert systems
Model solution	CAD-files and Database connected to product model of expert systems and also presented in design guides
Structured mapping	Unified Modelling Language (UML) for linking logical structures of power plant techniques and product model of expert system
Informal mapping	Mind mapping for logical structures of power plant techniques
Form	GUI dialogues and database
Document	Textual, spread sheet and CAD files in document management system
Interview	Discussions between experts of power plant techniques and knowledge engineering
New tacit knowledge	New data in databases, new documents, new ideas for products

Table 2 User groups of knowledge creation environment.

User group	Knowledge involved	Role
Engineering Procurement and Construction personnel	Equipment configuration, plant construction and purchasing knowledge	Entering knowledge within the responsibility area of the user group. Verification of the knowledge is mainly delegated to the leading expert of the technical area.
Sales personnel	Product structure and sales knowledge	
Product developers	Product structure and product constraint knowledge	
Knowledge creation environment developers (knowledge engineers)	Product structure, product constraint, equipment configuration, plant construction and purchasing knowledge	Browsing the entered knowledge and reformulation of the knowledge if necessary, user support and development of the knowledge creation environment
Expert system developers		Browsing the entered knowledge for implementation purposes and reporting of the implemented knowledge

Leading experts and knowledge engineers are mining knowledge so that they can describe the logical knowledge structures of product, design and project delivery. Knowledge engineers can also define the structure of the data warehouse and component structure in the expert system. "Active comments" under Knowledge Creation are also structured.

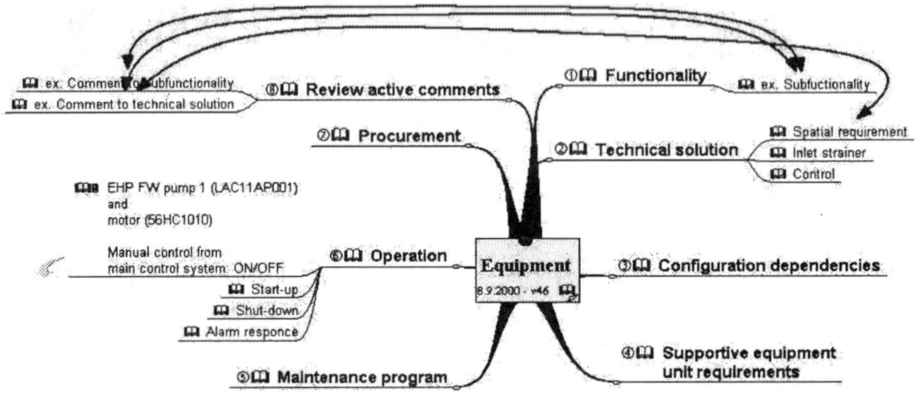

Figure 5 Informal mapping of knowledge by Mind Mapping

It is possible to work concurrently with different experts and at the same time to have joint evaluations in meetings. Furthermore, the mind mapping application may be used as a case-tool for WEB application development. This may then be used for browsing, proposing, entering, modifying, verifying and validating of knowledge.

5.2 Structured Mapping

In structured mapping, the collected and created rules and associated structures are stored in knowledge repository, which is a relational database. Work between informal and structured mapping is manually defined using UML. Figure 6 gives an example how formal mapping is supported. The knowledge is connected to a product model tree at the left-hand side of the dialog with the same semantics as that of the informal mapping (Figure 5).

6 CONCLUSION

This paper presented the underlying motivation and commitment of Fortum Engineering towards the development of its knowledge creation environment. The key point is the translation of scattered tacit knowledge to more meaningful and structured explicit knowledge for use in expert applications where possible.

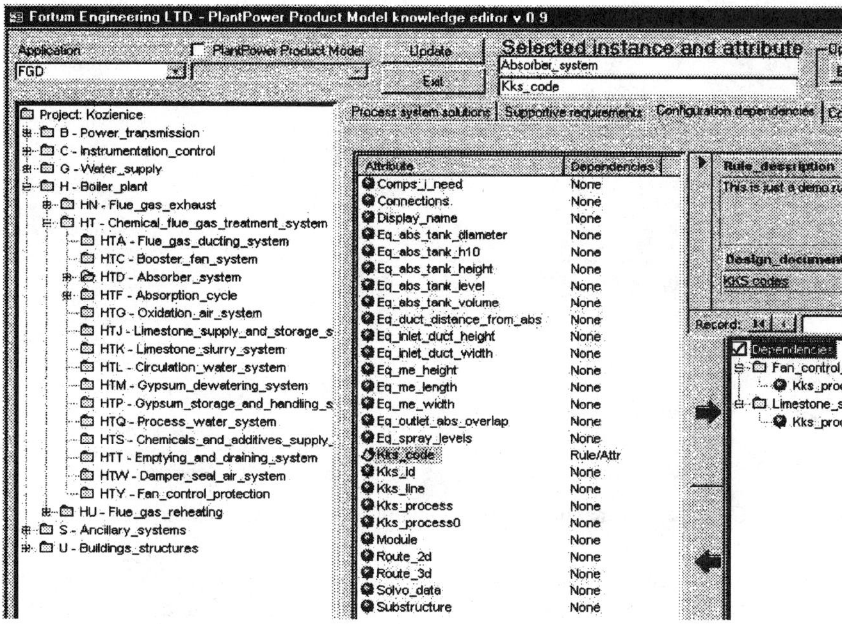

Figure 6 User interface to formal mapping of knowledge

This is an ongoing development work and only some initial lessons have been learned and insights into the future have been provided. More findings will be reported and published as they become available during the course of the development of the knowledge creation environment.

7 REFERENCES

[1] Kazi A.S., Puttonen J., Sulkusalmi M., Valikangas P., Hannus M. (2000 forthcoming) Knowledge Creation and Management: The Case of Fortum Engineering Ltd., *Knowledge Management in the Sociotechnical World.*

[2] Nonaka I., Takeuchi H. (1994). The Knowledge-Creating Company, *ISBN 0-19-509269-4*

[3] Davenport T. H., Prusak L. (1992). Working Knowledge, *ISBN 1-57851-301-4 (pbk)*

[4] von Krogh G., Ichijo K., Nonaka I. (2000). Enabling Knowledge Creation, *ISBN 0-19-512616-5*

Formal Ontology for Participative Simulation

J.B.M. Goossenaerts, C. Pelletier
Eindhoven University of Technology, Faculty of Technology Management, The Netherlands
Em: J.B.M.Goossenaerts@tm.tue.nl

C. Reyneri
DataConsulteam, Turin, Italy

R.J. van den Berg
Baan Development, Barneveld, The Netherlands

Keywords Ontology for manufacturing, simulation, participation, enterprise modelling

Abstract This paper reports on the first results of the Ontology Workpackage of the HUMACS/PSIM project which started on April 1st under the IMS/EU 5th Framework Program. After summarizing guidelines for ontology development, the PSIM choices regarding domain and related knowledge areas are presented. A draft ontology configuration schema is presented. It reflects findings from the area of enterprise reference architecture and integration methodology. The position and impact of other knowledge areas, i.e., ergonomy, socio-technical systems design, IT tools and procedures, are addressed informally.

1 INTRODUCTION

Today's manufacturing enterprises have to optimize their production in a highly competitive and global market place. Innovative prototypes must be turned into a manufacturable product, at a much higher speed than ever before. This trend accentuates the significance of knowledge and knowledge processing in organizations. *Participation* is the process which allows employees to exert some influence over their work and the conditions under which they work. Competence and capability are "both a requirement for and a consequence of participation" [13]: a requirement because participation needs a minimum level of skills in order to be effective; and a consequence because participation enhances the skills levels of those involved. Participation assumes a smooth mutual communication between management and employees. Information infrastructure services can support this communication by providing highly visual representations of abstract processes, that contribute to a common basis for discussions and suggestions. In this respect, *simulation* is the construction and use of a computer-based representation, or model, of some part of the real world as a substitute vehicle for experiment and behaviour prediction. It offers an attractive opportunity for engineers, planners, managers and teams to try out ideas in advance of commitment to a course of action [10].

The goal of *participative simulation*, is to enable workers to exert direct influence over the product and process designs by bringing in their tacit knowledge, to combine it with expert knowledge, and to put the blend of both insights to the test. When these activities are supported by an attractive ICT interface, the resulting continuous improvement process may become intrinsically motivating for the work force[1]. Besides, it also contributes to the competitive advantage of the organization. Unfortunately, several problems hamper the introduction of participative simulation in manufacturing enterprises. The total number of users of simulation tools is still low. Another problem is how to generate a common description and let people with various experiences participate in the analysis. In addition, the tools or content-bases are not connected to the real world, and therefore, often reflect a state of the business, that is outdated.

In order to overcome these problems and accomplish that several tools and content bases are combined with the transparency of "a single multi-user tool", the PSIM project needs to develop several aspects: an *extensible and modular ontology* that will serve to facilitate interoperation and communication between applications and people; *navigation functions* in support of problem solving and decision scenarios, these functions must often be adapted to the changing needs of each company; and *integration services* for tools and systems including ERP (enterprise resource planning), and MES (manufacturing execution systems), ergonomic and sociotechnical applications. In this paper only the ontology is addressed in detail. The second Chapter describes the common process of ontology development. The third Chapter describes which additional aspects have been considered in the PSIM ontology development, and how.

2 ONTOLOGY DEVELOPMENT

An ontology corresponds in practice to a set of formal terms, usually with a hierarchical organisation, with associated formal definitions that specify their relationships with the other formal terms, and a set of constraints about their use in the knowledge representation of the domain studied. Each term can be seen as a knowledge category that can be instantiated[14]. General guidelines have been stated to help the ontology builders in their task. An ontology has to be rooted in a broad mutual understanding and agreement of the different domain stakeholders. The stakeholders have to agree on the general hypotheses made on the knowledge domain. These hypotheses deal with the different high level categories or concepts, which are used to express the objects or object types and their relations [12].

In a first step, an informal ontology is developed. Uschold [18] proposes to use two complementary processes to identify scope, terms and concepts of an informal ontology for an application domain: (1) identify *motivating scenarios* that can serve as the basis for defining a complete set of *informal competency questions*; the competency questions express different reasoning problems that must be supported; and (2) Use *brainstorming* to identify as many potentially important concepts as possible, and use *trimming* by grouping the concepts into various more or less distinct work areas such that terms within an area show more similarity in meaning, than between different areas (eg. terms related to activity, versus terms related to organisation); within each area, and for each term indicate the importance of including it in the ontology; remove the less important and/or the duplicated terms.

The resulting informal ontology serves as a specification for the subsequent formal code [18]. Setting a definition of terms used in an ontology and the choice of them is the most difficult task in the building of the ontology. Guidelines for choosing terms and defining them in a suitable way have been addressed by several authors [11,14,17]. *Clarity* is achieved by using examples to illustrate what is and what is not intended, and by stating all underlying assumptions, especially where they are not explicitly formalised in axioms. *Consistency and coherence* is achieved by avoiding circularity, by using terms that best conform to the common usage, by avoiding the introduction on unnecessary new terms. Dictionaries, thesauri, and technical glossaries should be consulted. *Extensibility and reusability* can be achieved by getting the right balance between being specific enough to perform the required tasks, but not so specific that it will be of little use to others; also by avoiding several terms that mean roughly the same thing. The authors recommend to work in a middle-out fashion rather than top-down or bottom-up: a bottom-up approach results in a very high level of detail and an increase of overall effort, it also makes it difficult to spot commonality between related concepts, and increases the risks of inconsistencies which leads in turn to re-work and yet more effort. To reach agreement when terms are used ambiguously, one should concentrate on the underlying ideas first, ignoring the terms, define each related idea, then decide on the most important ideas and lastly choose appropriate terms.

3 ONTOLOGY FOR PARTICIPATIVE SIMULATION

In addition to using the common ontology development techniques, an ontology for participative simulation in manufacturing enterprises, needs to leverage available bodies of generic or scientific knowledge "around" the domain. In the PSIM project, the following knowledge areas are addressed explicitly: reference architecture and methodology in enterprise modelling and integration; socio-technical systems design; ergonomy and the reduction of physical and mental workload; the improvement process and intrinsic motivation of employees; the development and deployment of ICT tools, especially ERP software packages, and their use of enterprise modelling techniques (IT vendors and IT consultants).

The PSIM ontology development gives each of these knowledge areas its due attention, as is explained below. An enterprise meta model based on CIMOSA[3] and European pre-standards [5,6,7] was taken as the source for the first core terms and their definitions. The completeness of this model was then checked by confronting it with the pertinent concepts related to the domains of socio-technical systems design, ergonomy, the PSIM procedure, and the AS-IS analysis at three pilot-sites. If a concept can not be explained or expressed with the help of the concepts in the model, it has to be added to the core. In a first step concepts have been formulated in natural language and translated to a semi formal one using the proposed model. By this way, we are building the first version of the ontology, the informal one.

3.1 Reference Architecture and Methodology

The relevance for participative simulation of reference architectures and methodology is derived from their relevance for integration projects as expressed in the report of the IFIP-IFAC task force[21]: "*... a large part of integration projects **[participative simulation projects]** is in fact similar and common in every type of enterprise. Thus it could be captured, standardized and utilized instead of developing it again from scratch. Once standardized, generally accepted architectures can be supported by tools, methodologies, and a range of compatible products thus making the entire endeavour[**participative simulation]** efficient in time and cost.... Methodologies usually are developed in conjunction with* reference architectures..*A methodology (for enterprise integration **[or participative simulation]**) is a set of proven guidelines, techniques, procedures which the* *end-user can follow in order to implement an enterprise integration policy as well as carry out integration* ***[and change]*** *projects within that policy."*

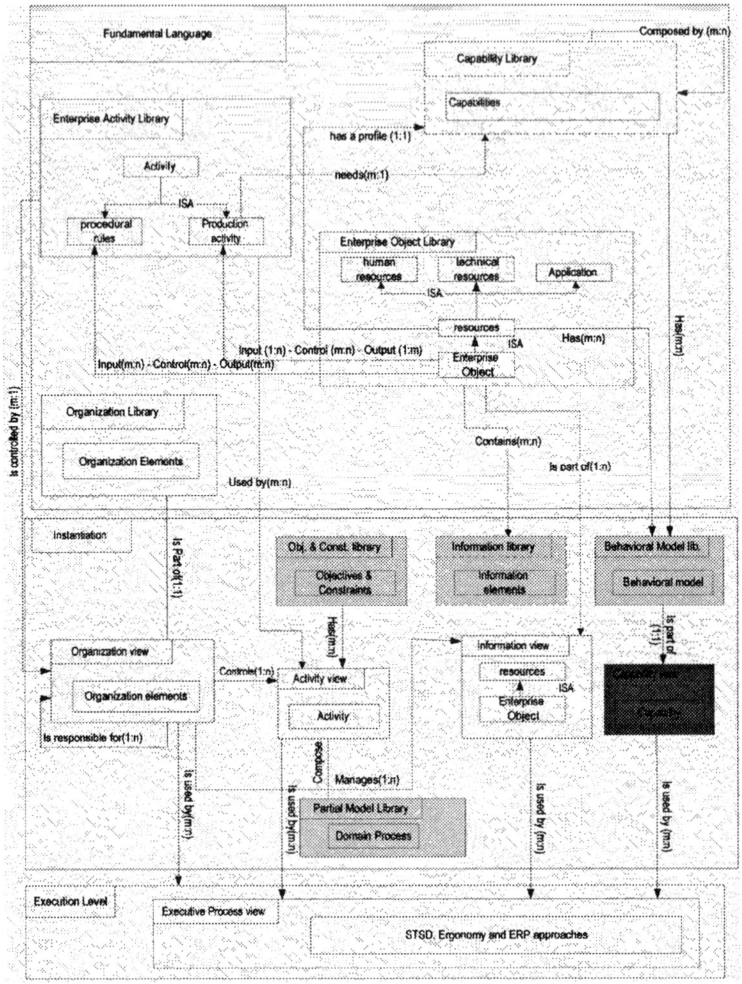

Figure 1. PSIM Draft Ontology Configuration Schema

The PSIM ontology draws on existing enterprise integration knowledge (CIMOSA [3], ENV 40 003 [5], ARIS [17], DEM [19]), as well as on concepts for design and maintenance of enterprises for their entire life history (GERAM or Generalized Enterprise Reference Architecture and Methodology" [2,8]). The backbone of the participative simulation environment is an "ontology configuration schema" with a hierarchical structure that reflects the ENV 40 003 step wise instantiation process. The ontology configuration schema is depicted in figure 1. Its aim is to support the initial configuration and further modification of the ontology for any particular enterprise.

The fundamental constructs at the "generic" level – the **fundamental language** (figure 1) – are three: activities, organization elements and enterprise objects. Their definition draws on the constructs defined in ENV 12204 [16]. Instances of these meta-constructs are created as template elements in libraries. Capabilities, at this level, serve as attributes, to support evaluations of whether or not production activities can be executed by or for the enterprise objects envisioned.

At the **instantiation layer** (figure 1), entities – organisation elements, activities and enterprise objects – are instantiated within three enterprise views: the organisation view, the activity view, and the information view covering the enterprise objects. When configuring the particular level model of the enterprise at the instantiation layer -- compare with the ENV 40 003 particular level -- use is made of a number of "partial level" libraries: the partial model library, the objectives and constraints library, an information library and a behavioral model library. The capability view is not configured as such, since it results from choices made in the activity and information view. At this level, plenty of cross-view rules on enterprise configurations can be stated and enforced.

At the ***execution layer***, the particular enterprise model configured at the instantiation layer, is blended with operational data, work orders and emergent problems to be solved. At this layer the enterprise model execution and integration services matter, as described in ENV 13354 EMEIS (Enterprise Modelling Execution and Integration Services [7]). Decision scenarios might cover the allocation of work to team members, the definition of the process plan for a new product, and so forth.

3.2 Socio-technical Systems Design

The social subsystem of worksystems has remained largely unrecognized in the modelling of manufacturing systems and the development of ICT applications. Reasons for this include: the overemphasis on computer technology to integrate enterprises; the lack of knowledge among ICT professionals of the scientific basis on which to model social and human factors; and the inadequacy for modelling social and human aspects of existing modelling techniques and methodologies.

Socio-Technical Systems Design (STSD) is concerned with the optimization and integration of the human factor in manufacturing systems, predominantly at the work group, departmental and organizational levels. It aims at improving the quality of work and organisation simultaneously through adaptation or fundamental redesign of the contents and composition of technology and human tasks [4]. Within the sociotechnical framework, a method was developed that specifically addresses the issue of allocating tasks between humans and technology, i.e. defining the degree of automation. Key to this method are design criteria on the level of work system, individual task, and human-machine interaction, which can be used also in system modelling and simulation [9].

3.3 Ergonomy

While the focus of the sociotechnical framework is the human-human-technology interaction, the more specific aspects of fitting tasks and technology to human operators is dealt with by an ergonomic approach concerned with optimizing the tasks, technical systems and work stations in order to improve human performance and to reduce mental and physical workload. Data from the European Foundation for the Improvement of Living and Working Conditions indicate that a rise in 'time pressure' has taken place throughout Europe. Approximately 30% of the workers in the European Union are involved in painful and tiring postures for more than half of their working day and 40% of the workers are exposed to short repetitive tasks, which often lead to reduced quality, productivity, complaints or even sick leave. A recent survey reports on the work-relatedness of drop out from work due to psychological disfunctioning. Some important aspects in the reduction of workload are the good fit between task and personality, possibilities to develop and regulate your own work[20]. Therefore an important function in a participative simulation environment is envisioned that will warn users when unacceptable workload for humans and teams is anticipated in a particular work system design.

3.4 Procedure

Today improvements in companies are usually achieved, when problems occur and are successfully solved. The employees who deal with this task, are often motivated by the fear, that they will be called to account for the losses a problem causes to the company. Thus, the motivation for problem-solving activities is rather extrinsic with all its negative effects. Since it is not the attractiveness of the task but a coercion which induces the employees to solve problems, this activity lasts only as long as a pressure exists. In order to achieve continuous improvement the problem-solving has to motivate the employees intrinsically: the improvement process has to enrich the job contents and animate the work force to deal with it because of its attractiveness [15]. PSIM needs a generalized procedure for a situation-independent application. This procedure has to map the steps of an ideal improvement process considering an involvement of employees on all hierarchical levels.

The PSIM procedure contains the following steps: (1) *Selection of the topic*: Every employee can initiate an improvement process. That means that not only the worries of the management are dealt with, but even the concerns of the employees such as work overloading or occupational safety. (2) *Selection of participants*: a problem classification scheme supports the identification of concerned company units. (3) *Ontological convergence*: During team sessions the current situation is described and it is stated why this situation is a problem. This step is supported by several tools, for instance process and product simulation. (4) *Definition of goals*: Based on the previous step the problem-solving team defines the goals. The goals are recorded by PSIM. (5) *Idea generation*: PSIM assists the idea generation by creativity conducive tools such as morphological matrix or Ishikawa diagrams. (6) *Simulation of the ideas*: relevant simulation tools are selected from those available in the environment, to simulate the effects of the different ideas. (7) *Evaluation*: results of the simulations, both improvements and possible negative effects, are evaluated and presented. (8) *Selection of solution*: based on the results of the previous step and relevant financial criteria the team selects a solution in cooperation with supervisors. (9) The *implementation* of the chosen solution is supported by PSIM.

The above description of the procedure implies a further important component of PSIM: the moderator function. The simulation platform has to support the problem-solving process by informing the participants about meetings, coordinating possible dates, sending reminders about workpackages which have to be done *et cetera*. The moderation is necessary to keep the process going on, even when the work load endangers the progress.

3.5 Tools

A final challenge is the integration of various simulation and other tools in the participative simulation environment. On the one hand these tools have to be connected to the different steps of the procedure, as it is applied in variable work contexts. On the other hand the tools have to be linked to the users with a user-friendly interface. Also it must be possible for the tools and systems to exchange data, results of their applications, as well as data on the state of the enterprise. These tasks should not be underestimated, because of existing tools and systems have considered software ergonomic aspects and data exchange to a very various extent, because of the short cycle time of some software tools, and the frequent occurrence of in-house developed IT systems.

PSIM will provide a structure which facilitates an easy plug-in of other than the originally integrated tools. For certain tools, this plug-in has to be relative to the particular ontology that has been configured for the enterprise deploying the tool.

4 CONCLUSION AND FUTURE WORK

To develop an ontology for participative simulation the common ontology development techniques have been combined in order to leverage available knowledge in areas such as enterprise architectures, ergonomy, socio-technical systems, change procedures, and IT tool development and integration. The PSIM approach and domain choices have been presented in this paper. The draft ontology configuration schema that has been presented will be subjected to depth, scope and particularization tests during the coming months, so that it evolves towards the kernel of a participative simulation environment that can easily be expanded and deployed in multiple organisations, for multiple user profiles and use-scenarios.

Further developments also include capture of the ontology in UML/XML schemas, and the building of interfaces for navigator, tools and users. Towards the end of the PSIM project, the participative simulation environment will be tested at three pilots.

5 ACKNOWLEDGEMENT

This research has been carried out in the EU IST-IMS PSIM Project (Participative Simulation environment for Integral Manufacturing enterprise renewal (IMS 1999-00004). PSIM is part of HUMACS, a project within the international IMS research program.

6 REFERENCES

[1] Berg, R.J. van den, Eijnatten, F.M. van, Vink, P., & Goossenaerts, J.B.M., *Leveraging human capital in assembly organizations: The case for participative simulation.* Proceedings IST conference. Helsinki, 1999.

[2] Bernus, P. and L. Nemes, editors. Modelling and Methodologies for Enterprise Integration. Chapman & Hall, London, UK, 1996.

[3] CIMOSA, ESPRIT Consortium AMICE, editor. CIMOSA: Open System Architecture for CIM. Springer Verlag, Berlin, 2nd, rev. and ext. ed edition, 1993.

[4] Eijnatten, F.M. van, & Zwaan, A.H. van der. The Dutch IOR approach to organizational design: An Alternative to Business Process Re-engineering? *Human Relations,* 1998, *51* (3), 1-30.

[5] ENV 40 003. CEN/CENELEC/IT/WG ARC. Computer integrated manufacturing -- systems architecture -- framework for enterprise modelling. European prestandard, CEN/CENELEC, 1990.
[6] ENV 12 204. CEN TC310 WG1. Computer integrated manufacturing -- systems architecture -- constructs for enterprise modelling. European prestandard, CEN/CENELEC, 1996.
[7] ENV 13354. CEN TC310. Computer integrated manufacturing -- systems architecture – Enterprise Modelling Execution and Integration Services, European prestandard, CEN/CENELEC, 1999.
[8] IFIP-IFAC Task Force. GERAM: Generalised Enterprise Reference Architecture and Methodology. Version 1.6.1, May 1998.
[9] Grote, G. A participatory approach to the complementary design of highly automated work systems. In: G. Bradley & H.W. Hendrick (Eds.), Human factors in organizational design and management - IV (pp.115-120). Amsterdam: Elsevier, 1994.
[10] Groumpos P, Krauth J. Simulation in Industrial Manufacturing: Issues and Challenges, In: Fichtner D, Mackay R. Facilitating deployment of Information and Communications Technologies for competitive manufacturing. Dresden: The European Conference on Integration in Manufacturing - IiM ,1997.
[11] Gruber, T. Toward principles for the design of ontologies used for knowledge sharing. (ed. N. Guarino; Proc. of the International Workshop on Formal Ontology, Padova, Italy, 1992.
[12] Guarino N., M. Carrara M., P. Giaretta, Formalizing Ontological Commitments. In Proc. Of AAAI'94, Seattle, Washington, 31 july-4 august 1994.
[13] Heller, F., Pusic, E., Strauss, G., & Wilpert, B., 'Organizational participation: Myth and reality'. Oxford University Press, 1998.
[14] Martin P. (1996) Exploitation de graphes conceptuels et de documents structurés et hypertextes pour l'acquisition de connaissances et la recherche d'informations, Ph.D., Nice-Sophia Antipolis , France.
[15] Orban, P. & Völker, S., How to Improve Working Processes through Collaboration in Cross-Departmental Teams. In: P. Vink, E.A.P. Koningsveld & S. Dhondt (Eds.), Human Factors in Organizational Design and Management – VI, Den Haag: Elsevier Science 1998.
[16] Reyneri, C., Operational Building Blocks for Process Modelling, Enterprise Engineering and Integration Standardisation Workshop, Berlin, May 24-26, 2000.
[17] Scheer, A.-W.. Business Process Engineering -- Reference Models for Industrial Enterprises. Springer Verlag, Berlin, 1994.
[18] Uschold M., M. King, S. Moralee, and Y. Zorgios (1998) The enterprise ontology. The Knowledge Engineering Review, 13.
[19] Van Es, R.M. and Post, H.A. 1996, *Dynamic Enterprise Modelling: linking business and IT*, Kluwer, Deventer, the Netherlands.
[20] Vink P, Peeters, M. Balancing organizational, technological and human factors: the vision of production management. In: Vink P, Koningsveld EAP, Dhondt S, eds. Human factors in Organizational Design and Management VI. London: Elsevier Science Ltd, London, 1998:7-11.
[21] Williams, T.J., P. Bernus, J. Brosvic, D. Chen, G. Doumeingts, L. Nemes, J.L. Nevins, B. Vallespir, J. Vlietstra, and D. Zoetekouw. Architectures for integrating manufacturing activities and enterprises. *Computers in Industry*, 24/2-3, 1994.

Managing Technical Documentation for Large Defence Projects: Engineering Corporate Knowledge

William P. Hall
Naval Projects and Support, Tenix Defence Systems, Australia

Key words: document management; content management; SGML; XML; HTML; database

Abstract: Documentation activities may account for 10% of the acquisition cost of major Aerospace and Defence systems. Document content requires strict management, yet needs to be replicated (and be kept identical across many uses of the same information). Tenix encountered these issues with support documents for the 10 ANZAC Frigates it is building for the Australian and New Zealand Navies, and the situation was growing more difficult with each Ship delivered. Tenix contracted Aspect Computing to implement RMIT's SIM (Structured Information Manager) to manage documents and content. Compared to the word processing system it replaces, SIM reduced the number of maintenance routines needing management for the class of 10 ships from around 20,000 to less than 2,000. Reuse of redundant texts within procedures will provide another 50% reduction, for an overall 95% reduction in text requiring management. With comparatively little additional development effort SIM technology can substantially improve a wide range of documentation processes and reduce costs throughout the project cycle.

DOCUMENT MANAGEMENT ISSUES FOR LARGE PROJECTS

Felix Litman (1997) of Oracle Corporation's ConText Group, estimated that 90% of the value of corporate knowledge resides in textual documents. Documentation is particularly critical for most large A&D projects. The Australian Defence and Industry Strategic Policy Statement (DOD 1998), estimates that the cost of tendering (a documentation activity) for Defence projects is around 3% of the total acquisition cost. This probably does not amortise costs for unsuccessful tenders by suppliers and subcontractors.

Adding costs for documentation deliverables (operating, maintenance and technical manuals, etc.) raises the overall documentation cost to something around 10% of total acquisition cost.

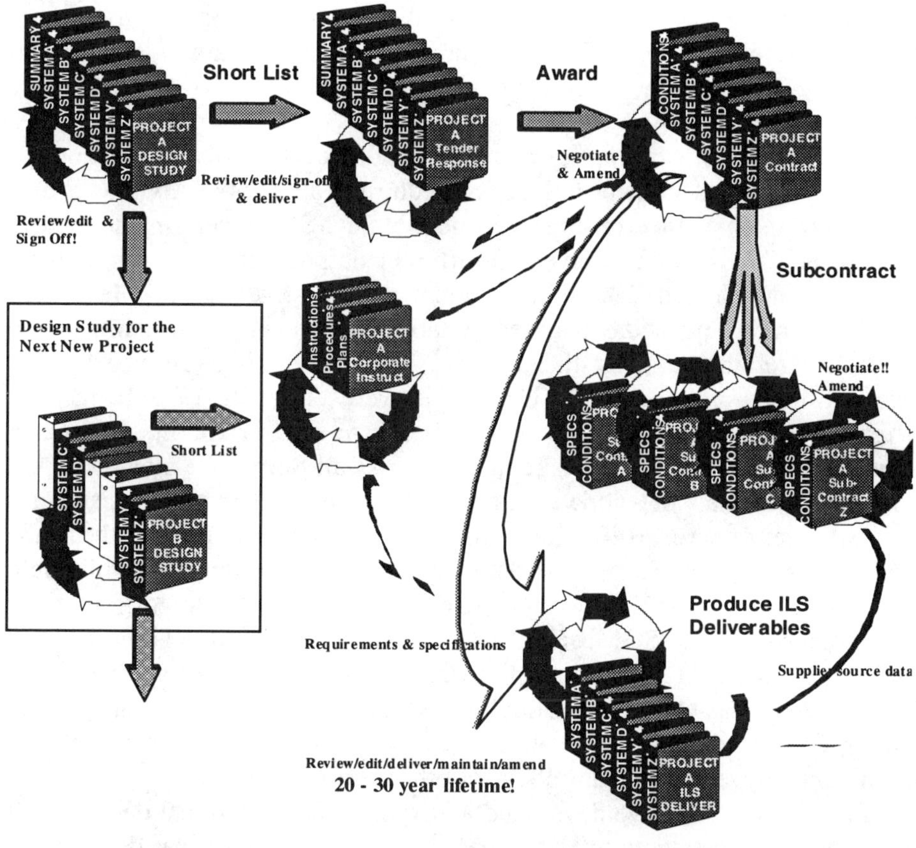

Figure 1. Major project documentation cycle

Project documentation is highly redundant (Fig. 1). Text is duplicated horizontally, across similar kinds of system-related documents produced at a given stage of the project; and vertically through time, as information flows down from the design, through tenders, contracts and subcontracts, to the support documentation deliverables. The safe identification and reuse of duplicated content could make major savings in authoring and management costs and schedules.

Most corporate documents are produced with word processors. However, word processors are inappropriate for long or complex documents like contracts and maintenance manuals. Word processors mark up documents with proprietary codes to format text for printing, so major problems are often encountered converting documents between different systems. Word processors are also hugely complex, offering multiple ways to produce the same visible format. They often crash in use or corrupt documents. The likelihood of problems increases substantially where different authors (using different features or the same features differently) collaborate on the same document or try to combine separately produced texts into one document.

Tenix has been working for more than eight years to solve these kinds of problems with its document production requirements for the ANZAC Ships. From 1992 maintenance routines for the ANZAC Ships were maintained in WordPerfect merge tables. Over 20 different outputs, some with automatic validation of selected data items against authoritative master files, were "single sourced" from the one set of tables. However, even using merge tables to structure text, by the time we began preparing routines for our second Ship, word processing limitations were clear. To reliably manage differences between Ships, we needed separate sets of routines for each, even though 95% of the data is identical across all Ships. We also could not control text formats or validate data as entered, or to detect inconsistencies in similar texts across different maintenance routines or across the same routines for different Ships. The data was also at increasing risk while held in a proprietary format depending on ever more obsolete technology.

We began to use SGML (ISO 1986) as a non-proprietary authoring and publishing format in 1994 for equipment overhaul and repair specifications documents. It was clear this technology would provide content management capabilities we needed for maintenance routines, but the cost would be significant, so we continued to look for other alternatives.

In 1996 and 1997 we developed an experimental relational database for authoring and managing maintenance routines for the Australian Navy's Landing Platform Amphibious ships. The project provided management for more than 100 routines (including reuse of text). However, compared to databases for tabular information, maintenance routines have complex and hierarchically deep data structures. Elements to be managed and reused (e.g., paragraphs, notes and other standard texts) are deep in the hierarchy, and complex sequences of join tables need to be traversed to decompose and re-link these elements. Also, by contrast to SGML databases that parse DTDs to build tables more-or-less automatically, in a generic RDB the table building process is manual. Consequently, besides being slow, the relational application was restricted to a single document type and was a nightmare to maintain. The experiment was abandoned and the data was delivered to the

Client as MS Word text files. SGML offered the only route for a long-term solution.

PROCURING AN SGML DATABASE

By the end of 1997, we knew there was no choice for our document management needs but to implement an SGML database. A case was made to complete a detailed requirements analysis and product review as an R&D study to identify a solution able meet our needs. We reviewed all document management systems we could obtain information on. Applications without SGML content management abilities were not considered further. Allette Systems and our initially preferred system vendor (Texcel - now part of BroadVision) were hired to help develop detailed specifications. Xyvision provided similar analysis and specifications development services at no cost. CSIRO Mathematical and Information Science's Text Information Management group helped to review technical aspects of the RFQ development and bid evaluation. An RFQ package, including draft DTDs for maintenance routines and repair specification data structures, was prepared and circulated four suppliers. The principal requirements were:

- hold data in SGML/XML
- convert existing data (~8,000 routines) to SGML/XML
- validate critical data against master tables on edit/delivery
- manage text applicability to Ship configuration items
- link document change effectivity to engineering change orders
- maintain multiple client-specific languages (RAN, RNZN) in one doc
- register source documents
- 2-way links between deliverable text elements and source documents
- manage workflow processes
- manage and reuse content objects (e.g., graphics)
- manage and reuse document components (e.g., text)
- meet delivery requirements for Client's maintenance management system
- fixed price bid.

The initial two bids (Xyvision and Texcel) were technically satisfactory, but neither offered convincing local support for implementation and data conversion. Other options were reviewed again, emphasising local support. CVSI bid to implement XyEnterprise's mature Parlance and provide local support with their own SGML-literate resources. Aspect bid to implement RMIT's SIM. SIM's object oriented repository was designed from first principles to index, store and deliver SGML/XML data over the World Wide Web and includes no third party products, but did not yet have work flow or element re-use capabilities. These required additional development. Both bids involved substantial code development to migrate data from WordPerfect into SGML, for data validation and to deliver data to the Client's maintenance management system (AMPS) via a unique data transfer format. Both bids were technically and commercially acceptable.

We accepted Aspect's bid to implement RMIT's SIM (see http://www.simdb.com). SIM includes no third party code, and the greater development risks with SIM compared to the more mature Parlance solution were mitigated by the fact that SIM's underlying IP expertise was 100% based in Melbourne. Development cost risks were mitigated by the fixed price agreement.

IMPLEMENTING SIM

Mitigating Project Risks

None of the existing SIM users reused document components deep in the document hierarchy. To mitigate the cost risks to develop this capability in SIM, we negotiated for the required generic component management and workflow capabilities to be developed at RMIT risk and for these to be included as an integral part of the "COTS" product being provided within the agreed license fee. This gave us the technology we needed at fixed price less than the actual development costs; but gives us no rights in the intellectual property developed from our specifications. The implementation project began in June 1999.

Converting Data from WordPerfect to SGML

We anticipated major difficulties converting WordPerfect tables into the SGML "CALS" tables as required by the draft DTD. We even anticipated that most or all tables might have to be manually re-entered after the files were converted to SGML. Within three weeks we had demonstrated the critical steps in an end-to-end conversion. Most tables converted perfectly, and needed only minor intervention to improve display formatting. The full conversion involved five steps, which benefited greatly from SIM ACE's ability to recognise and parse Microsoft's RTF formatting codes:

1. A simple WordPerfect macro converts the WordPerfect files into Microsoft's RTF format and replacing WordPerfect's proprietary merge table field delimiters with identifiable ASCII tags.
2. Writing an ACE script to convert the RTF format into SGML conforming to a "loose" DTD, allowing files to be stored temporarily within SIM.
3. Manually cleaning up "loose" text in an SGML editor to ensure full conformance with the controlled authoring DTD for the final product.

4. Collapse 4 Ship-sets of documents into a single class-set by incorporating system applicability and dual language information in one file. Because this task involves subject matter expertise, it is also largely manual and provides the opportunity to add value to the data in other areas.
5. Eliminate redundant text elements from the class set. (Depends on SIM Release 2).

Conversion Steps 1 and 2 successfully converted around 97 percent of some 8,000 records for the first four Ships to a "loose" DTD in the first attempt, with about 70% of the converted data fully compliant with the authoring DTD. Conversion failures were due to author faults like illegal paragraph number sequences, using space characters rather than tabs to indent text, and setting tables with tabs or spaces instead of using table functions. Such errors are common in a word processing environment where authors are not trained and disciplined technical writers.

Authoring Over the Web (Intranet)

SIM makes no demands on the user's computer beyond having the default Web browser (to IE 4 standard) and an SGML compliant editor. User passwords, sessions and activities are controlled via HTML forms. System administration, file and metadata management, review, signoff and delivery require no software on the user's desktop system beyond the default Web browser. Authors who edit procedure text are able to do so with any SGML editor (as we have confirmed with ArborText's Epic Editor, FrameMaker+SGML, and XMetaL). SIM is not customised in any way for a particular editor or vice versa. On checking out a document for editing, SIM downloads the SGML text as an SGML mime-type and the DTD and associated object files to default directories. The SGML editor maintains text compliance with the authoring DTD. On check in or check out, or whenever the author requests an update, SIM validates all aspects of the content being edited against both the DTD and master data files to ensure that the file held within the SIM repository complies with all established rules. All that would be required to enable distributed authoring over the World Wide Web would be an ACE script to zip and send the downloaded files as a self-extracting package, and to include in web pages sent to the user a Java script able to repackage and return the changed file(s).

Managing Workflow Via the Web

All interactions with the workflow system are via browser forms. SIM's work flow architecture is based on document status and roles rather than explicit routings. User roles are determined by their logons, and only work items eligible to be worked on by that user are displayed in the list. SIM maintains work-in-progress lists and displays status indicators against items as these are tracked through the predetermined workflow. Although the workflow is rigidly determined for a document type, the system is quite flexible in practice because it is based on roles, not individuals. Authors may check out items for editing, check them back in and release them for peer review. On release for review, documents appear instantly on the worklist in a review status. Other authors may view items with a review status and record comments against them. After peer review, a supervisor releases the work item for QA review. QA may release the document for delivery or send it back for redrafting. The delivery process again validates the contents and produces the delivery formats. Documents with 'fatal' faults cannot be delivered. All other inconsistencies are logged for annotation to the Client.

Managing Configuration

WordPerfect offered no facility to relate a document to more than one configuration item in a single Ship even though one routine could apply to many different configuration items across several ships. Also, the two Navy's often use different materials or refer to different documents in otherwise identical routines. Our DTD allows all instances of a routine's use to be encoded in one record: SGML "language" elements allow RAN and RNZN versions of the text side-by-side within a single container element. On delivery, SIM splits the output into RAN and RNZN specific sets of routines. Changes relating to engineering changes (ECs) are linked to the appropriate EC number. An effectivity field determines for AMPS whether the document changes are effective immediately or as when engineering change takes effect. Even without reusing elements of text, for the first four Ships, SIM Release 1 reduced the number of individual routines to manage from 7753 to ~ 1800 - a 75% reduction. For the full class of 10 ships, the text management requirement will be reduced by more than 90%.

Tracking and Managing Source Documents

With WordPerfect we had no practical solution but to rely on our SME's personal knowledge to recognise how changes in a source document would impact deliverables. Our SIM implementation includes a source document register for recording source details (e.g., titles, supplier, version, publication date, etc.), and a linked repository where electronic copies (in any format) of the source material can be held for archival purposes. It is then possible to create two-way hyperlinks between the register entry and any text created using information from that source, whether supplier documents, client letters, etc. The links may be document references delivered to the client or an author's working references. Thus, when a source document changes, we can easily and exactly determine the potential impact of those changes on our deliverable documents. This substantially reduces the work required to complete an impact analysis and improves its reliability.

SIM Release 2

Release 2 completes the initial implementation project, to provide re-use and configuration controls for the lowest level document elements requiring it. Release 2 allows standard text elements within the routine such as paragraphs in tasks, steps, warnings, cautions, notes, and standard phrases to be shared and managed from single locations ("write once, use many times"). Based on the large amount of redundant text in routines, this reduces the net volume of text requiring management by at least another 50%. *This amounts to no more than **5%** of the text that would have been required to be managed for the full set of 10 Ships by comparison to word processing!*

Release 2 also includes an annotation capability, allowing authors, reviewers and users to record comments and notes against particular elements of text or hyperlinks. This is considered to be a valuable source of information documenting the reasons for change or how the text was derived from a particular item of source data.

BUILDING A CORPORATE INFRASTRUCTURE

Our corporate use of SIM system is being extended to other document classes. These extensions can be developed at low cost by using capabilities developed for the ANZAC project as models. Combined with our existing document analysis capabilities, and the commonality of the ACE code across the whole system, much of the work can be done in-house.

We are currently building work flows and templates for the DEF AUST 5629A Core DTD. This enables us to move the ANZAC Ships' Damage Control Books (DCB) and Ship's Information Books (SIB) into SIM. Although to now the DCBs and SIBs have been treated as ship-specific documents, like maintenance routines were, most of the text is duplicated across the entire class. Configuration management, dual language and effectivity management capabilities will reduce the amount of text requiring management - especially if the client can be persuaded to accept electronic delivery of the products .

We are also targeting contracts and tender documents as a major area for process improvement based on management in a SIM environment.

CONCLUSION

Based on our research and development in document engineering and our implementation of SIM, Tenix is well down the track in implementing state-of-the-art document and content management. SIM clearly has the capacity to be extended to form a core corporate nervous system for capturing, managing, querying and acting on most kinds of business knowledge. If we continue to exploit this head start, we should be in a position to establish the kind of virtual corporation that James Martin (1996) believes can beat the world. Martin understood the theoretical requirements for the agile nervous system such a cybercorp would require, but it is probably fair to say that when he wrote his book, the technology was not yet mature enough to work the way he proposed. Today, we are building this nervous system in Tenix, and are well on the way to effectively managing and efficiently reusing the knowledge contained in the overall project documentation cycle (Figure 1).

REFERENCES

[1] Department of Defence (1998). Defence and Industry Strategic Policy Statement. Defence Publishing and Visual Communications, Department of Defence, Canberra 2600 - http://www.dao.defence.gov.au/ipi/IPP/Industry_Policy_Statement/PDF/Policy.pdf.

[2] International Standards Organisation (1986). ISO 8879 - Information processing; Text and office systems; Standard Generalized Markup Language (SGML). International Standards Organisation (Fri, 31 Oct 1986) - http://www.isostandards.com.au/Script/Details.asp?DocN=isoc000516637

[3] Litman, Felix (1997). Unlocking the Value of Text with Oracle ConText Cartridge. *Oracle Magazine*. (September 1997).

[4] Martin, James (1996). Cybercorp: The New Business Revolution. American Management Association. 326 pp. - http://www.jamesmartin.com/

[5] Science Applications International Corporation. (1996). Task Group B - Establish Metacomputing Testbed: Task B4 - Analysis of Alternate Document Management Systems: Sub-task B4-1 - Selection of Fine-Grained Document Management System. Distributed Object Computation Testbed (DOCT) Technical Report (31 October 1996) submitted to San Diego Supercomputer Centre. - http://www.sdsc.edu/DOCT/Publications/b4-1/b4-1.html

AGORA: An Integrated Knowledge Management Environment

P.M. Chrissohoos, M.P. Anastasiou, I.P. Kouranos, N.N. Kalogeropoulou, M.P. Aslani
INTRACOM S.A., Greece
Em: phry@intracom.gr

Keywords Knowledge management, Knowledge awareness, GLOBEMEN

Abstract Telecommunication Industries in their efforts to ensure long-term survival in today's rapidly changing business environment, they should increase their responsiveness and be both pro-active and re-active to challenges driven from their environment. In an effort to address this issue, INTRACOM will demonstrate the use of an integrated knowledge management environment called "AGORA" aiming at increasing its responsiveness of the Sales and Marketing business process and support the execution of this process in an inter-company environment.

1 INTRODUCTION

In today's global business environment, telecommunications industries face new challenges. Market liberalisation, privatisation, deregulation, and the rapid deployment of new information and communication technologies have lead to a business environment that changes often, irregularly and almost unpredictably. In order to ensure long-term survival, telecommunications industries should develop and deploy new services rapidly, in order to differentiate themselves from their competitors. Key factors of differentiation are the external and internal responsiveness, and knowledge awareness. The external awareness of an enterprise refers to its ability to understand its external environment, while its internal awareness is the enterprise's ability to understand itself, its strengths and weaknesses [1]. Being aware of its competencies and skills does not provide the enterprise with a clear path to successful performance, unless it is combined with its ability to efficiently manage internal change and leverage internal competencies and resources to meet a market opportunity or respond to a customer need [2]. This enterprise ability is called internal responsiveness [2]. In that way the enterprise is able to rapidly and efficiently respond to challenges driven from the external environment and outplay its competitors [1].

In this context, knowledge management is becoming of increasing importance to industry and organisations, in order to improve organisational efficiency, competency and innovation. Knowledge management involves the identification and analysis of available and required knowledge assets and their knowledge related processes, as well as the subsequent planning and control of actions to develop both the assets and the processes so as to fulfill organisational objectives [3].

2 OBJECTIVES

In its effort to address the above mentioned issues, INTRACOM, the largest Hellenic industrial group in the field of Telecommunications, Electronics and Information Technologies, aims at increasing the responsiveness of the Sales and Marketing business process, so as to be able to rapidly and successfully respond to global market opportunities, and to achieve what is called "global reach".

From this perspective, the above-mentioned objective could only be achieved through the company's acquisition of knowledge awareness of the Sales and Marketing process. This business process is a very complex and critical one, and its main activities are:

- Selection, analysis and refinement of information related to market trends, market place competition, identified opportunities and potential problems in the marketplace.
- Monitoring of the technology advancements, aiming at identifying key opportunities of technological growth.
- Identification, contact establishment and acquisition of new customers.
- Identification of customers' requirements for product enhancement and/or new products and services development.
- Management of the existing company's business network, aiming at ensuring its responsiveness to future market opportunities. This business network consists of the company's partners including suppliers, contractors, sub-contractors and engineering offices, which may all be spread over different sites and/or geographical regions.
- Response to Request for Tenders by preparing a bidding document in compliance with the Tendering organisation's criteria and requirements.

The efficient performance of the above activities presupposes the effective management of a large amount of information and knowledge. Despite that, there is an inherent complexity in managing this knowledge that stems from the fact that various relationships and dependencies exist between the various knowledge categories. Moreover, this knowledge is globally distributed in different company departments, as well as in different organisations within the business network, thus creating communication difficulties and sharing inefficiencies. It is often the case that the required knowledge resides in many different places and sources, such as the WEB, press releases and existing databases in the business network. In addition, the individuals involved in the process have important process – relevant knowledge that could be helpful to their colleagues if it could be derived and be available to them in some way.

Therefore, INTRACOM aims to implement a prototype of the "AGORA" and demonstrate its use as an integrated knowledge management environment. This prototype will be implemented throughout INTRACOM's involvement in the IMS project "GLOBEMEN" (Global Engineering and Manufacturing in Enterprise Networks), aiming at supporting and enhancing the performance of the Sales and Marketing inter-enterprise business process.

3 THE AGORA ENVIRONMENT

AGORA environment will consist of practices and applications that will enable the efficient acquisition, use, organisation and transfer of globally distributed implicit and explicit knowledge, related to the Sales and Marketing process.

3.1 AGORA modules

"AGORA" currently undergoes the phases of requirements and functional specification definition. The study of the Sales and Marketing process performed until this stage, indicated that "AGORA" should provide the process actors with timely access to valuable knowledge at anytime and anywhere in a fast and secure way. Such being the case, AGORA will operate over the INTERNET providing all users with access to the required knowledge via their web browser.

Based on the preliminary requirements derived until this stage, AGORA should incorporate the following four basic modules:

- **A Knowledge Repository.** This repository will be populated with information and knowledge related to the Sales and Marketing process.

The development of the repository should cater for the efficient management of its content, as well as for security issues. Furthermore, the user will be provided with efficient support to access and update the required information/knowledge.

- **The Knowledge Explorer module**. Since the information required to Sales and Marketing process is subject to irregular and unpredictable change, this module will provide the process participants with the required functionality to search for valuable information in various sources, such as WEB, local drives and CDs. The user will be able to select the most valuable parts of the information found and store it in the knowledge repository.
- **The Knowledge Organiser module**. This module will cater for role-based and personalised presentation of the available information and knowledge. This means that the users will be able to quickly and easily obtain the information and knowledge that is relevant to their task and responsibilities. Moreover, the users will be facilitated to organise existing knowledge and information, as well as to view it from different perspectives, depending on their role in the process.
- **The Knowledge Report Generator module**. This module will provide the actor of the Sales and Marketing process with functionality to aggregate existing knowledge, synthesise different pieces of information and generate reports on process related issues. An indicative list of such reports is: technology trends, existing and new products or services, market trends in a specific region.

3.2 Technical Solutions

The development of the AGORA environment will exploit productive Internet technologies and tools that support the rapid design, refinement, extension and integration of the various applications. In the following paragraphs the selected technical solutions are briefly described.

- **The 3-tier Architecture.** The three-tier architecture platform is today the dominating Internet platform. Software vendors, such as Oracle, IBM, HP and Sun are using it, in order to provide a business framework that Internet applications can rely on. This architecture consists of three tiers. The Web Server, the Application Server, and the Database Server (Figure 1).

The Web Server "listens" requests made up by web browsers through the World Wide Web. A client's request goes to the Web Server, where authentication issues and problems are solved. This request is then passed to the Application Server, which actually manages all the requests that the Web Server passes in.

The Application Server as a management console allocates time and space, in order to successfully let the client execute the application making use of the data stored in the Database Server. The whole platform runs with accuracy, enabling safe, complex and even parallel transactions to be completed fast.

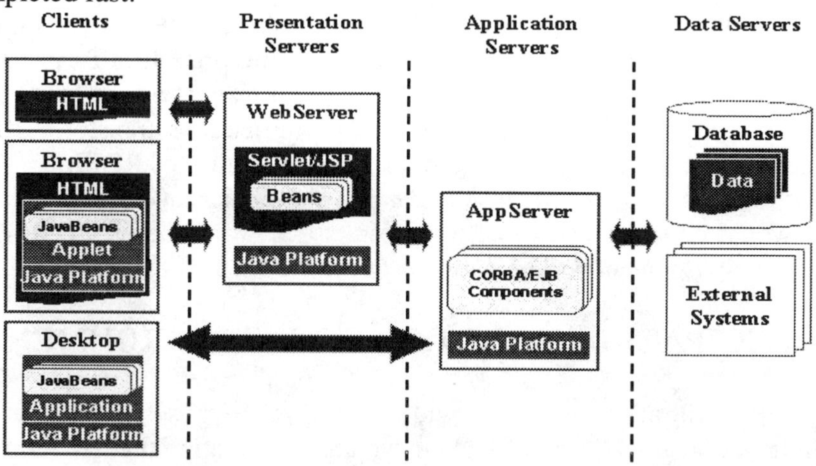

Figure 1: Business Logic in the 3-tier Architecture

The above architecture makes use of state-of-the-art technologies:

- **Java**. The Java programming language is the language of Internet. New programming features have been proposed by Sun, such as Java Server Pages (JSP), Enterprise Java Beans (EJBs), and Java Servlets, which are used by many vendors in their platforms to achieve what is called business logic. The biggest advantage by using these programming techniques is that a Java-enabled browser can run such an application with no additional requirements (space, memory, add-ins). Moreover, the Java language provides a very strong framework in which security and communication issues are being addressed.

- **XML.** The eXtensive Markup Language (XML) provides capabilities of efficient data exchange, in cases where the content of the information can be stored and further retrieved with ease and accuracy. The HTML language is the best language known so far for the presentation of information and data that resides in a RDBMS system to an Internet site. However, it could not address problems where the content and the information itself is so valuable and critical that need to be stored with accuracy, and further retrieved or updated if needed. XML can now address these problems [4,5]. Furthermore, Java provides the framework to parse, change, exchange and even store that information in a RDBMS system.
- **Oracle.** Oracle 8.i was the first xml-enabled database. With many capabilities such as backup and recovery, Oracle can be customised and optimised according to user needs, providing a very accurate, powerful and user friendly RDBMS system. It works ideally with Java (JDBC drivers, SQLJ) and even its own language (PL/SQL) can be used to obtain the best results.

The implementation of the "AGORA" environment will be based on this 3-tier architecture, consisting of a collection of Enterprise Java Beans, Java Servlets and Java Server Pages, each of which will play its own role towards the application development. Java APIs will be used to transform XML documents into valid ones (capable to be stored in the RDBMS). The RDBMS system will be Oracle 8.i. Finally, the choice of the Application Server would be based upon real-time results that would be obtained when AGORA's environment will be published over the Web.

4 ANTICIPATED USAGE OF AGORA ENVIRONMENT

AGORA will allow enterprise wide collection, organisation, navigation, searching and dissemination of knowledge and information that are required by Sales and Marketing process owners, in order to perform their every day activities tasks.

Potential users of AGORA environment are personnel from the Sales and Marketing headquarters, sales representatives in a region, regional partners, and R&D personnel.

Sales representatives will be responsible for enriching the content of the knowledge repository with information and knowledge about:

- Market trends in the region of their responsibility
- Potential new customers in the region
- Customers requirements either for new products / services or for product enhancements
- Anticipated upcoming requests for tenders
- Competitors activities in the region

An operational scenario of the anticipated usage of AGORA environment is depicted in the above figure – Figure 2.

Regional partners will have access to the Knowledge Repository according to their role in the Business Network and in that way they will be able to find in it information and knowledge concerning their region and activities of the Business Network. In addition, they will be responsible to update the Knowledge Repository with their new R&D plans, and possible activities in the region.

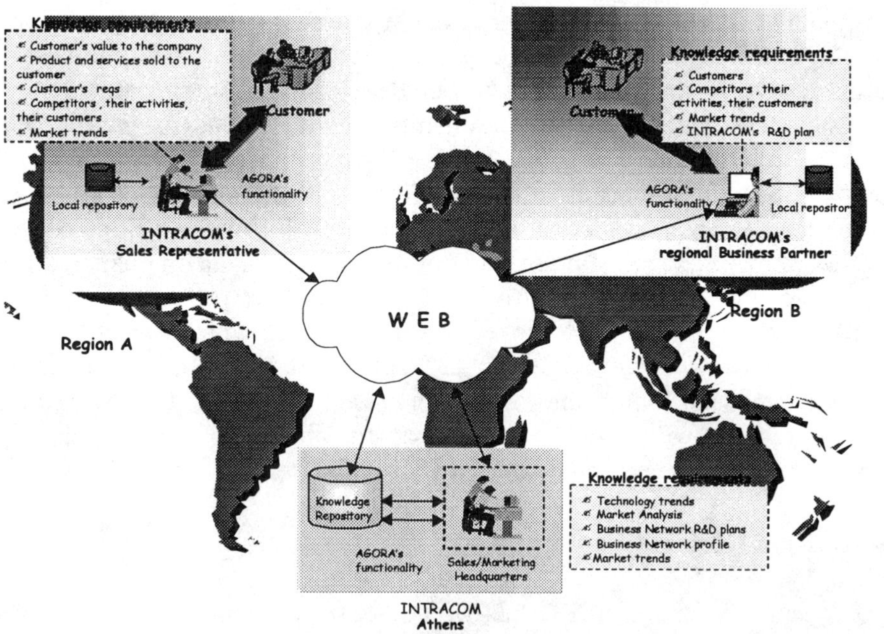

Figure 2. Anticipated usage of AGORA environment

R&D personnel use the environment to search for information about emerging technologies and will have to enrich the content of the knowledge repository with reports on technology innovation and key opportunities of technological growth.

Personnel from Sales and Marketing headquarters will examine the knowledge generated from the above mentioned actors, as well as information about existing competencies in the Business Network, in order to define product and market strategic plans, R&D plans and requirements for new business alliances.

Furthermore, it is anticipated that the use of AGORA for fulfilling Sales and Marketing actors' information and knowledge requirements will cultivate working practices that promote knowledge sharing and as a result continuous organisational learning. In that way, it is anticipated that in the long term AGORA will facilitate the development of the internal and external knowledge awareness and responsiveness of both INTRACOM and it business partners.

5 CONCLUSION

This paper presents through the "AGORA" environment, INTRACOM's approach and practice towards the management of knowledge related to the Sales and Marketing inter-enterprise business process. Throughout INTRACOM's involvement in the GLOBEMEN project, it is anticipated that a prototype of the "AGORA" will be developed and demonstrated at its pilot site. . The preliminary requirements that the AGORA environment should fulfil in an inter-enterprise environment have been currently derived. Moreover, a study of the state-of-the-art development tools and technologies was performed, and the most appropriate ones for the development of AGORA were selected. Future work will be focused on implementing a prototype of the "AGORA" environment and demonstrating its practical use in a real case scenario. Moreover, during the following project phases, methods of using such environment will be determined, in order to efficiently manage Sales and Marketing knowledge. In that way new working practices will be defined and modeled, which will promote learning and knowledge management sharing in a virtual enterprise network.

5 REFERENCES

[1] Thomas Koulopoulos, The Delphi Group, (1997). Creating Corporate Instinct, *BusinessWeek*

[2] Carl Frappaolo, What's Your Knowledge IQ?, ", Intelligent KM, http://www.intelligent km.com/feature/08/feat1.shtml

[3] Yogesh Malhotra (1998). Knowledge management for the New World of Business, @Brint.com, http://www.brint.com/km/whatis.htm

[4] Magnus Lonnroth, (1999). Tech Brief: Introduction to XML, Oracle Corporation, http://www.oracle.com/oramag/webcolumns/TB03xml01.html

[5] Rohib Khare, Adam Rifkin (1997). Capturing the State of Distributed Systems with XML, *World Wide Web Journal Special Issue on XML*, **2**, 207-218

PART EIGHT

Information Technologies for Manufacturing

A Multi-agent Based Information Infrastructure for Manufacturing

Manas R. Patra, Richard Moore
United Nations University, International Institute for Software Technology, Macau, China
mrp@iist.unu.edu, rm@iist.unu.edu

Keywords Multi-agent, Information Infrastructure, Formal Model

Abstract In the present competitive market environment manufacturing enterprises cannot function as monolithic entities. Rather they have to collaborate with other enterprises both for raw materials as well as for component fabrication. In this paper we propose a generic information infrastructure for manufacturing enterprises. The rationale behind building such an infrastructure is to facilitate information access which can help manufacturing enterprises of all sizes to collaborate. The model presented here is based on a multi-agent paradigm.

1 INTRODUCTION

Today's manufacturing companies are faced with global competition and fast changing customer requirements. In order to cope with this challenge they introduce variations in their products and at times completely new products. But this involves large investments both for building the necessary infrastructure and for other operational costs. Thus, manufacturing companies tend to collaborate with other enterprises both for raw materials and for component fabrication. This helps them concentrate more on innovative product design and quality considerations. But, with the mushrooming of small and medium-size enterprises it is difficult to gather information about the ones which can provide the required services reliably and at a competitive cost. In addition, even enterprises which build components rather than finished products need to search for bigger enterprises which require their components. This observation has motivated us to propose an information infrastructure to support manufacturing enterprises in their search for information about enterprises for possible collaboration.

Currently, a number of research projects are underway to develop infrastructures for supporting different functionalities of manufacturing enterprises [1,6,4]. Most of them emphasize the automation of internal processes, the management of internal activities, workflows, supply chain management, etc. and to our knowledge little attention has been paid to developing an information infrastructure that can help manufacturing enterprises collaborate in an effective way. We believe manufacturing enterprises of all sizes can greatly benefit from this kind of infrastructure.

Our model is based on a *Multi-agent paradigm* which is appropriate for inherently distributed application domains. We formally specify the requirements of such an infrastructure using the RAISE (Rigorous Approach to Industrial Software Engineering) Specification Language, RSL [5].

2 A MULTI-AGENT MODEL

In recent years, *Agent technology* has been propounded as a promising approach to handle applications that are information-rich and involve repetitive information processing. The *Multi-agent computing paradigm* makes use of this technology for modelling application domains which are inherently distributed. The basic agent architecture for the proposed model is depicted in Figure 1.

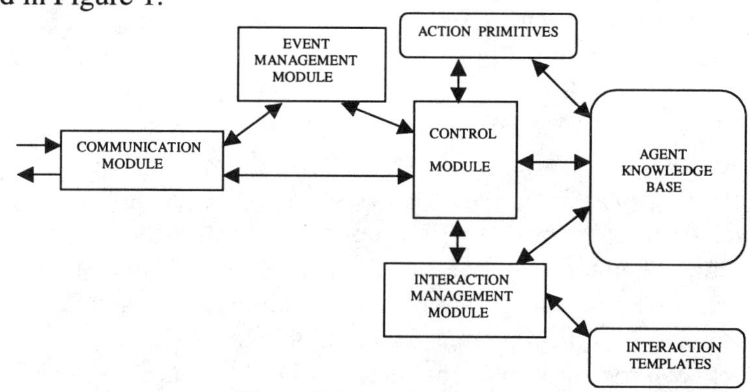

Figure 1 Agent Architecture

The agent software is conceptually organised in seven major components: *communication module*, *event management module*, *control module*, *interaction management module*, *agent knowledge base*, a library of *action primitives*, and a repository of *interaction templates*. The communication module, which is augmented with a *message-in-queue* and a message-out-queue, manages all incoming and outgoing messages. The behavior of an agent is modeled in an *event-driven* manner where events are triggered by the arrival of messages. The event management module identifies events, sequences them (based on priority) before passing them to the control module, and signals errors in the case of unknown events. At the core of an agent software is the control module which is in fact an inference engine responsible for the control of the software. When an event is signaled the control module triggers the appropriate action. The actions are mostly predefined at the time of agent design and are maintained as a library of action primitives. The interaction management module monitors all ongoing interactions at an agent. The agent knowledge base is the information repository for an agent which encompasses knowledge of its own capabilities and of other agents with which it interacts.

2.1 The Agents

The information infrastructure that we model is depicted as a multi-agent environment. The environment is populated by enterprises of varying sizes involved in some form of manufacturing activity, namely fabricating components or making finished goods. The enterprises interact with each other through their personalized software agents.

We identify three different roles of agents: *buyers*, which represent enterprises intending to purchase components or products; *sellers*, which represent enterprises which are offering to supply components or products; and *brokers*, which serve as intermediaries between buyers and sellers to facilitate the buying and selling process. Any enterprise can of course adopt one or more of these different roles at any given time.

Both buyers and sellers can register with brokers, who maintain information about them and the products they are interested in buying or selling. Brokers may also pass requests for products they do not deal in themselves on to other brokers. Both buyers and sellers thus effectively have access to all the enterprises which are registered with the same brokers as themselves. This eases the search for information about smaller enterprises from whom a required component can be obtained.

Brokers can be useful to buyers for both *product brokering*, which addresses the problem of *what to buy*, and *merchant brokering*, which helps to determine from *whom to buy*. When sellers advertise their products to brokers, this information can be passed on to prospective buyers, and when buyers contact a broker regarding products or components they are interested in buying the broker can pass this information on to prospective sellers. Brokers may additionally act as mediators for any purchase deal that they arrange in this way. We include all these aspects in our model.

3 THE FORMAL MODEL

We now give an overview of our formal specification of the proposed information infrastructure using the RAISE specification language. Full details, which are omitted here for brevity, can be found in [3].

The abstract types *BuyerId*, *SellerId* and *BrokerId* are used to uniquely identify the buyer, seller, and broker agents respectively. Similarly *ProductId* uniquely identifies products, the details of each product (its category, brand, features, etc.) being specified by the record type *ProdDesc* with which the *ProductId* is associated in the map type *Products*.

```
BuyerId, SellerId, BrokerId, ProductId,
ProdDesc ::
    prodName : ProdName
    prodCategory : ProdCat
    brand : Brand
    feature : Feature ↔ chg_feature,
Products = ProductId  m→  ProdDesc
```

The information related to a seller includes the brokers with which it is registered together with details about the products it can supply, including their basic cost, discount rates, and the quantity of stock available. Since a seller can deal with many products we use the map type *MyOffer* to record this information for all the seller's products. The map type *Sellers* records information about all sellers in the market. Buyers are specified similarly, though we omit the details here for reasons of space.

```
SellerAddr, SellerBankDetail,
SellerInfo ::
        sbrokers : BrokerId-set ↔ chg_sbk
        sAddr : SellerAddr ↔ chg_sAddr
        sBank : SellerBankDetail ↔ chg_sBank
        canSell : MyOffer ↔ chg_canSell,
ProdOffer ::
        basicCost : Nat ↔ chg_cost
        discount : Nat ↔ chg_disc
        stockQty : Nat ↔ chg_qty,
MyOffer = ProductId  m→  ProdOffer,
Sellers = SellerId  m→  SellerInfo
```

In order to implement the required functionality of a broker, the broker must maintain information pertaining to its registered buyers and sellers as well as other brokers which it deals with. These form three fields of the record type *BrokerDB0* defined below. For the remaining two fields, *notavailProd* records information about products about which buyers have enquired but which the broker cannot currently supply - this information is retained while the broker attempts to find a broker or a new seller who can deal with the request. Finally, the field *trans* records information about transactions which the broker is mediating. Each transaction is allocated a unique *TransId* and the details of the transaction - the buyer, seller and product involved together with the current status of the transaction - are recorded in the associated *Transac*. Various consistency or well-formedness conditions on brokers, for example that transactions can only involve registered buyers and sellers, are embodied in the function *is_wf_broker*, details of which are omitted here for brevity.

```
TransId,
Transac ::
  prod : ProductId
  buyer : BuyerId
  seller : SellerId
  status : TransState,
TransState == orderRecvd | shipmentConfd,
BrokerDB0 ::
  buyerList : BuyerId-set ↔ chg_buyerList
  sellerList : SellerId-set ↔ chg_sellerList
  notavailProd : ProductId  m→  BuyerId-set ↔ chg_notavail
  brokerList : BrokerId-set ↔ chg_brokerList
  trans : TransId  m→  Transac ↔ chg_trans
  no_of_quer : Nat ↔ chg_quer
  no_of_purs : Nat ↔ chg_purs,
BrokerDB = {| bdb : BrokerDB0 • is_wf_broker(bdb) |},
Brokers = BrokerId  m→  BrokerDB
```

The information infrastructure as a whole is viewed as a market place populated by buyers, sellers and brokers dealing in particular products. We specify this using the record type *Market0*, which is composed of each of these four components. Again, consistency conditions, for example that the buyers and sellers registered with a broker must belong to the market, are captured by the well-formedness function is_wf_market.

```
Market0 ::
  buyers : Buyers ↔ chg_buyers
  sellers : Sellers ↔ chg_sellers
  brokers : Brokers ↔ chg_brokers
  products : Products ↔ chg_products,
Market = {| mk : Market0 • is_wf_market(mk) |}
```

3.1 A Dynamic View of the Model

Agents join and leave the market, new products are introduced, existing products become unavailable, and so on. The possibility of such changes is incorporated into the model by defining functions which act on the data types defined above. Each function corresponds to one of the possible actions that can be performed in the market and has an appropriate precondition to ensure that the consistency conditions are preserved when it is executed.

We have defined a set of basic functions for updating information available in the electronic market. These include functions for registering a buyer or a seller with a broker; adding a new product to or removing an existing product from a seller's product range or a buyer's interests; recording a request from a buyer to a broker for a product which the broker cannot supply directly (by storing the request in the broker's *notavailProd* field; changing the details (address, bank details, etc. relating to buyers and sellers; and changing a seller's offer price, discount, etc. for a product. We explain one of these functions here. Other functions can be found in [3].

The function *add_Seller* is used to register a new seller with a broker. The seller is added to the broker's *sellerList* and the broker is added to the seller's *sbrokers*. The precondition ensures that both the broker and the seller belong to the market and that the seller is not already registered with the broker.

```
add_Seller : BrokerId × SellerId × Market ~→ Market
add_Seller(kid,sid,mk) ≡
 let nbinf =
   chg_sellerList(sellerList((brokers(mk))(kid)) ∪ {sid},(brokers(mk))(kid)),
   nkinf = chg_sbk(sbrokers((sellers(mk))(sid)) ∪ {kid},(sellers(mk))(sid)),
   cs = chg_sellers(sellers(mk) † [sid → nkinf], mk)
 in
   chg_brokers(brokers(mk) † [kid → nbinf], cs)
 end
 pre is_valid_broker(kid,mk) ∧ sid ∈ dom sellers(mk) ∧
      sid ∉ sellerList((brokers(mk))(kid)) ∧ kid ∉ sbrokers((sellers(mk))(sid))
```

3.2 Seller Selection

In a similar way, we have specified interactions between buyers, brokers and sellers, beginning with the request from a buyer to a broker regarding availability of a product through to the brokering of the sale. When a buyer contacts a broker for information about sellers of a product, it may wish to know all the sellers who are registered with the broker with a view to select a particular seller from this list itself, or it might provide the broker with some criteria for selecting an appropriate seller and ask the broker to select a seller based on these criteria. The appropriateness of a selection would depend on how closely it satisfied the criteria. In the following specification the variant type *SelMode* captures these two selection options. The record type

SellerReport gives the format of the response to the buyer's query.

```
SelMode == giveAll(ProductId) | preferential(ProductId, Criteria),
SellerReport ::
        prodId : ProductId
          sellerid : SellerId
          prodinfo : ProdOffer,
                    Reply == replyFormat(SellerReport-set) | no_seller_found
```

The function *prodSellers* finds all the sellers of a specified product that are registered with a particular broker, while the function *applyCriteria* takes the buyer's specified criteria as an additional input parameter and selects only those sellers that satisfy the criteria.

```
prodSellers : BrokerDB × ProductId × Market → SellerId-set
prodSellers(db,pid,mk) ≡
 {sid | sid: SellerId • sid ∈ dom sellers(mk) ∧ sid isin sellerList(db) ∧
                                pid ∈ dom canSell((sellers(mk))(sid))},
applyCriteria : Criteria × SellerId-set × ProductId → SellerId-set,
```

The function *select_seller* uses these functions to determine the appropriate set of sellers which match the request from a buyer.

```
select_seller : SelMode × BrokerDB × Market → AgentId-set
select_seller(sm, db, mk) ≡
 case sm of
         giveAll(pid) → prodSellers(db,pid,mk),
         preferential(pid, cr) → let f = prodSellers(db,pid,mk)
                                  in applyCriteria(cr, f, pid)
                                    end
 end
```

4 AGENT COMMUNICATION

In order for the information infrastructure to be effective, agents must communicate with each other in an efficient way. We enforce structured communication among the agents by defining a set of communication primitives. These primitives are similar to the ones used in the KQML [2]. Some of the primitives that we define to facilitate agent communication include: *subscribe* and *unsubscribe* for agents to subscribe to or unsubscribe from services offered by a broker; *advertise* for agents to advertise their capabilities, e.g. the products they deal with; *enquire* for agents to submit a query about a particular product; *reply* to respond to queries; *request* for agents to request a particular service; and *commit* for agents to commit to provide a requested service. The interaction of the agents is governed by a set of protocols such that the agents adhere to a well-defined pattern of communication. By well-defined we mean that agents know *what to communicate, when to communicate*}, and *with whom to communicate*. We do not address the issue of *how to communicate* because this as a lower-level network communication problem which involves message communication through Internet or Intranets whereas our emphasis is on the high-level communication which can be built on top of these lower-level details.

5 CONCLUSION

In this paper we have proposed a multi-agent based information infrastructure which can be used for the benefit of the manufacturing enterprises who would like to collaborate with each other. The formal approach adopted in specifying different aspects of the infrastructure ensures conceptual clarity in the design level and avoids inconsistencies during implementation.

REFERENCES

[1] Bussmann S. Daimler AG. (1998). An Agent-Oriented Architecture for Holonic Manufacturing Control, Ist Open workshop IMS Europe, Lausanne, Switzerland

[2] Finin T. et al. (1992). Specification of the KQML Agent Communication language, The DARPA Knowledge Sharing Initiative, External Interfaces Working Group

[3] Patra M.R., Moore R. (2000). A Formal Model of an Agent-mediated Electronic Market, UNU/IIST Tech. Report No. 211, Macau, August 2000

[4] Shen W., Xue D., Norrie D.H. (1997). An Agent-based Manufacturing Enterprise Infrastructure for Distributed Integrated Intelligent Manufacturing Systems, In Proc. of the 3rd Int. conf. on the practical application of Intelligent Agents and Multi-agents, London, UK, March 23-25, 533-548

[5] The RAISE Language Group (1992). {\em The RAISE Specification Language}, Prentice Hall International (UK) Ltd.

[6] Vision of the National Industrial Information Infrastructure Protocols (NIIP), http://www.niip.org/vision.html

A Cost Estimation Tool Integrated into FIPER

David Koonce, Robert Judd, Thomas Keyser
Ohio University, Athens OH, 45701 USA
Em: koonce@ohio.edu.

Keywords FIPER, Cost Estimation,

Abstract The estimation of a part's manufacturing cost in all phases of the design stage is crucial to concurrent engineering. To better estimate the cost for a product, data must be available from both engineering systems and business systems. This paper presents a cost estimation system being developed to support design time cost estimation using FIPER; the Federated Intelligent Product Environment, which is being developed as part of the NIST Advanced Technology Program. The FIPER project will be developing an architecture that interconnects design and analysis software tools in a peer level architecture using JAVA to support Multidisciplinary Design Optimisation, Design for Six Sigma and Robust Design.

INTRODUCTION

One key constraint in engineering design is product cost. However, estimation of the cost of a yet to be produced part or product is a difficult process. The NIST ATP sponsored FIPER project allows for a new and truly integrated cost estimation process for aiding design time cost analysis. FIPER will unify design tools for optimisation across multiple analytical disciplines. For cost to be included in this optimisation, a new and highly integrated cost estimation tool must be available. This tool will include the capability of developing cost estimates of a product from whatever data is available in the environment. In the early stages, simple parametric cost relations, like weight-based relations may be used. As more information is developed, cost estimates will be developed on design features, manufacturing features or even a developed process plan will automatically be substituted for the parametric relations. All estimates will also be adjusted for bias using a scaling factor determined by comparing the estimated cost of a similar existing part to the known actual cost of that part.

This paper presents the design for a new highly customisable cost integration tool to support design time optimisation that considers cost as a constraint.

FIPER

FIPER, an acronym for Federated Intelligent Production EnviRonment, will utilize a technology called Intelligent Master Modelling (IMM) to allow design engineers to reduce the time to evaluate potential designs across all analytical disciplines. FIPER will support advanced design methodologies such as DFSS (Design For Six Sigma), MDO (Multidisciplinary Design Optimisation) and robust design. While the IMM helps coordinate design and analysis, a supporting architecture will be built upon the concepts of federated information systems. The FIPER infrastructure is being developed entirely in JAVA to support the mixed computing platforms typical in product design.

FIPER is being developed as part of a four-year National Institute of Standards and Technology (NIST) Advanced Technology Program (ATP). The development team consists of the General Electric Corporation, BFGoodrich Aerospace Aerostructures Group, Engineous Software, Parker Hannifin Corporation, Ohio University, Stanford University and the Ohio Aerospace Institute. In specifying the capabilities and services in FIPER, the development team addressed five problems common to many businesses in the United States:

- the need to reduce time to market,
- the need to reduce design cycle time,
- the need to reduce product costs,
- the need to improve product performance, and
- the need to improve product quality and reliability.

As a consequence of the need to reduce production costs, FIPER must have the ability to accurately predict the cost of a potential design. The software tool that produces the estimates must operate in the FIPER environment and be able to, with no user interaction, generate a cost estimate as a service to a calling program.

MANUFACTURING COST ESTIMATION

It is often posited that the major portion of a product's cost, as much as 80%, is determined early in the design process. Decisions like material can easily be seen to impact cost. However, decisions like a radius or blend may result in the need for a tool change, new setup or even a processing machine change, adding to the manufacturing cost of a part. As such, producibility often figures into the estimation of a part's cost. Regardless, functional specifications usually drive the design process [1].

Despite the importance of the design details to product cost, a recent study found the delay between the design decision and cost determination hindered the designer's ability to learn about the process implications of design decisions [2]. In addition, the consequences of the costly decisions were often not fed back to designers at all [2].

Numerous commercial cost estimation tools exist and many organizations have developed proprietary cost estimation systems. The sophistication of these tools ranges from spreadsheets to multi-user mainframe database systems. The capability of these systems ranges from the ability to estimate costs for highly specific parts to generic systems which can be used to estimate costs for virtually any manufactured part. Regardless of the sophistication or size of the system, the manufacturing cost of a part is estimated using one or more of four basic methods: intuitive, analogous, parametric and analytical [3].

The intuitive approach relies on the experience of the estimator to predict the cost. An analogical estimate is essentially a variant estimate using similar parts, often matched using a group technology code. Parametric estimates use the values of key part attributes to determine the cost. A parametric estimate may rely on very high-level parameters of the product's performance or use detailed geometric data. Lastly, an analytical estimate relies on a summation of the steps in the production process; and as such, can only be done late in the design process.

A newer trend in cost estimation is the inclusion of manufacturing system costs in the estimation of a part's cost. Various methods in this area include: activity based costing, throughput accounting, target costing, life-cycle costing and strategic accounting [4]. For example, traditional activity based costing (ABC) can be further decomposed into activity based costs such as processing cost and non-activity based costs such as inventory holding costs [5].

Costs such as these are usually buried in an overhead factor. Overhead is also an area of cost estimation research with the notion that product complexity is the primary driver for overhead [6]. This study indicated that production volume and the number of transactions like engineering change orders strongly correlated with manufacturing overhead costs. For example, production volume may necessitate a need for increasing capacity or even process capability in the manufacturing system.

COST TOOL ARCHITECTURE

The architecture of the FIPER cost estimation tool (FCET) is largely influenced by the cost estimation methodologies employed. FCET will use a combination of generative and variant costing, with designs being evaluated using either a work breakdown structure or a parameter based estimation from a similar part. This dual estimation approach will require that the tool maintain a repository of existing parts, with costs and match parameters, as well as traditional cost estimation equations and associated data files.

The FCET will consist of three separate user interfaces, a cost estimation engine and an associated database for storing estimates, and operational data. The three interfaces are the module builder, the template builder and the cost estimator. The end user interface, the cost estimator, will allow the estimation of parts using predefined structures to produce high-fidelity estimates. Figure 1 shows a simplified architecture on the tool.

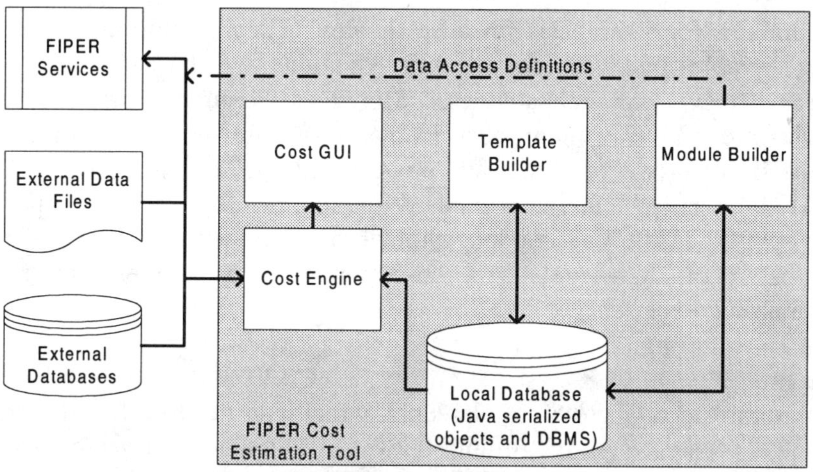

Figure 1. Tool Architecture

The Development Environment

The developer is presented with two tools: the module builder and the template builder. Dividing the development environment allows the separation between how costs are estimated from the view that the user is presented. This is analogous to database design, where the design of the schema is separate from the design of the end user data access forms. This allows for different views of the estimate, based on user needs. (e.g. an engineer may need a different view than a manager)

User Environment

The user is presented with a tool that can call up any template. The costs in the template are divided into elements with focus on a work breakdown structure. For detailed designs, a work breakdown structure (WBS) can be utilized to generate detailed cost estimates. The WBS is a hierarchy consisting of three classes of modules: operations, components and assemblies. The WBS is generated for a specific estimate with pre-defined template.

Associated with each element is a group of attributes that are set by the user. These attributes may also be a default value or take their value from an element higher in the WBS. The template specifies a default value or relationships. For example, a high-level element may specify a material attribute, which is used for the material attribute in all machining sub-elements.

Modules

Modules form the basis for all estimates. They contain the definitions of the variables necessary for the cost equations and the definitions to the linkages to external data sources. Stored as executable JAVA objects, modules are usable in many cost estimates and are the instantiation of an organization's product families and processing logic.

Modules can use any cost method: parametric, feature based, operation based, ABC, etc. The module builder allows the developer to create modules in a high-level GUI environment, which does not require programming experience.

Templates

A template is a group of elements, which can be used to estimate the cost of a product. At the simplest level, a template is a set of wrapped modules that are linked with defined relationships, defaults and mappings that enable the costs to be rolled-up into a compete cost estimate. Elements can be configured with:

- a set of input attributes (with defaults),
- a set of buckets (for summing calculated values),
- a set of fixed sub-elements,
- a set of optional sub-elements,
- a set of modules,
- mappings between element variables and sub-element variables, and
- matching rules for scaling an estimate.

Templates allow for interaction between the variables and equations in the modules that make up a cost estimate. For example, a material variable in one element (like a machining operation module) may be linked to the material of a higher-level element. The linkage then allows a material change to cascade through the cost estimate.

Templates are not generated at the time of estimation. They are a library of products and processing steps predefined by cost estimation experts. While a basic set of process templates may be applicable to many organizations, it will be necessary that the FCET be customized for each implementation.

Templates can also be used to present the user with a view different from than what is provided in the constituent modules. For example, the element attributes may present a geometric feature with the element attributes while the modules are operation based. All necessary mappings and translations are maintained in the template structure.

Risk Analysis

The cost estimation tool will have the ability to perform risk analysis. The risk analysis will be done using Monte Carlo simulation. Given a set of inputs that may be any combination of deterministic and probabilistic values the Monte Carlo simulation will generate minimum, maximum and average cost estimates. Additionally, the variance will be reported as well as a graphical representation of the resultant distribution. The cost estimates will be based on user defined confidence interval levels (e.g. 90%, 95%...) or a user defined run length (number of simulated trials).

Scaling

Key to cost estimation is the minimization of error in the equations estimating the cost. One method for minimizing error is to use cost estimates of existing parts. By comparing computed cost with actual cost, a measure of the error can be determined and applied to a new estimate for a similar product. Figure 3 shows the equation for scaling a cost estimate using a closely matched part. The keys to scaling are one, having a complete and accurate database of current part costs and two, determining which existing part most closely matched the planned part.

$$C_S = C_U \frac{M_A}{M_E}$$

Where,

C_S Scaled cost estimate

C_U Estimated cost of the part

M_A Actual cost of matched part

M_E Estimated cost of matched part

Figure 2. Cost Scaling Equation

While the calculation for scaling is not complicated, determining a closely matching part may be difficult. A set of matching parameters must be defined for each element. For scaling to work, the error in the equations must be duplicated with both sets of calculations. For example, if two parts with very similar parameters use significantly different materials, then the method for estimating the costs may be entirely different. In this case, scaling the new part against the old part could add only random error. Beyond scaling processing cost, labour scaling is possible if data is available. However, for purchased parts, material and labour are combined which makes decomposition for scaling more difficult.

Trade Studies

The cost estimation tool will also store and track trade studies. Engineers will be able to set a reference point and then perform what-if analysis using multiple scenarios. At any time the user may reset the cost of any element or sub-element to a stored value in the trade study.

CONCLUSIONS

This paper has presented an architecture for a new cost estimation tool capable of generating estimates at all stages of the design process. FIPER presents an excellent opportunity for the creation of a design time cost estimation tool. With FIPER providing the necessary design data and a directed search toward optimality with respect to cost and other constraints, the FIPER cost estimation tool will help designers reduce costs while meeting performance constraints.

REFERENCES

[1] Locascio, A. (2000). Manufacturing Cost Modelling for Product Design, *Journal of Flexible Manufacturing Systems*, **12**, 207-217

[2] Busby, J.S. (1997). The Limited Informativeness of Resource Discrepancy Feedback to Designers, *Journal of Operations and Production Management*, **17**, 6, 630-646

[3] Duverlie, P. and Castelain, P. (1999). Cost Estimation During Design Step: Parametric Method versus Case Based Reasoning Method, *Journal of Advanced Manufacturing Technology*, **15**, 895-906

[4] Boons, A. (1998), Product Costing for Complex Manufacturing Systems, *Journal of Production Economics*, **55**, 241-255

[5] Park, C. and Kim, G. (1995). An Economic Model for Advanced Manufacturing Systems Using Activity Based Costing, *Journal of Manufacturing Systems*, **14**, 6.

[6] Banker, R.D., Potter, G., and Schroeder, R.G. (1995). An Empirical Analysis of Manufacturing Overhead Cost Drivers, *Journal of Accounting and Economics*, **19**, 115-137

Collaboration and Application Integration: Distributed Design with Virtual CAD

P. Bertok
Department of Computer Science, Royal Melbourne Institute of Technology, Australia
Em: pbertok@rmit.edu.au

J.P.T. Mo, S. Woodman
CSIRO Manufacturing Science and Technology, Australia

Keywords CAD, Application integration

Abstract The pervasiveness of computer networking and rapid progress in computer performance have made global cooperation of designers a real possibility. While these technologies provide the hardware infrastructure, software development is struggling to keep pace with these developments. There are many legacy applications and competing new developments, but unfortunately most of those operate on specific platforms in isolated environments. Cooperation of designers working on those systems is very difficult; a significant problem being that different data converter facilities are needed to exchange, interpret and combine designs components produced on different systems. The system presented in this paper aims at integrating different CAD systems and achieve true interoperability, where components designed in one system can be easily viewed, modified or integrated into other designs. The approach taken was creating an integration platform, and different CAD systems are integrated into a virtual CAD system in a seamless manner. The original CAD systems store objects designed locally on that system, while remote access is provided via an integration layer.

1 INTRODUCTION

Traditionally design has been a cooperative activity, but early CAD systems were designed for single users and without sharing information with other systems. Later CAD systems became more open as they provided interfaces for data exchange to other programs and utilities, and multiuser operation has also become possible. While these developments already allowed designer cooperation, system heterogeneity still remains a problem. As there are many CAD systems in use, it is quite common in large scale global projects involving several design teams around the world that different groups use different CAD systems [1]. Solving the problem of designer cooperation is becoming more and more urgent. While global connectivity is already a reality, the infrastructure to connect any computer to any other computer in the world exists; its use in cooperative work is still in its infancy. This paper addresses this issue by proposing a common user interface that can access different CAD systems and can use their native CAD model data. In essence, the system described here was developed to facilitate unambiguous information flow in engineering projects.

2. THE VIRTUAL CAD SYSTEM

The primary objective of the "Virtual CAD" project is to develop a system that allows remote users around the world to access CAD models the same way as local users do, even though the design data may be stored in different CAD systems in different formats. The Virtual CAD system was to offer a unified interface to the user and add a convenient remote design facility while retaining useful CAD features.

Additional functionality was implemented in the unified interface, and retaining the original features was achieved by integrating the original CAD systems without any modifications. All components were integrated by using special interfacing, or wrapping software, and common utilities provided a uniform appearance. In the process existing software was used wherever possible, which included the integration platform as well. Several methods and systems were considered as integration platform, and the selection was based on user requirements, which included the required services and functionality, the preferred user interface and the user environment.

First a world-wide-web based solution was considered. Web-browsers are available under most platforms, provide convenient easy-to-use user interfaces and can display arbitrary data types by using plug-ins, hence enabling component based integration. A web-based model viewer could be easily implemented and could provide an easily expandable platform by adding new plug-ins for new model data types. The real shortcoming of a web-browser-based solution was manifested in handling modifications to design data. While data of significant complexity, e.g. graphical or even audio data, can be transferred from server to client, transferring data from client to server is restricted primarily to textual information e.g. via CGI scripts. The Virtual CAD system was intended to provide an active front-end at the user sites and allow users to modify different design data, however, changing design data via textual data input is very cumbersome. Based on this evaluation, the idea of web-based integration was abandoned at a very early stage.

Next, platforms providing integration on the procedure level were examined. Procedure-level integration enabled a more versatile architecture that included a user interface providing design data manipulation. This meant accessing procedures in the individual CAD systems, and also adding procedures for remote data display and manipulation. An aim was to integrate existing CAD systems, so that the Virtual CAD system will include their capabilities and enhance them.

Distributed object technology (DOT) has been widely accepted as enabling technology for distributed software improvement and re-use [2]. DOT has been used to create middleware, a middle layer software between components such as clients and servers. Middleware offers an integration platform for diverse applications, including legacy systems. DOT also uses encapsulation technique to preserve the original essence of the legacy system and to provide an external interface. The encapsulation approach has become a widely applied method and is known as *wrapping*.

When selecting a platform, CORBA and the Distributed Computing Environment (DCE) were looked at. CORBA soon became the final choice, as CORBA offered a real object oriented framework with many services available. In addition, the implementation chosen (Orbix) was available under the operating systems in the user environment, making the implementation work clear and straightforward.

3. CORBA

The Common Object Request Broker Architecture (CORBA) is a one of the most commonly used middleware. It provides an object-oriented framework that provides a universal communication infrastructure supporting a variety of object mechanisms. CORBA was designed to deal with heterogeneous components and to integrate them into cohesive distributed systems. CORBA suits client-server architectures in particular, clients being the consumers of resources provided by servers. In CORBA, as in any object-oriented system, every entity in a running program is viewed as an object with a message-handling interface. The interface specifies the behaviour of the object and it is kept separate from the implementation. Objects are encapsulated and the implementation is usually hidden from the user/client, which enables modifications or replacement of the implementations behind the interfaces.

CORBA achieves interoperability through language independence. CORBA defines a special interface definition language (IDL) that has mappings into every major programming language. Object interfaces describe operations and associated attributes in IDL. They translate functionality offered by the server into object-oriented specifications required by the object management system. An interface described in IDL constitutes a contract between client and server, both client and server use the same interface definition to construct the executable code.

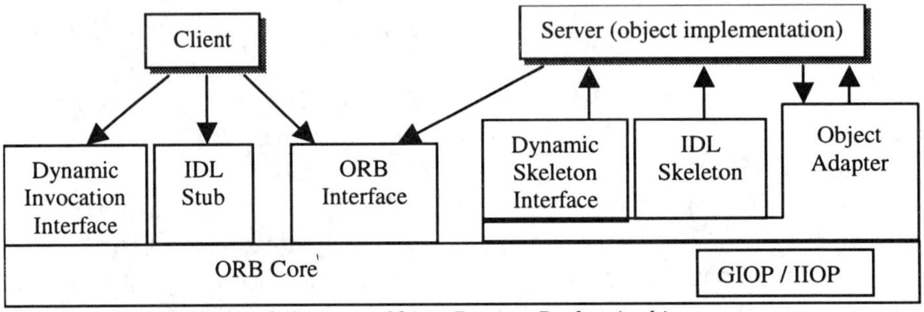

Figure 1 Common Object Request Broker Architecture

The most essential components of a CORBA system are the Object Request Brokers (ORB), which provide a communication bus connecting the different components, as shown in Figure 1. ORBs provide a mechanism to invoke a method on a remote object: an ORB locates the object, activates it if necessary and communicates the client's request to the object. ORBs support static and dynamic invocation interfaces. With static invocation the client knows the server's interface beforehand, while with dynamic invocation the client learns the server's interface through inquiries during run-time.

Interoperability of ORBs is supported by inter-ORB protocols. The General Inter-ORB Protocol (GIOP) has been developed to allow communication between independently developed software modules, without explicitly identifying the underlying network. GIOP works over any transport protocol that satisfies a minimum set of requirements. The Internet Inter-ORB protocol (IIOP) uses GIOP with TCP/IP protocol stack for transport, which is more restrictive but has become very popular due to the pervasiveness of the Internet.

To support different object domains CORBA defines the interoperable object reference (IOR). IORs are used when crossing object reference domain (ORB) boundaries, as there are no general rules for object implementation, every ORB can implement an object reference in a way that is the most convenient locally.

CORBA was designed with interoperability in mind, and provides an excellent platform for integration. Independently developed components can be easily connected into one system, and it is especially suited to the integration of legacy systems into a collaborative system.

4. DESCRIPTION OF THE VIRTUAL CAD SYSTEM

4.1 System Architecture

The task described in this paper consisted of integrating a commonly used CAD system, AutoCAD, into a larger, CORBA based system, as illustrated in Figure 2. Other CAD systems, such as Pro-Engineer, SolidWorks, IDEAS, are also being considered for future integration. The clients need not be co-located, they can be distributed around the world, and each of them is running the same code, albeit with possibly slightly different local configuration data.

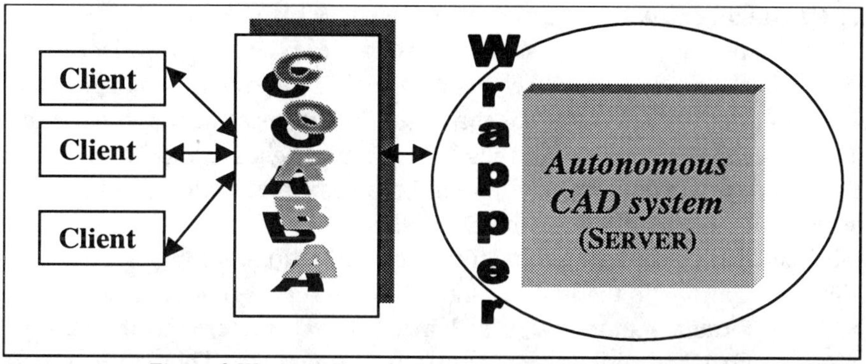

Figure 2 Wrapper Facade Pattern Applied to Virtual CAD

As different clients need access to different CAD functions and data, the so-called Component-First based wrapping approach was used [3]. This approach employs the legacy system by using its modules and application logic in the new environment. System architecture, functionality and interfaces, however, are following the new system design, so the existing logic and code are seamlessly integrated into the new environment.

The work involved a feasibility study that also examined the appropriateness of messages and data transfer. The aim was to make best use of the legacy CAD system's capabilities and of the available bandwidth in the communication network (Internet or Intranet). There was no clear one-to-one correspondence between the services/data provided by the legacy system and those required via the new client interface. However, since developing the client was also part of the project, it was possible to find a quasi-optimal mapping of objects between the old and new systems, which did not require too complex conversions of functions and data; in mathematical terms a many-to-many mapping system was established.

The extensibility of the wrapper was also considered, as modifications, such as including additional functions in the new interface and including other CAD systems, may need to be incorporated at a later stage. Different wrapping patterns were examined, and a Wrapper Façade [4] was eventually implemented, as it provided a simple, extendable structure that implemented many-to-many mapping.

4.2 Implementation

The implementation was divided into two parts, the first one laying down the class foundations and the second one addressing error handling and exceptions.

4.2.1 Clustering functions and data structures into classes

Within the wrapper class several classes were defined to reflect the new user interface and functionality. In many cases, the client was specified to ease the burden of interface conversion, and the local server functions are replicated at the client side. In these cases simple message passing between the two sides was sufficient. Great help was the Java 3D based client interface that could emulate many of the CAD server's 3D-manipulation capabilities [5]. Classes representing these functions simply forward their method invocations to the underlying low-level server functions.

An important feature of the Virtual CAD system is its ability to manipulate CAD primitives and modify CAD models. This is illustrated in Figure 3: the class `BasicPrimitive` defines common interfaces required by a CAD primitive. It associates with the `TransformGroup` and `BranchGroup` classes defined in the Java 3D library. It defines basic interfaces to transform the geometry of a primitive. In addition, this class defines two methods (post-order and pre-order) to traverse this structure. The class `LeafPrimitive` can define a primitive such as standard shapes and CAD models represented in neutral formats. The class `CompositePrimitive` can define a primitive that contains other primitives (defined as children) such as `CompositePrimitive` or `LeafPrimitive` objects. The class `CompositePrimitive` provides interfaces to add and delete children.

These basic primitives support the fundamental user operations in the Virtual CAD environment. For more complex operations, such as assembling a few components in the CAD server, additional levels of indirection are required. A possible way to implement additional indirection is dynamical dispatching of wrapper facade method implementations, which can also increase extensibility. Aggregated functions need to be developed and wrapped in the server by using languages associated with the CAD system.

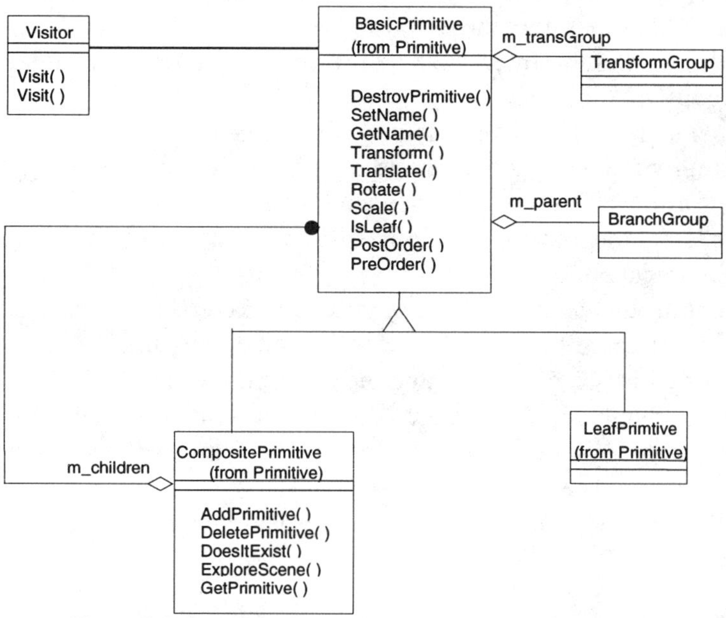

Figure 3 System Structure with CAD Object Manipulating Primitives

4.2.2 Establishing error reporting mechanisms

Handling errors is essential in any application. It can be kept simple by reporting back to the caller only, or can be quite complex when corrective actions may need to be initiated, and the pattern used may need to accommodate that.

In this project it was kept fairly simple. The options were the following. Errors could be communicated back to caller via standard error codes, as done on many conventional platforms. Using exception-handling mechanisms provided by CORBA was another option, but it was more complex and needed to be justified. Currently the exception handling facilities in Java on the client side are used, which provide a comprehensive set of functions for contingencies. In future the error may be communicated back to server. The error handling capabilities of the server will also have to be considered, as the server may be coded in a language different to the client.

5. CONCLUSION

This paper presented a case of a virtual CAD system. The most important features of the system were the following:

- It can integrate different, common-off-the-shelf (COTS) CAD systems into one large virtual CAD system, and provides a seamless data flow from one CAD subsystem into another.
- It provides a common interface to all different CAD systems integrated into virtual-CAD.
- It can operate in a globally distributed environment.

The system was implemented using distributed object technology on a CORBA platform, with the following characteristics.

- Procedures in the individual CAD systems were accessible from the virtual CAD system.
- Implementing the Wrapper Façade software pattern allowed a simple, extendable structure that suited the application environment well.
- The program classes reflected interface and functionality.
- Message passing between client and server was the main form of communication, albeit direct remote method invocations were also necessary in a few cases.
- Error reporting was kept simple, and was implemented in form of messages sent back to the client.

6. REFERENCES

[1] Fowler S. and Karinthi R. (1996). Remote access to CAD databases using an information sharing system, Computers in Industry, Vol.29, 1996, pp.117-122.

[2] Mowbray T.J. and Zahavi R. (1995). The Essential CORBA-Systems Integration Using Distributed Objects, John Wiley & Sons Inc. 1995

[3] Morin T. (1999). Migrating Legacy Systems to CORBA, http://developer.netscape.com:80/viewsource/morin_corba/morin_corba.html, 1999

[4] Schmidt D. (1999). Wrapper Facade, C++ Report, February 1999

[5] Chan P. and Lee R. (1998). The Java Class Libraries. Second Edition, Volume 2, Addison-Wesley, 1998

World Wide Web Adapted Geometric Model in the Context of Functional Design

F. Danesi, C. Dartigues, Y. Gardan
CMCAO, IFTS, Charleville-Mézières, France
Em:{danesi, dartigues}@netcourrier.com, gardan@infonie.fr

E. Perrin
CMCAO, Metz university, Metz, France
Em: perrin@lrim.univ-metz.fr

Keywords World Wide Web, Functional Design, Geometric Modelling

Abstract Our goal is to create a CAD system dedicated to WWW. Offering a CAD system over the WWW is a viable topic that is still too difficult to handle. The main problem is to transfer CAD system data, which are classical huge. Unfortunately this transfer is time and web-line consuming and no specific method exists. This paper begins with a survey of the current field of algorithm reducing data volume, in order to transfer less data on the web. Since they are not dedicated to CAD model, which are still huge after treatments, we propose three transmission methods specific to each main model used in CAD: CSG, octree and B-Rep. We show their limits, which can be pushed back by a functional model.

INTRODUCTION

Current CAD-CAM systems are efficient in their favourite field: geometric modelling. They propose designed interfaces to build parametric objects and to act on shapes. If present systems help the designer when the shape of the object is known, it is regrettable that only parametric and variational designs take into account the designer's know-how. In practice, design intent and methodology are totally given up at the beginning of the computer use. This arises numerous basic drawbacks. The user has to start with geometric constraints and shape's creation and nothing shows that the geometric model is adequate to functions and constraints of the specifications. No interface software tool is available to capture user's know-how and functions of the specifications.

In that context, a tendency rises in order to define less monolithic systems that work with interoperability: virtual companies have emerged due to the World Wide Web extend and communications between server and host (or client) are numerous. For a time, only bitmaps of a 3D scene have been transmitted to the client through compression algorithms. In that case, the server made computer calculations and the client was only showing the results. Nowadays, technique changes: it may be interesting that the client could make specific operations by its own and relieve the server. This means that a geometric description of the 3D scene is then necessary to the client. Unfortunately, geometric description is huge. Thus the entire description can not be transmitted to the client because of communications time. If the image compression has been largely studied, geometric compression research (including algorithmic geometry and data compression) is recent and available papers deal with triangle mesh compression.

FUNCTIONAL WWW CAD SYSTEM

The will to capitalise the know-how and to reduce times of production make that one now truly approaches the CAD-CAM systems in the optics of a functional modelling. Our goals are to assist the designers during the earlier stages of the product design and to create a system dedicated to any kind of firm. The latter needs to take into account the level of computer knowledge of all the employees, who are potential users of the system. The interface will have to integrate intuitive notions in order to capture the user's know how. However the current systems are still based on geometry and in order to achieve the wished goals, it is necessary to delay computations and to introduce higher semantic level concepts [1].

Some commercial firms try to use WWW to distribute their system but commercial CAD systems are heavy to manage, tedious to understand and to learn and require powerful computers. We want to develop a new approach to implement more ergonomic CAD systems, using WWW as a tool. The firm will not have any longer the responsibility of the system management. This will be done by the company that distributes the system on internet; the user will just have to load software components he needs to work; computations will be done by the server computer so time responses will be related only to network and not to host processor's power. Thus, the small companies will be able to reach technologies that were hitherto inaccessible for lack of financial, material and intellectual means.

All of this implies to create a functional model and an adequate interface, to manipulate the model, to translate it into geometrical data to obtain a shape and to transfer it from the server to the host computers. As it is difficult to detail functional and WWW problems at the same time, we will highlight problems resulting from the use on Internet with a geometrical model in the next section. For further details on functional modelling and functions to shape translations, see [2,3].

TRANSFER OF GEOMETRICAL DATA FOR WWW

Our goal is to create a CAD system available on internet with the same characteristics as a classical one, with powerful tools (for example to draw up an estimate of mechanical parts) and with more specific ones useful to a given profession (such as the one to calculate the shape of a mould in foundry [4]). Obviously, the system responses to the user's interactions must be similar to a classical CAD system ones. This problem relies on current network flow, computer memory space and huge size of used geometrical data. This study consists in managing the transfer of geometrical information between the host and the server as quick and transparent as possible. In the next sections, we will show that the methods encountered in the bibliography can not solve problems coming from these constraints and we will suggest solutions for the main CAD models: historic construction, octree and boundary representation.

Overview

In order to work, the CAD system user on a distant host has to receive some topological and geometrical data contained in the model. The product model may become rapidly consequent and it can not be sent as it is.

The compression methods [5,6,7,8,9,10,11,12] permit to reduce the data volume. They are usually used in virtual reality and in medical imaging but they can be adapted for CAD because they commonly work with a mesh. In that case, the CAD model must be translated into a mesh before the method runs [9], which is often realised in a CAD system. Compression methods proceed in two steps. First, they code the topology of the mesh, which gives a list of vertices sorted by proximity. Secondly, for each vertex of the list, they predict its position according to the previous vertices. The vertex code is obtained by subtracting the real position with the predicted one. It can be noted that good results have been obtained in finite element meshes to reduce time calculus by reducing the mesh. In our case, compression methods are not adequate because CAD model must be translated into a mesh, which spends time. Moreover data volume remains large after compression and must be sent once: the user has to wait end of transmission to work and he receives the entire model even if he does not need it.

The simplification methods [13,14,15,16,17,18] send the model in several times and then give an approximate view of the model from the first transmission. The model is then refined step by step. As compression methods, they work on meshes. Before transmission, the mesh is simplified recursively until it reaches the desired level of approximation that is fixed at the beginning of the process. Each iteration gives a level of simplification and affects the model homogeneously. In that case, the user's waiting is compensated by a progressive view of the model but he still can not work before the end of the reconstruction.

The selective simplification methods [19,20,21,22] are more convenient because they adapt the model reconstruction to the user's needs. Parts of the model are chosen according to the user point of view and are firstly reconstructed. Unfortunately modifications he may realise can alter parts of the model that have not been yet transmitted. Model maintenance problems may occur between the host and the server computers. It is also damageable that selective simplification methods rely on meshes.

Nevertheless, selective simplification methods give a way of thinking to an intelligent system taken into account the user's will, that is here reduced to point of view. However, this intelligence may increase if more information on the user's know-how are given to the interface. The latter is then able to interpret and control the will instead of only capturing it: if a modification alters a part of the model that has not been yet transmitted, then the interface can detect it and anticipate the transmission. In that case, the interface has no longer this passive role of translator it used to have between user actions and system orders. This purpose would not be done without a functional model. As we have mentioned it, we only focus at the moment on geometrical models. So we restrict our next study to the three mains geometrical model used in CAD in order to choose the most convenient.

Dedicated methods to geometrical CAD models

The methods of the section 0 work with meshes. However, the three main geometrical models used in CAD are constructive solid geometry, octree and boundary representation. They could be translated into meshes but it is time consuming and more important, information losing. Because of the lack of study about transfer with the three main models, we then propose three dedicated methods to them and show their limits.

Our method to transfer Constructive Solid Geometry

CSG (Constructive Solid Geometry) is a graph, whose nodes are Boolean operations and leaves are simple geometric primitives. CSG represents the model implicitly. Historic construction is the same way of representation as CSG but it is used with form features (like slots, bosses or blind holes): instead of primitives, features are used. The order of features defines the order of design. Its advantage is then to contain higher level information that corresponds to the user's design intent. That is the reason why we prefer studying historic construction rather than CSG.

To add a feature, the user has to view the object he is constructing in order to position the feature. This means that the historic construction has to be translated into an explicit model, in general into a boundary representation. The position of the feature is, for example, relative to a face of the object that is selected by the user. Then the form feature is instantiated by a generic procedure, whose parameters are defined by the user and the new object is created. Only the kind of form features (which can be coded by a number), its relative position and their associate parameters are necessary to transfer. The relative position is related to the explicit model. As the transfer must be as quick as possible, we want to avoid sending the explicit model. To realise this, the relative position has to be coded as an absolute position, which is totally defined by the viewing parameters (eye's co-ordinates and sighting axis co-ordinates) and the mouse co-ordinates.

The historic construction is compact and is obviously rapid to transfer. Nevertheless, the explicit model has to be reconstructed just after the transmission, which costs time. As seen in the overview (in section 0), simplification and above all selective simplification can be used to reduce this problem. Selective simplification is totally adequate to our case because it takes into account the user point of view, which has been saved for the transmission of the model. The idea is then to select only the form features with a close point of view of the host current one. This operation raises to a major problem: selecting the form features according to the point of view neglects interference between features. In that case, adding a new feature may perturb the design intent and then arises to a wrong solution.

Our method to transfer octree

Octree is not an exact model but it is often used when it facilitates graphic treatments. It is a space recursive decomposition into black, white or grey cubes. The colour of the cube expresses its belonging (black) or not (white) to the object. A grey cube has to be subdivided. So, its main drawback is that it is memory space consuming. To reduce this problem, we propose a method to calculate the biggest black boxes (instead of cubes) at each decomposition step. The principle of this method is to subdivide a box according to [Ox) axis, then to subdivide independently the two resulting boxes according to [Oy) axis and to iterate with the four resulting boxes according to [Oz) axis. Subdivision according to an axis is realised by choosing the best section plan in order to obtain the biggest black box. Subdivision can be avoided when it may start with a black or white box.

The created structure is then a degenerated octree. It is more compact than a classical one and then is quicker to send even if it is more time consuming to build. The degenerated octree is naturally adapted for selective simplification. At the beginning of the session, a tree corresponding to a low level of decomposition is sent to the host. In visualising it, the host is able to detect which boxes are visible and then asks the server their next level of decomposition in order to obtain a more precise visualisation.

However, even if the user views exactly his object, he could not modify it before he receives the exact model too, which could not be reconstructed from the octree without losing information. So it has to be transferred.

Our method to transfer Boundary representation

B-Rep (Boundary representation) is a structure, which contains all the vertices, edges and faces of the object that it represents exactly (when it is polyhedral) and explicitly. We have shown, in the two previous sections, that an exact representation is required in a CAD system. Unfortunately, B-Rep is often a huge structure and it is inconceivable to send it once. We have overviewed that the methods in section 0 are not adapted to our problem. To reduce the user delay time and to let it work as soon as possible, only the visible faces can be sent to the host. The number of visible faces is relative to the complexity of the object according the user point of view. In general, the number of visible faces is very less than the number of total faces but, for the reason mentioned above, this can be consequent anyway.

To avoid this problem, we propose to send first an image to the host instead of the model. This principle can be used each time the user changes of point of view. This can be done in three steps as following: first, the server asks the point of view to the host. Secondly, the server calculates the corresponding image: this can be done actually fast in using appropriate processors. Finally, the server sends the image and the host shows it.

The needed part of the model can be sent later during the time that the user is watching to the object. If this delay is too short, the user may interact before the transmission ends. In that case, the needed faces, that are dependent of the interaction, may have been received or not. If not, the appropriate faces must be selected by the server and sent to the host. A local image can be then calculated and the global image can change following the user interaction. But finding the adequate faces (and then finding the parts of the image that must be recalculated) is a main problem because the faces are not only the ones directly concerned by the interaction (the user shows a face) but also the ones modified by the interaction (some faces may become visible because of the modification of another). In a B-Rep, only topological and geometrical information is saved and this problem concerns higher level information and calls for a functional model: for example, it would be easier to know that a given face is behind another face [23].

CONCLUSION

Our goal is to create a CAD system dedicated to WWW with the same properties as a classical one. Such an aspiration needs to solve data transfer time, quantity of transferred data, etc. The current methods in the state of the art dedicated to rapid data transfer concern compression and medical imaging. Unlucky, no method is specific to CAD system because this problem raises with particular models such as CSG, octree and boundary representation that are not used for compression and medical images. We have shown that CSG and octree, even if they can be compact, are inapt to let the user works because they are not exact representation. Boundary representation, that is an exact representation is unfortunately huge and can not be send in totality. If images are used to reduce data quantity in that case, this treatment needs higher level information, that boundary representation does not possess. Such information would be saved in a functional model. However, even if only geometric models are used, we have shown that by taking into account the inherent properties of each one, it is possible to obtain good results using, basically, the three following directions. First, modify the model in order to adapt it to the future treatments. Secondly, take the user point of view as a fundamental information to know what to send in priority. Finally, send information step by step, enriching the details concerned by the user interaction.

Using a functional model may permit to compress naturally data contained in a geometrical model and perhaps to realise actions corresponding to the user interactions. In that case, we could neglect totally geometrical model in order to use only the functional one. Unfortunately, the latter is still too conceptual to be used in a current CAD system. Thus we are still working with geometrical model in a first time, and our future work will consist in implementing it in our WWW-CAD conceptual system.

REFERENCES

[1] Johnson A. L.,Thornton A. C., Fong C. F. (1993). Modelling functionality in CAD: implications for product representations, *9th international conference on engineering design, Hubka eds., The Hague - Netherlands*, August 17-19, 1610-1617

[2] Y. Gardan, C. Minich, D. Pallez, E. Perrin, (1999) From Functions to Shapes, *Proceedings of EDA'99*, Vancouver

[3] Y. Gardan, C. Minich, D. Pallez, E. Perrin (2000) Towards a specifications-to-shape translation tool , *Proceedings of TMCE 2000*, Delft, April 18-21, 373-382

[4] Gardan Y., Lanuel Y., Pallez D., Vexo F. (2000). Methodology for functions to shape translation tool in foundry, *Computers in industry*, to be published

[5] Allier P., Laurent N. (1999). Compression et Représentation Echelonnable de Maillages Triangulaires. *Proceedings of AFIG,* Reims, November 24-26, 89-102.

[6] Chow M. M. (1997). Optimized Geometry Compression for Real-time Rendering, *Proceedings of IEEE Visualisation*

[7] Hayley M. B., Blake E. H. (1996). Incremental Volume Rendering Unsing Hierarchical Compression, *Proceedings of EUROGRAPHICS,* Poitiers, 45-55.

[8] Li, J., Kuo C.C. J. (1997) Progressive Compression of 3D Graphic Models, *Proceedings of The International Conference on Multimedia Computing and Systems*, 135-142.

[9] Masuda H., Ohbuchi R. (2000) Coding topological structure of 3D CAD models, *Computer Aided Design*, **32**(5-6), 367-375.

[10] Taubin G., Rossignac J. (1998) Geometric compression through topological surgery, *ACM transactions on Graphics*, **17**(2), 84-115.

[11] Touma C., Gotsman C. (1998) Triangle mesh compression, *Graphics Interface*, 26-34.

[12] Rossignac J. (1999) Edgebreaker: Connectivity Compression for Triangle Meshes, *IEEE Transactions on Visualisation and Computer Graphics*, **5**(1), 47-61.

[13] Algorri M.E., Schmitt F. (1996) Mesh Simplification, *Eurographics'96*, 77-86.

[14] Belblidia S. (1998) Modélisation et visualisation par niveaux de détail de scènes architecturales complexes, *PhD Thesis* , Institut Polytechnique de Lorraine INPL.

[15] Belblidia S., Perrin J.P. (2000) Simplification de surfaces polygonales complexes *MICAD2000*, 161-167.

[16] Cignoni P., Montani C., Scopigno R. (1998) A comparison of mesh simplification algorithms, *Computer & Graphics*, **22** (1), 37-54.

[17] Fischer A. (2000) Multi-level models for reverse engineering and rapid prototyping in remote CAD systems, *Computer Aided Design*, **32**(1), 27-38.

[18] Rossignac J., Borrel P. (1993) Multi-resolution 3D approximations for rendering complex scenes, *Conference on Geometric modelling in Computer Graphics* 453-465.

[19] Ciampalini A., Cignoni P., Montani C., Scorpino R. (1997) Multiresolution Decimation based on Global Error, *The Visual Computer*, **13** (5), 228-246.

[20] Cignoni P., Puppo E., Scorpino R. (1997) Representation and visualisation of Terrain surfaces at variable resolution, *The Visual Computer,* **13**(5), 199-217.

[21] Hoppe H. (1998) Efficient implementation of progressive meshes, *Computer & Graphics*, **22** (1), 27-36.

[22] Reinhard K. (1998) Multiresolution representations for surfaces meshes based on the vertex decimation method, *Computer & Graphics*, **22** (1), 13-26.

[23] Eustache J., Lanuel Y., Vivian. R (1999) Utilisation de la fonctionnalité d'un objet pour optimiser le calcul d'une image de synthèse, *12èmes journées de l'AFIG, Reims, France.*

Moving XML to a Manufacturing Enterprise

Paul Lau, Jasper Wong, Edward Cheung
Industrial Centre, The Hong Kong Polytechnic University, Hong Kong
Email: icpaul@polyu.edu.hk

Keywords Information system reengineering, XML, SGML, Metadata, Data Source Object, Active Serve Page, Servlets, ODBC

Abstract Information generation, representation and control in a manufacturing enterprise are often done through interactions of a variety of application software. Application software such as resource planning, electronic design automation, computer aided design and manufacturing, product data management, various database and supply-chain management applications are essential for an enterprise to be successful in this information age. For an enterprise to operate effectively, individual application must communicate and interact with each other, and user groups can share and exchange pools of information. Although client/server technology can break some of the barriers, with applications being shared among common databases and open system platforms, it does not imply that the applications are capable of inter-operating data and document can be efficiently managed and distributed over the manufacturing enterprise at large. XML provides a tool to integrate data from different sources and eliminate client-extensions to achieve universal access. In this paper, we will incorporate XML in a manufacturing enterprise to provide a cross-organizational support for a board range of users, across hardware and software boundaries conveniently and cost effectively.

1 INTRODUCTION

Information management is a key factor for the success of a manufacturing enterprise. The enterprise resources have unleashed the current technological revolution that has fuelled this Information Age. Internet technologies applied to build corporate intranets and also extranets between customers, suppliers, and business partners have opened up heterogeneous access leading to manufacturing intelligence. An effective manufacturing enterprise needs a modern and flexible data delivery system so as to access both structured and unstructured data in the enterprise databases. Most of these databases exist as unstructured data that include textual documents, reports, graphics, audio and video resources. Developing after the successful application of Hypertext Markup Language (HTML), Extensible Markup Language (XML) is an Internet technology that can be applied to document metadata used between different systems. It helps in organising and locating data [1]. XML is a data format for information interchange on the Web [2]. In addition, it acts as a bridge between structured and unstructured data. Metadata can be published in a Document Type Definition (DTD) file, which defines the structure of an XML file. XML can be used for describing data of virtually any types. It is a language for creating mark up. It is more flexible and powerful than HTML. It is supported by most of the productivity tools and office suites [3]. It offers an effective way to provide communication between different databases systems and platforms.

The XML technology offers two structuring methodologies, Heavy-weight structuring and Light-weight structuring, to define information structure of documents [4]. Heavy-weight structuring applies in a way similar to designing data structure for programming. It consists of complete sets of syntax rules, constraints, attributes, tags and semantics. The structure of the documents plays an important part in the delivery of document information. Light-weight structuring applies when information is best expressed in natural texts. The markup of a particular key-sentence in a content page helps readers to identify vital information and it is the key to success for the delivery of information in the enterprise. Instead of defining a complete document model, markup identify elements, add sentinel marks or express semantic links with the documents.

2 XML DATA MAPPING

XML-data is closely related to the MCF specification [5], and can be applied for structured data and metadata exchange on the Internet. It describes an XML vocabulary for defining and documenting object classes. It can be used for classes that are strictly syntactic or indicate concepts and relations among concepts as in relational databases [6].

An XML document is primary concerned with data and the document is a data source. XML is a universal data format for structured document. It is text-based and platform independent. Different applications and HTML pages can access and manipulate a relational database once it is converted into an XML data source. One can convert a Microsoft Access table into a XML based data by accessing the data source through scripts or Data Source Objects (DSO). Another method is to transform the table, using the

Active Server Pages (ASP) technology, into an XML data source. In this way, Web or Intranet users can access this information more quickly. By using this method, XML data source can be generated and updated dynamically. This data conversion can be achieved through some simple mapping methodologies between XML and objects or between XML and tables [7]. Figure 1 shows that there is a one-to-one relationship between the object model and the XML model.

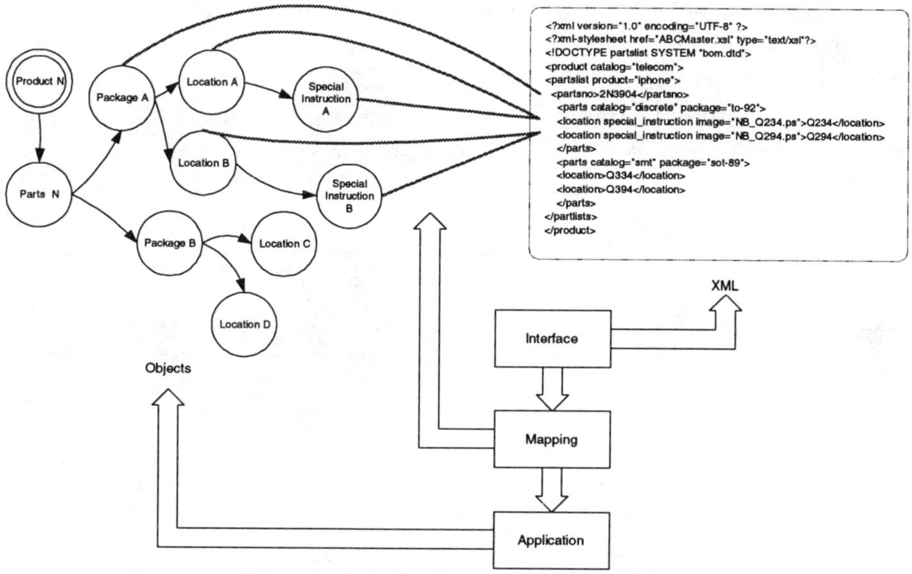

Figure 1 Mapping between XML and objects

As the object model and the XML model are similar, the Mapping between the XML and objects are simple. If you parse XML, you can obtain a tree of common data structure for the Objects. Figure 2 shows a Mapping between XML and tables. Since the two models are different, and it is not a one-to-one relationship, the Mapping between XML and the relational tables is more complicated.

3 THREE TIER MODEL FOR XML

3.1 XML and Databases interface

Figure 3 shows a Java based Three-tier Model which consists of a Web Browser Client as tier1, a Web Server as tier 2 and a Backend Server as tier 3. The most common Backend Servers are Relational database management systems (RDBMs). A Backend Server is the data source for enterprise legacy applications. XML web applications exchange information between organizations through this database access architecture. The RDBMs are very efficient in dealing with large amount of data. It is difficult to exchange data between RDBMs. XML offers a flexible and standard way for data exchange. The middle tier is a web server to provide connectivity for web clients and business logic support in the manipulation of data from the databases. In Java enabled system, a Servlets enhanced Web Server can be used. Servlets are effective in providing secured Web-based application which normally requires interaction between databases and clients. It can also generate dynamic and generate custom HTML documents for display while maintaining unique session information for each client. Using Servlets, input/output streams and Java Database Connectivity (JDBC), robust multi-tier client-server applications that access databases can be built.

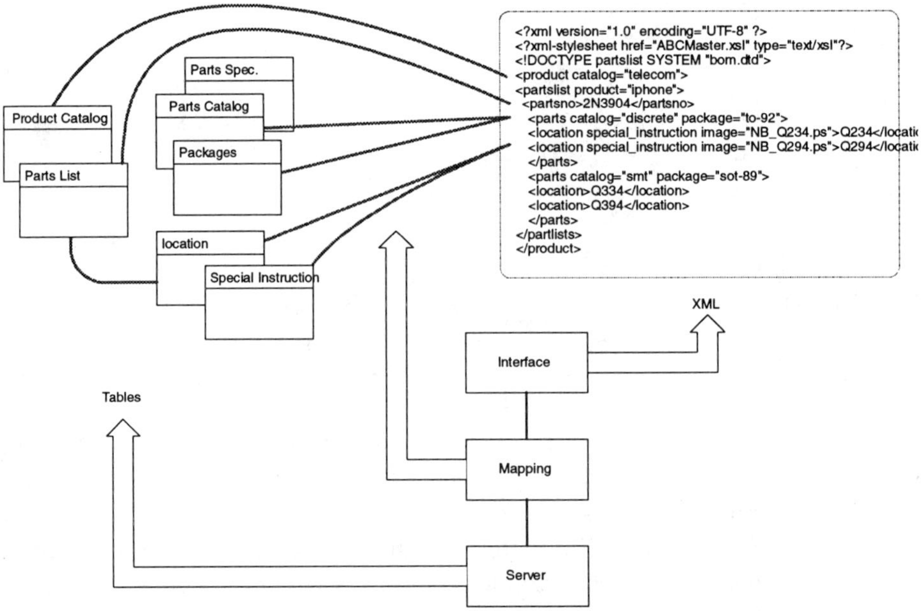

Figure 2 Mapping between XML and tables

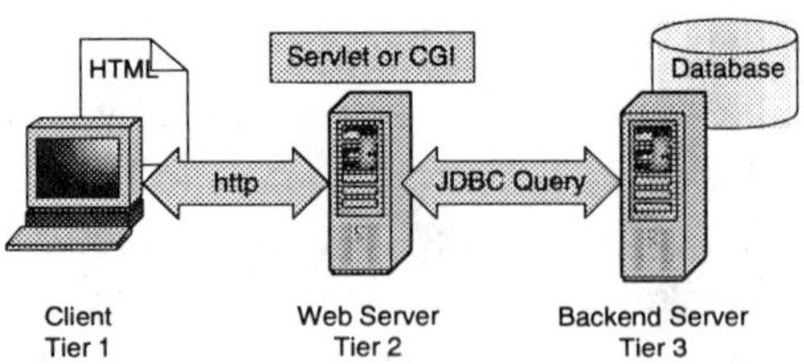

Figure 3 XML and Database Interface

3.2 Application Model

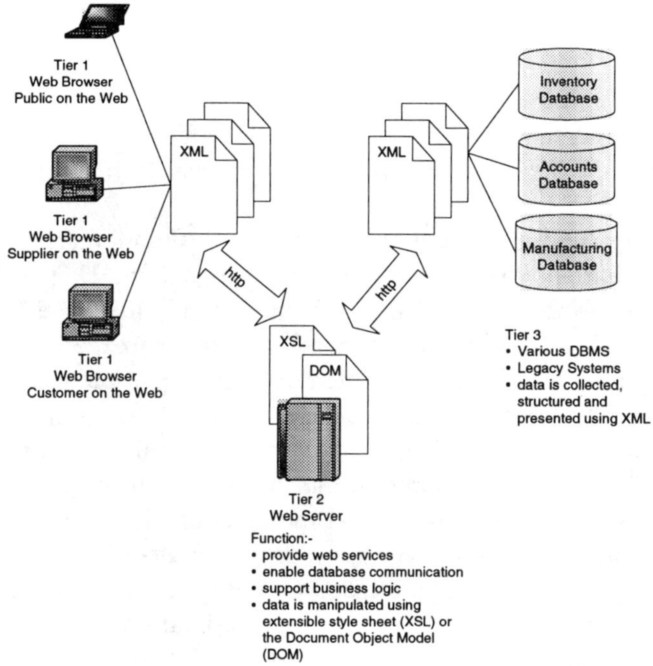

Figure 4 XML Three-tier Application Model

In most organizations, management database is normally stored in a proprietary file system. It doesn't make sense to convert all the existing data to XML format because individual application has its uniqueness and schema. For security reasons, most of the manufacturing enterprises do not want to expose unnecessary information on the Internet. Hence, it is necessary to isolate the database from the Internet clients. Figure 4 shows a typical XML three-tier application model. Tier 3 consists of a set of legacy systems and databases. Structured data are entered into or retrieved from these systems by the enterprise users. The existing data in the databases stays in its current format. In tier 2, structured information is compiled or converted to XML raw data. The XML data will be exchanged over HTTP to the Web client in the tier 1. Together with the XML data, a style sheet can be created for the Web client to download and display information in the original format from the content provider. The Web client can then perform editing, rendering, calculations, graphical interpretation, statistical processing and other manipulation on the XML data independently. In this way, web client can take advantages of their hardware rather than depending on the processing power as distributed by the server.

4 INFORMATION MANAGEMENT SYSTEM FOR A MANUFACTURING ENTERPRISE

The use of information system has been increasingly playing a critical role towards the demand on productivity and throughput in a manufacturing enterprise. Figure 5 shows an information process within a manufacturing enterprise.

The enterprise, in general, consists of sales and marketing, design, process planning, production, material control and purchasing functional units. Manufacturing information and documents are managed in traditional databases, drawing files, office application, bitmap images, vector images together with different databases, which cater for a particular manufacturing system or process. Integration for these manufacturing information and documents are necessary. The primary drive for this integration is the demand in efficiency, collaboration and quality in an organisation. These demands cannot be met by partial automation in individual process. Collaboration tool is essential because it can integrate all documentation and databases. With a good user interface, the collaboration tool can provide a convenient environment to improve co-ordination between business functions and improve decision-making. Many problems and issues relating to the design, development, integration, evolution and maintenance of Information System in a large scale and complex plants have become apparent and are not adequately addressed by the traditional process. A sophisticate information system makes a manufacturing enterprise ideal for manufacture-to-order, assemble-to-order, configure-to-order and engineer-to-order. This flexibility is essential for a manufacturing enterprise to success in the Internet business.

A collaborative information system enables the management team to track and report on all activities, integrates materials and costs which associated with individual products with entire manufacturing functions or business function. Through this information system, the enterprise management makes comprehensive order processing and inventory information available for immediate analysis. The management can analyse customers' buying patterns, track the performance of items and identify market opportunities that can shape future marketing activities. It supports international quotes and sales orders and allows the management of sales contracts anywhere anytime. Purchasing is fully integrated with all manufacturing and financial function so as allows the enterprise to reduce its inventory for a more profitable operation. This system can also provide formatted information for the suppliers. This will enable collaboration not only limited within an enterprise itself, it virtually extends its business boundary. The enterprise can set up strategically supply chain or partners without worrying the incompatibility of systems. Manufacturing enterprise is normally supported by different systems introduced during various stages in its development. These may include resource planning, electronic design automation (EDA), computer aided design(CAD) and manufacturing (CAM), product data management, decision support and data warehouse and supply-chain management. By integrating with different objects or DTD, the enterprise information system provides the capability to integrate information reside in different RDBMS such as engineering drawings in CAD/CAM systems with resources management systems such as MRP.

5 AN IMPLEMENTATION EXAMPLE

ABI is a manufacturing enterprise. It has more than twenty sales offices and a factory over the world. Manufacturing information such as cost, shipment, product specifications, production rate, stocks and parts need to exchange between the factory and sales offices frequently. There are over a hundred products for the ABI enterprise. It is very common to produce over twenty different products in a day. The factory implements a Just-in-time inventory system and multi-vendors policy. Hence, a tight interaction between the factory, vendors and the sales offices is very important. ABI enterprise has a long history. Its information systems and databases have evolved over many years. Consequently, it suffers from loose coupling in its information system with isolated or proprietary processes that operates in different platforms and legacy DBMS. Management information exists as word processor files, spreadsheets, legacy relational databases, engineering drawing, sales and suppliers catalogues that have different formats and metadata. Under the current environment, information is neither coalesced nor intelligent.

Figure 6 shows a proposed Information Architecture for the ABI enterprise using XML technology. It is based on the Three-tier Model that we previously discussed. It consists of three parties: vendors, sales offices and the factory. The clients are the end users from the vendors, sales offices and the factory. One the client sides, data is formatted by the Data Source Object (DSO) scripting. Information is sent to the Internet Explorer 5 browser from the factory's Internet Information Server.

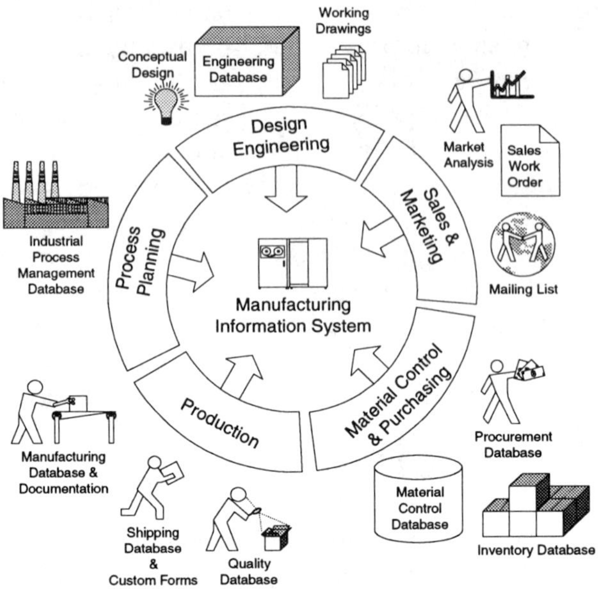

Figure 5 Information Process in a Manufacturing Enterprise

In addition to a Web server, the middle tier comprises an INSIGHT XML database and the INSIGHT software system [8], which automatically combines document and graphic-based information from databases and file systems into intelligent Web applications. Whenever a customer places an order to a sales office, a salesman can issue query to the Business Server from the browser. Figure 7 shows part of the XML sample codes implement in this application.

6 CONCLUSION

In this paper we have discussed some key features of XML as a tool to integrate data from different sources as well as to achieve a universal data and document access. It shifts the process of data visualisation from a server to the client and allows various data manipulations at the client side. It is a convenient way to publish document and exchange data either internally or externally in a manufacturing enterprise context including Internet access for its strategic partners. The XML technology offers Heavyweight and Lightweight structuring methodologies through data mapping from data sources. This facilitates the creation of XML documents from various data formats.

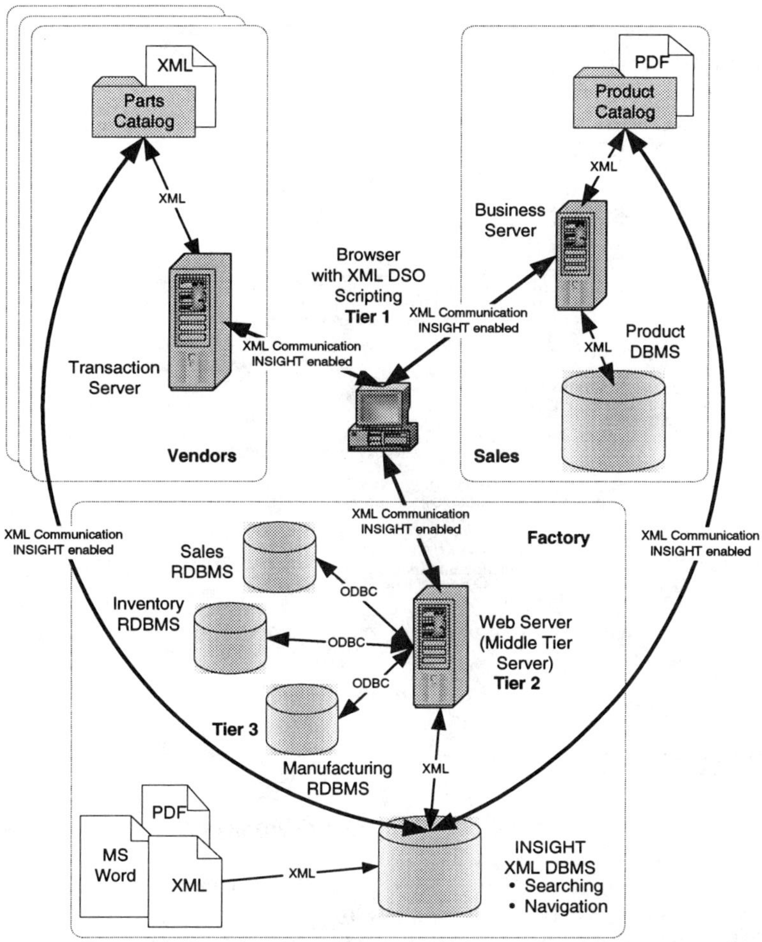

Figure 6 Manufacturing Enterprise Information Architecture

A Three-tier model was introduced in this paper. The model can be applied in any manufacturing enterprises. As information flow and system is typical complicated and important in such environment, it is essential to move the XML technology to the enterprise safely and seamlessly with minimal disturbance. In our application, we have illustrated that moving the XML technology to the manufacturing enterprise can seamlessly improved its information system and hence the productivity and the profitability of the enterprise. An example has been discussed to demonstrate the implementation of a XML enabled information system in a manufacturing enterprise. By incorporating XML in a manufacturing enterprise, cost effective information flow can be achieved across various hardware and software platforms. A manufacturing enterprise can run ahead of its competitors in this information age.

```
Product Information
<?xml version="1.0"?>
<PRODUCT>
    <PRODUCT_LIST>
        <AUDIO_CAT>
                <PROD_NUM>PA010</PROD_NUM>
                <PROD_NAME>IMAGE MP3</ PROD_NAME>
                <PROD_COST>US200</ PROD_COST>
                <PROD_DES>Silver, metallic finished<PROD_DES>
        </AUDIO_CAT>
        <TOY_CAT>
                <PROD_NUM>PT020</PROD_NUM>
                <PROD_NAME>REMOTE CAR</ PROD_NAME>
                <PROD_COST>US110</ PROD_COST>
                <PROD_DES>Red, ABS plastic </PROD_DES>
        </TOY_CAT>
    <PRODUCT_LIST>
</PRODUCT>
Factory Information
<?xml version="1.0"?>
<FACTORY>
        <PROD_NUM>PA010</ PROD_NUM>
        <PRODUCTION_RATE> 2000 per day </PRODUCTION_RATE>
        <INVENTORY>
                <PART_NUM> IC123</ PART_NUM>
                <STORE_INVENTORY> 2000</ STORE_INVENTORY >
                <VENDOR>
                        <NAME> XYZ SEMICONDUCTOR</NAME>
                        <TRANSACTION-SERVER-URL>
                        http://trans.xyzsemi.com
                        </ TRANSACTION-SERVER-URL>
                <VENDOR>
        </INVENTORY>
        <FINISHED_GOOD_QTY> 5000</ FINISHED_GOOD_QTY>
        <SHIPMENT_DATE> 12/12/2000</ SHIPMENT_DATE>
</FACTORY>
```

Figure 7 XML Sample Code

7 REFERENCES

[1] Scott Mace, Udo Flohr, Rick Dobson and Tony Graham(1988). Weaving a better Web: Byte Magazine, March 1998, pp. 58-68

[2] Extensible Markup Language (XML) 1.0W3C Recommendation, February 1998. http://www.w3.org/TR/1998/REC-xml-19980210

[3] Widergren S., deVos A., X J. (1999). XML for Data Exchange: Power Engineering Society Summer Meeting, IEEE Vol. 2, 840-842

[4] Bapst F., Vanoirbeek C. (1999). XML documents production for an electronic platform of requests for proposal: Reliable Distributed Systems, Proceedings of the 18th IEEE Symposium, 330-308

[5] Meta Content Framework using XML (MCF), Specification (1997) http://www.textuality.com/sgml-erb/w3c-mcf.html

[6] Dianne Kennedy (1999), "Introducing XML-Data", XMLperts Articles library http://www.xmlxperts.com/xmldata.htm

[7] Barry D. (1999). XML Data Servers: An Infrastructure for Effectively Using XML in Electronic Commerce, Barry & Associates, Inc. http://xmml-data-servers.com

[8] INSIGHT: "White paper: Enigma's e-Publishing Product Suite" http://www.enigma.com

Manufacturing Enterprise Integration Using Simulation Software to Coordinate Budget Planning

E.J. Colville
Hon. Research Associate, School of Engineering, University of Tasmania, Australia
Email: respub@access.net.au

Keywords Budget simulation

Abstract An examination of departmental planning and communication needs within an organisation, and in particular in manufacture, reveals the major importance of planning policy making and staff motivation. Clearly these are at the heart of business management and its infrastructure. A method of tackling these needs is presented using budget simulation. This involves preparation of budgets in a way which encourages participation in decision making and more effective implementation of the agreed budgets.

BACKGROUND

Recent research studies have defined the principal tasks undertaken by manufacturing managers, in addition to a need to understand the technology of their industry. These studies have outlined the principal needs for further knowledge by many engineers to assist them to take on the combined technology and business management responsibility our industries urgently need. From board membership through to the chief executive officer and the departmental managers of finance, production and marketing we need managers who have a technology background but also have skills additional to traditional engineering ones. [1,2]

Inquiries overseas in this year 2000 confirm a further extension of global subcontracting and international investment. They also reveal the effect that currency variations and drive for market share are having on manufacture, product design and productivity as well as weakness in the employment opportunities for traditional engineers in some age groups.

The information technology revolution has reduced some of the routine tasks of professional groups, for example, engineers, accountants and lawyers, and this leaves room for this necessary broadening and an overlapping of professional activities [3,4,5]. Unless we as technical people now take up this opportunity to broaden our professional scope, others, who do not have the important and essential technology skills, will become the predominant management group. This will be to the detriment of our global competitiveness.

Incorporation of specific and clearly defined management and business skills for engineers in industry and in our university engineering courses is now essential for the future prospects of our engineers and manufacturing industries.

Not only is training and guidance required when a management position is achieved by engineers but more management training is needed during the earlier formative undergraduate years of the engineer. To assist both these needs, i.e., both plant management and education needs, a systems approach to management using a simulation of an organisation as a model for management action has been developed and successfully applied to manage a firm. Successful use of this process is reported for the management of small to medium size manufacturing operations and in manufacturing education [6,7,8]. It encourages participation by senior executives in the forward planning of a business, develops staff and stimulates implementation of budget plans.

Wider use of budget simulation to assist management and in particular to augment the business skills of technically trained people is therefore recommended.

SIMULATION AND TECHNOLOGY TRAINED EXECUTIVES

Industry now needs engineers and technology trained people who have expanded their horizons beyond the traditional professional engineering scope of the past. This is clear from personal observations in Australia and overseas, comments from industry development groups and manufacturers battling the combined effects of "level playing fields", and global competition.

Our manufacturing industries in particular need leaders on their boards and in senior management positions who thoroughly understand the technology of their industries and also can positively liaise with and direct the marketing, subcontracting, financial and political aspects of their firm.

To achieve this we need to become more outward and widen our focus. We now need to apply our ability to calculate and understand probability to such items as finances, people development, customers and suppliers. The analysis [1] of these additional specific skills required by technology trained people to provide the balanced senior management responsibility so urgently needed in our manufacturing industries has already received comment

If we don't take up this challenge it will be to the detriment of our industries as they are forced to rely on such approaches as short term returns, dangerously high gearing, market share, supply side economics and movement of investment against the national interest. Many non-technical executives acting without the knowledge to cope with the rate of change in technology, international communications and social structure change are unable to manage efficiently.

Adding to this weakness has been the abdication of professional engineering institutions, in both Australia and U.K., of their responsibility to specifically prescribe subjects in our university engineering faculties. They need to specifically include financial management, business knowledge and people development subjects in their courses by people who have experienced this need. [9] In Australia the view appears to be "we will not prescribe, it is up to the universities to interpret what we want" but what if many faculty members have little, if any, recent experience of current manufacturing industry needs. In U.K. the view was presented that " it is up to industry and universities to jointly provide the management training required for young engineers" so that the stimulus needed at the university undergraduate stage is still only halfhearted or even negligible in some cases.

A primary purpose of this paper is to present a method using simulation of a manufacturing firm's overall activity which can be used to train both young and more mature engineers in company management and later be useful in developing their overall management skills when managing an enterprise. As emphasised by Williams and Johnson [10] in a keynote address to an international manufacturing conference in Manchester in July 2000, there are two sides to manufacture in a global economy of innovation and regulation. One a world of mechanisms, machines, materials and facilities, the other that of overall systems, management, finances, people, customers and subcontracts. As most engineers in the first category think in terms of systems, an overall systems and simulation approach to both technology and business management will bring together the many interacting groups needing the coordination of a manager. It joins the two sides of manufacture. It promotes improvement and team activity and places the accent on implementation and results. It makes senior staff part of a continuous planning and budgeting process. It promotes the manager as a people developer and at the same time facilitates the transition of the traditional engineer to that of manager so that he/she can take the wider responsibility so needed in our current economy.

SIMULATION AND MANUFACTURING MANAGEMENT

Five important stages in preparing a simulation [11] are:

- Entry of data describing the forward plans of the company using a user friendly Excel spread sheet allied to a description of each group of data. Almost 100 matrices covering around 1000 numerical entries are employed to describe a small to medium size enterprise. Production, supply and marketing units of activity are translated using dollar values per unit to provide a common language throughout the simulation. However as this process is reversible as outlined in figure 1 it subsequently allows transfer of budget answers, which are in financial terms, back into production unit terms to give production planning and pricing information.

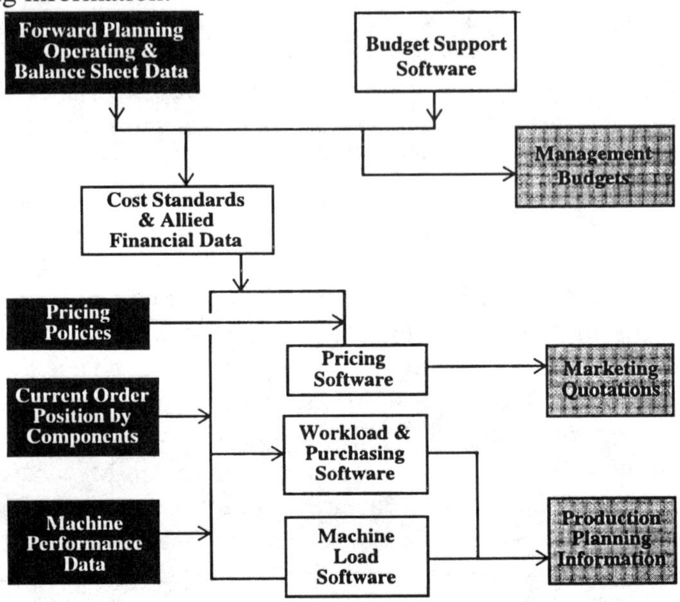

Figure 1 Coordinated Budgeting and Planning Information

- The data allows for 12 periods ahead, for example months, quarters or years. A macro operates on the above spread sheet to automatically produce a compact numerical data bank applicable to a company's total financial affairs and forward plans.

- Software programs of two types are utilised to process the data as set out in the pull down menu of figure 2. The first program is the basic mathematics interconnecting all the operating inputs and outputs applicable to the company and its components. Some 30 groups of equations cover balance sheet and operating criteria as well as matrices defining overhead distribution by component and department. The second program, in visual basic, presents the results in a readily accessibility compact form so that executives can bring up budgets, standard costs and financial ratios quickly, on their screens to check proposals for improvement.
- A manual describes the purpose of the program and guides users in data and budget production. Suggestions for management of the budgeting process using the programs and maintenance of the data and the budgeting process as the firm develops are provided.
- Fractional increases based on previous performance are used to align budget data predictions with executive thinking
- Ancillary programs may be prepared based on the cost standards developed by the master budget simulation program. These programs can provide a basis for pricing by the marketing department, production planning by the production planning department and purchasing implications for the finance department, all directly related to a company's overall policies.

IMPLEMENTATION

The initial task of data entry follows detailed investigation of the firm's activities. Clarification of the terms describing the firm's products, subassemblies and accounts categories is followed by meetings with the department heads and CEO to consider their needs and views to be fed into the budget. Trials follow using the simulation and meetings take place to coordinate plans and iron out anomalies and differences of opinion. A joint plan is agreed which the key executives are prepared to positively implement. At this time alignment of the company's book categories with the simulation descriptions allows later comparison between budget and actual performance as the budget unfolds.

The next phase of operation is that of managing using the budget process to guide the company's activities. Responsibility is defined as to who should call budget meetings in which progress is reported and minutes taken. The person responsible for the maintenance of the system also needs to be defined. The key to this process is that if people own the plan and believe in it, they are more likely to push it through despite obstacles encountered on the way.

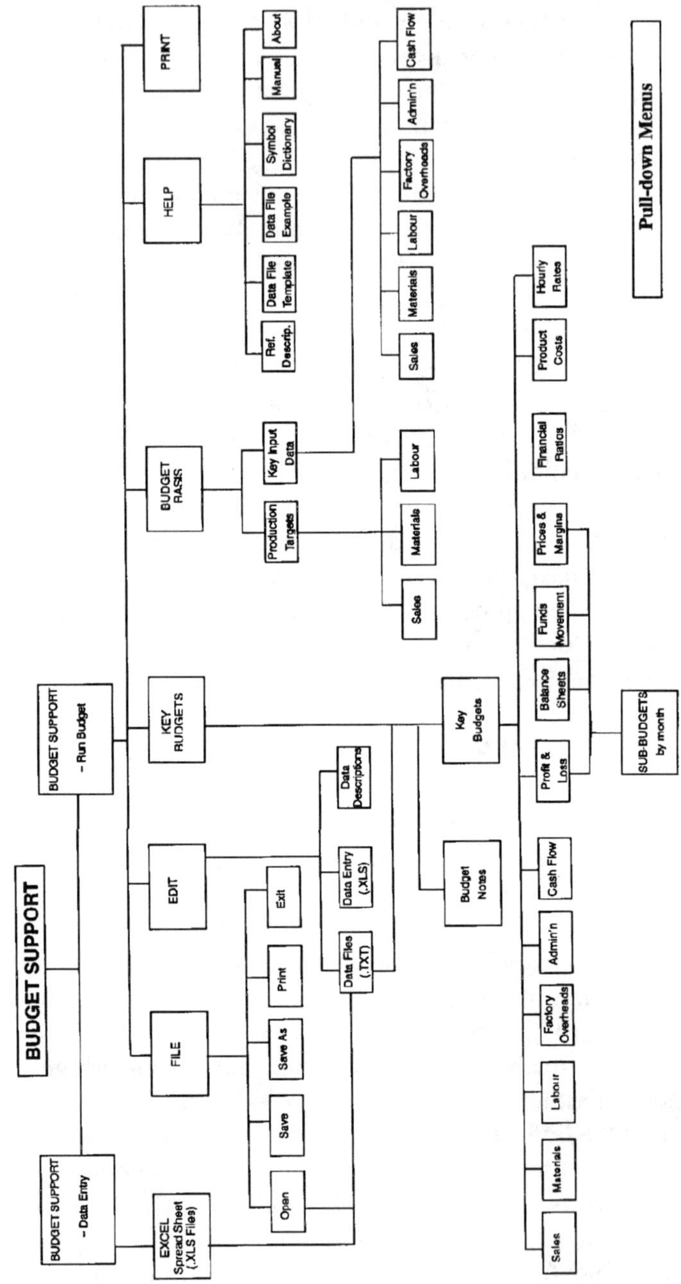

Figure 2 Pulldown menu of main Budget Support Program

Some points to watch, applicable to most innovation, are the following:

- Patience with introduction may be needed as this process provides long term benefits and security while many managers are primarily interested in short term returns.
- Managers who operate in a hierarchical fashion can be frightened of sharing information and improving communication which is fundamental to this improvement process. They sometimes resort to the "our business is different" attitude.
- The coordinator needs to have an insight into the potential of computer programming as well as management needs.
- The production of internal budgets, involving full participation as outlined above, may be resented by those external professionals such as accountants, previously solely responsible for the budgeting process.

CASE STUDIES AND RESULTS.

This simulation developed from a need to produce budgets and test them economically and rapidly in real time. Tests in industry and in particular in printing and publishing have been carried out over some twenty years since the research was initiated.

At the same time there has been the stimulus to provide management education for mechanical and manufacturing engineers at the University of Melbourne. More recently a consolidation of industrial and educational experience applicable to practicing engineers using simulation has been undertaken with the University of Tasmania.

During this period the extraordinary development of computer software and hardware has played a significant part in advancing simulation. Results have been increased profits, better liquidity and wiser general management. This has been the result of a saving in infrastructure costs, such as accounting, estimating and liaison costs between supply production and customers, and realignment of business structure when predictions from the simulation dictate this.

The ability to meet the needs and timing of macroeconomic change and revise standards by testing the effect of radical technical, financial and process changes has been a feature of this process. An example is movement from vertical to horizontal integration as buyer- sellers relationship change. Using this process the traditional accounting and skills needed to manage people, customer relations and subcontracts in a manufacturing enterprise can be embraced by engineers.

CONCLUSION

The budget support system described meets two important needs of current management. First the provision of economic budgets quickly as business climates change. Second it encourages a high degree of participation by senior executives and their staff. This results in efficient implementation of forward plans as a result of team work. The procedures outlined shorten the lines of communication and promote flexibility, vital to being competitive in the current global market.

As a result of an increased awareness in UK and Australia of the need for more competitive manufacturing management, current activities include promotion of this simulation approach for manufacturing industries in Australia, United Kingdom and South East Asia.

REFERENCES

[1] Colville E.J., (1999). Advancing Medium Size Manufacturing Enterprises through Management by Engineers. Doctoral Thesis, Univ. of Tas. (Inc. 73 refs. & software).

[2] Colville E.J., (2000). General Management Opportunities for Maintenance Engineers. ICOMS 2000, 4th International Conference of Maintenance Societies, Wollongong, N.S.W. May. Proceedings paper 076.

[3] Colville E.J., (2000). Management by Engineers. EIS 2000, 2nd ICSC Symposium on Engineering of Intelligent Systems, Paisley, Scotland, June Proceedings p. 334

[4] Colville E.J., (2000). Strengthening Production by Extending Engineering Skills. Special ICPR 2000, International Conference on Production Research, Bangkok, Thailand Aug.

[5] Colville E.J., (2000). Engineers Managing Businesses. ICME 2000, 8th International Conference on Manufacturing Engineering, Sydney, N.S.W., Aug paper 157.

[6] Colville E.J., (1999). Manufacturing Management Improvement through Rapid Production of Budgets. IPMM 99, 2nd International Conference on Intelligent Processing and Manufacturing of Materials, Hawaii, USA, July Proceedings p. 649.

[7] Colville E.J., (1986). Management Information Systems for Small Manufacturers. Society of Automotive Engineers International Conference, Auckland, New Zealand. P. 156

[8] Colville E.J., (1992). Mathematical Simulation of a Manufacturing Concern. An Important Part of Engineering Management Education. ACEME, Sydney, N.S.W. April Proceedings.

[9] Personal communications, (2000), I.E.Aust , Canberra, May, and I.Mech.E. London, July.

[10] Williams D.J., and Johnson.W., (2000). Hard and Soft approaches to Manufacturing: Which is the most important? Keynote Address - MATADOR Manufacturing Conference, Manchester UK July.

[12] Colville E.J., (1997) Budget Support – A Management and Development Program. *Research Publications, Vermont, Victoria.*

PART NINE

Computer Integrated Manufacturing

An Expert System for Plasma Cutting Process Quality Prediction and Optimal Parameter Suggestion

Sam Y. S. Yang
CSIRO Manufacturing Science and Technology, Australia
Em: s.yang@cmst.csiro.au

Keywords Plasma cutting, CNC, Information Infrastructure

Abstract An hybrid expert system, which uses both rule-based reasoning and statistical modelling, is developed for prediction of cut quality and suggestion of optimal process parameter settings for CNC controlled plasma metal-plate cutting machines. A scheme for process parameter classification has been developed. They are classified according to whether they are fixed, operator selectable or adjustable. A scheme for cutting quality definition and practical method of assessment have been developed which incorporate both the quantitative requirements for each quality attribute and their respective relative importance. When process parameter values are entered, the system gives the expected quality outcome. When the quality requirements are specified, the system outputs the optimal parameter setting values. The prototype of the system has been tested and the predictions are consistent with experimental measurements.

1. INTRODUCTION

Plate cutting is one of the most important manufacturing processes for metal components making. The quality and the cost of cutting process is often critical for final product quality and cost. It is estimated that the plasma cutting machine with a computer numerical controlled (CNC) torch movement is the optimal choice for 80% of the metal plate cutting processes [1]. However, the quality of the plasma cutting process is often unpredictable largely due to the unknown effects of the different process parameters. Difficulty in quality control is one of the factors which affect the wide spread use of the plasma cutting process in industry.

Requirements for quality control and prediction are quite common in other manufacturing industries as well. Various expert system approaches have been used to tackle them [2-3]. The *hybrid expert system* [4], which combines both the shallow-knowledge (rule based) approach with the deep-knowledge (modelling and simulation) approach, has been used in various industrial processes [5].

This paper outlines the design and formulation of an expert system for the metal plate plasma cutting process. The system predicts cutting quality for a given process parameter setting, and provides advice on the optimal process parameter setting for a given quality requirement. The major tasks include definition and classification of process parameters, definition and representation of cut quality, knowledge capturing, data analysis and modelling, representation and delivering of knowledge in a form understandable by the users.

The software modules described in this paper form parts of a larger system for Remote Operations, Diagnostics And Maintenance (ROSDAM). The modules to be discussed include quality prediction and optimal parameter values suggestion modules under operation support, and process knowledge database maintenance module under database maintenance.

2. PROCESS VARIABLES AND CUT QUALITY

The first task is to find manageable methods and definitions for describing the input and output parameters of the system. This involves the definition of a complete set of process parameters and the cut quality in a quantifiable and repeatable way.

The process parameters are classified into three categories: fixed, selectable and adjustable. A fixed parameter is the one which cannot be modified for a given cutting job, such as the model of the cutting equipment. A selectable parameter is one that the operator can select from a discrete range of options, such as the type of plasma gas. Adjustable parameters are those parameters which can be set in a range of numerical values, such as the contouring speed of the cutting torch. For the plasma cutting process, there are a total of 120 process parameters.

The process parameters can also be classified according to whether they have a natural numerical representation or not. The parameters which have such a representation include plate thickness, cutting current, consumable condition (a lower value represent a better condition), etc.. The non-numerical parameters, such as machine model, plate material and grade, are used to sub-classify the process. Mathematically, the process parameters are represented as a parameter vector

$$P^{(\alpha)} = \left\{p_1^{(\alpha)}, p_2^{(\alpha)}, \cdots, p_N^{(\alpha)}\right\} \tag{2.1}$$

where the subscript denotes the numerical parameters with N being the total number of such parameters, and α denotes the non-numerical parameters.

Cut quality not only includes measurable qualities such as cut angle and accuracy, but also has abstract entities such as dross removability and surface finish. As the system is intended for actual use in industrial environment, the measurement procedures to obtain some level of accuracy in these qualities has been considered.

Even if the individual quality attribute can be quantified, it is often the case that there is not a universal definition for "good" quality. Whether a particular cut is good quality or not relates to the part to be made. For instance, if the part is for a drop-in fit, the cut size cannot be larger than the design value. If the part needs be joined to other parts in a given angle, the "cut angle" is critical. If the plate need to be joined by welding, the "nitrition" is a important factor. If the cutting surface forms part of the product finish, the surface finish may be the main concern [6]. To describe this job-dependent quality definition, a weighting factor is introduced in our system.

For plasma cutting, a total of eight quality attributes are considered. They include part size accuracy, cut angle, cut surface finish, dross peripheral coverage and removability. They are defined by their averages and standard deviations, and a relative weighting factor for each quality attribute which are represented by the following three vectors

$$\begin{cases} Q = \{q_1, q_2, \cdots, q_M\} \\ \delta Q = \{\delta q_1, \delta q_2, \cdots, \delta q_M\} \\ W = \{w_1, w_2, \cdots, w_M\} \end{cases} \tag{2.2}$$

where q_i, δq_i and w_i $(i=1,2,...,M)$ are the average, standard deviation and weighting factor for the i'th quality attribute. M (=8) is the total number of such quality attributes. A "good" cut is the one with averages of quality attributes close to the expected values and the variations (proportional to the respective standard deviations) are within specified limits.

The expert system to be described in this paper is based on the above definitions for process parameters and quality descriptions. The expert system has a user-friendly GUI for plasma cutting process data collection. When a cut job is completed, the machine operator can easily input the process parameters and the associated quality into the system. In the following sections, we will focus on the principles of quality prediction and optimal parameter suggestions.

3. CUT QUALITY PREDICTION

One of the functions of the expert system is to predict cut quality when the process parameter values are specified by the user. The architecture of the system is represented schematically as in Figure 1.

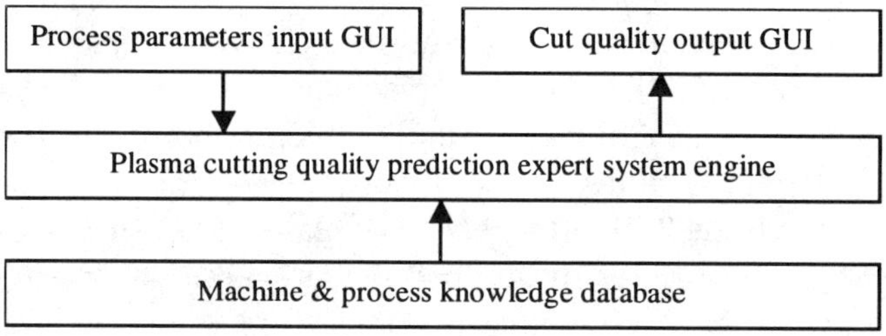

Figure 1: Block diagram of the expert system module for quality prediction.

Process parameter values are entered in to the system by the user through the process parameter input GUI. The plasma cutting quality prediction expert system engine uses the user input to search for the appropriate information in the machine and process knowledge database. The information enables the system to work out the expected quality. The predicted quality values are sent to the cut quality output GUI.

The quality prediction is based on the assumption that the change to quality vector ΔQ is small for a small change to the process parameter vector ΔP. Consequently, they can be related through a Jacobian or slope matrix S as in the following:

$$\Delta Q = S\Delta P \tag{3.1}$$

The matrix S in (3.1) is part of the process knowledge which is stored in the process knowledge database. The values of the matrix elements are obtained through a database maintenance process. The maintenance process can be represented by the block diagram as shown in Figure 2.

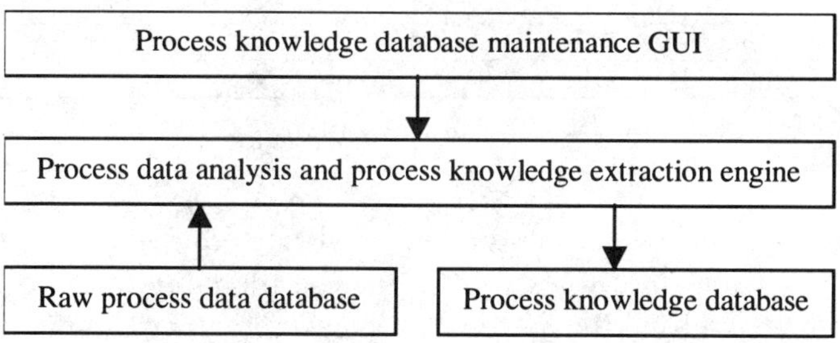

Figure 2: Block diagram for process knowledge database maintenance.

The process knowledge database maintenance process is initiated by the user when a batch of new raw process data has been accumulated in the raw process data database. The process data analysis and process knowledge extraction engine calculates the slope matrix for each entry of raw process data, and output the relevant process information and the slope matrix S to the process knowledge database.

The system has been populated and trained with 150 process data sets for a Hypertherm HT2000 plasma machine controlled by a FARLEY PDF32/Wizard II CNC. Another seven sample coupons have been cut and their quality attribute values were measured experimentally. All coupons were cut on 6mm grade 250 mild steel BHP plates. The cutting gas is oxygen and the shield gas is air. Dry cutting bed is used and no magnetic handling is used for the plates. New Hypertherm standard 200A consumables are used for the coupon cuts.

For the above cutting conditions, the HyperTherm manual recommended a cutting speed of 4.06m/min. The other cutting parameter values as recommended by HyperTherm are shown in the first row of Table 1. Due to software system limitations, the speed setting only preserves the first decimal place. In Table 1, each row corresponds to a test coupon. The parameters in Table 1 were so selected as they are commonly regarded in industry as the main influential plasma cutting parameters. It should be noted that these test coupons are used merely to verify the system predictions. They should not be mixed with the well-known design of experiments methodology for determining how quality depends on process parameter variations.

Table 1: *Cut process parameter set values for test samples.*

Sample number	Current (A)	Voltage (V)	Speed (m/min)	Cut flow (%)	Shield pressure (psi)	Kerf comp (mm)
1	200	120	4.1	64	60	1.25
2	180	120	4.1	64	60	1.25
3	200	130	4.1	64	60	1.25
4	200	120	4.4	64	60	1.25
5	200	120	4.1	51	60	1.25
6	200	120	4.1	64	40	1.25
7	200	120	4.1	64	60	0.00

The experimentally measured quality values and the values predicted by the system are listed in Tables 2. It is seen in the table that the predictions agree reasonably well with experimental measurements. The precision of the prediction will improve as the system is trained with more data sets.

4. PROCESS PARAMETER OPTIMIZATION

Ultimately, what the user want to know is about how to set up the machine parameters for a given quality requirement. The process parameter optimisation module of the system is developed to meet this user requirement. The module has an input GUI for the user to specify the cutting environment such as machine model, plate specifications, range of gas type selections, etc, and the quality expectation. The quality expectation by the user for each quality attribute is defined by its expected (average) value, a range which must be satisfied and a weight factor as defined in (2.2). The logical structure of the module is shown schematically in Figure 3.

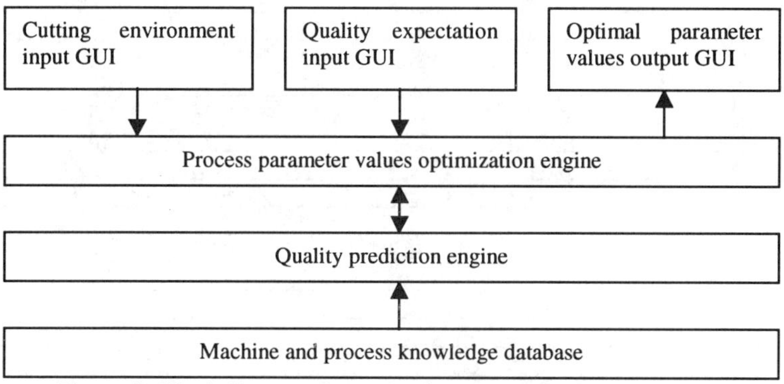

Figure 3: Block diagram for optimal parameter values suggestion

The process parameter values optimisation engine calls the quality prediction engine to calculate the expected quality output. By adjusting the parameter values, the process optimisation engine minimises the target function as defined in the following:

$$T = \frac{1}{M}\sum_{i=1}^{M} w_i \left| q_i - \hat{q}_i \right| / \delta_i \tag{4.1}$$

where w_i is the user supplied weight factor for the i'th quality attribute, q_i and $\hat{q}_i$ are the predicted and the user specified values for the i'th quality attribute, and δ_i is the typical amplitude of variation for the i'th quality attribute.

The optimisation is subject to the following constraints:

$$\begin{cases} q_i^{(\min)} \le q_i - \Delta q_i \\ q_i^{(\max)} \ge q_i + \Delta q_i \end{cases} \qquad i = 1, 2, \cdots, M \tag{4.2}$$

Table 2
Comparison of predicted and experimentally measured quality attribute values for the test samples. The lower bound (LB) and upper bound (UB) are average minus/plus standard deviations for predicted values and minimum and maximum for experimentally measured values.

Quality attribute	Sample number	Predicted quality values			Measured quality values		
		Averag	LB	UB	Averag	LB	UB
Top size error (mm)	1	-1.42	-1.81	-1.03	-1.40	-1.48	-1.32
	2	-1.36	-1.68	-1.04	-1.36	-1.53	-1.19
	3	-0.41	-4.22	3.40	-1.74	-1.45	-2.01
	4	-1.40	-1.79	-1.01	-1.51	-2.35	-1.07
	5	-1.51	-2.02	-1.00	-1.53	-2.11	-1.27
Bottom size error (mm)	1	0.24	0.01	0.46	0.39	0.20	0.58
	2	0.54	0.19	0.90	0.51	0.15	0.87
	3	-1.85	-8.15	4.45	0.17	-0.06	0.20
	4	0.33	0.01	0.66	0.41	0.16	0.65
	5	0.31	-0.38	1.00	0.34	0.27	0.54
Cut angle (degree)	1	6.27	4.89	7.65	6.40	4.63	8.17
	2	7.10	5.96	8.24	7.10	4.58	9.62
	3	1.20	-17.29	19.69	7.37	0.47	6.73
	4	6.41	5.03	7.79	6.67	5.37	8.31
	5	7.04	3.92	10.15	6.30	5.60	7.41
Surface finish (grade)	1	2	2	3	2	-	-
	2	2	2	3	3	-	-
	3	1	1	5	3	-	-
	4	2	1	3	3	-	-
	5	3	1	5	3	-	-
Dross removability (grade)	1	3	3	3	3	-	-
	2	3	3	3	3	-	-
	3	2	1	3	3	-	-
	4	3	2	4	3	-	-
	5	3	3	4	3	-	-
Dross peripheral coverage (%)	1	0	0	8	0	-	-
	2	9	0	30	0	-	-
	3	0	0	100	0	-	-
	4	0	0	9	0	-	-
	5	6	0	42	0	-	-
Top kerf width (mm)	1	3.37	3.28	3.47	3.35	3.21	3.49
	2	3.32	3.22	3.42	3.37	3.27	3.47
	3	4.17	2.24	6.10	3.63	3.58	3.78
	4	3.34	3.13	3.55	3.42	3.35	3.54
	5	3.35	3.02	3.69	3.41	3.33	3.47
Bottom kerf width (mm)	1	1.66	1.56	1.76	1.81	1.75	1.87
	2	1.54	1.47	1.60	1.61	1.56	1.66
	3	1.56	0.04	3.07	1.70	1.68	1.76
	4	1.53	1.36	1.71	1.63	1.61	1.64
	5	1.54	1.18	1.90	1.63	1.58	1.67

where $q_i^{(\min)}$ and $q_i^{(\max)}$ are user specified range and Δq_i is the predicted variation for the *i*'th quality attribute. The optimal cutting parameters are obtained through the above optimisation process. The output from the process optimisation module is the optimal choice of the process parameters for the given quality requirement.

As can be seen in the above discussions, the accuracy and reliability of the process parameter suggestion engine depends heavily on the performance of the quality prediction engine. Test with the existing small process data set produced output which agrees reasonably with opinions of the experienced machine operators.

5. CONCLUSION

This paper described the design of a hybrid expert system for plasma metal plate cutting machines. The system incorporates both the rule-based reasoning and the modelling/simulation. The quality predictions parameter suggestion from the prototype agree well with experimental measurements, even though the system is only been trained with a small number of process data sets at this stage. It is expected that the accuracy of the prediction will improve after the system has been trained by a larger number of process data sets. Use of the system is expected to reduce quality uncertainties and machine set up time. This will leads to improvements in cut quality and reduction in production cost.

The construction of the system and the methods involved are quite general and should be applicable to other industrial processes with appropriate modifications.

6. REFERENCES

[1] *FABCUT – The Fabrication and Cutting Bulletin* (July 1996) 1 "One in five" may regret buying laser

[2] J. P. T. Mo, C. Menzel, *Computers in Industry* **37** (1998) 171-183 An integrated process model driven knowledge based system for remote customer support

[3] W. B. Rowe, Y. Li, B. Mills, D. R. Allanson, *Computers in Industry* **31** (1996) 45-60 Application of intelligent CNC in grinding

[4] M. Kleiber, Z. Kulpa, *CAMES* **2** (1995) 165-186 *Computer-assisted hybrid reasoning in simulation and analysis of physical systems*

[5] Z. Kulpa, A. Radomski, O. Gajl, M. Kleiber, I. Skalna, *Engineering Applications of Artificial Intelligence* **12** (1999) 229-240 Hybrid expert system for qualitative and quantitative analysis of truss structures

[6] *Welding Design and Fabrication* (Dec 1998) 7-27 Cutting equipment

A Computation and Control Architecture of Virtual Manufacturing Shop

Zhiming Wu
Dept. of Automation, Shanghai Jiaotong University
E-mail: ziminwu@mail.sjtu.edu.cn

Keywords Virtual Manufacturing , FMS, Petri net modelling, Monitoring & Control

Abstract Virtual manufacturing (VM) is a model for an integrated application of manufacturing system. In this paper, only the information infrastructure of manufacturing shop is considered. The system architecture, modelling and analysis, include simulation and control software, can be tested and operated on a platform of distributed computer system (DCS). Generally, a client/server LAN with enough attached measurement/control terminals would be sufficient to imitate all the mechanisms and activities of such a manufacturing information system. A more extensive concept of VM that may include virtual product designing and virtual business operating is not covered in this work.

INTRODUCTION

Modern manufacturing system, such as flexible manufacturing system (FMS), has spent a large amount of enterprise budgets during its construction. Therefore, as soon as it is built, achieving a complete and efficient use of the whole system resources is especially important to quickly create more benefits and then compensate the huge investments.

Among all the technical resources of a manufacturing system, the information sub-system corresponds to the processes of message communication, functional computation and behavior control. They would not be the most expensive part in FMS, but they do give great influences to the normal and effective use of the whole facilities.

Since equipment and component of FMS are supposed to be stable and reliable, for instance the NC machines and tools are well developed nowadays, therefore, constructing high quality functional modules in the information system is of particular importance to make all facilities to be used in a coordinating/cooperating environment. Besides, approaches to check the correctness of all the operations in global FMS should be very well established.

ARCHITECTURE OF DISTRIBUTED FMS INFORMATION SYSTEM (IS)

Like many Communication-Computation-Control (C-C-C for short) Integrated Systems used in the application areas, FMS performs its activity within a fully distributed environment. Many of active resources in a FMS may play their roles independently in the real time. Different entities in a FMS can be classified by concepts of objects. For some active and autonomous ones among them, the name of actors or agents is always used.

Considering of information system (IS) in FMS simulation, although all the main C-C-C activities are able to completed by a powerful processor via a multi-thread operation system, a distributed computer system may describe the module structure and the inter-processes relations more accurately. In particular, when the message or event occurring concurrently in the real time, a fixed access policy of computer operation system thread might not be able to give the C-C-C time responses correctly because of competition of between them. Besides, as the simulation of each subsystem contains some more detail task to do, an independent processor may provide more neat software and I/O interface to handle quite a lot of requirements within a short time duration. Thus, a distributed computer system (DCS) should be introduced to represent all the system behaviours within such an information scheme.

Suppose the shop-controller, the material handling system (MHS) and each workstation (WS) are represented by a set of independent processors. Then, the DCS is given in Fig. 1.

From the idea of object-oriented programming (OOP) [Cor 1990, Boo 1994], it is able to form several real-entity classes to represent different kinds of real entities in the FMS. Among them, some active objects attached with their private methods, algorithms and data/knowledge may complete decision and control works separately in real time.

Work-pieces (WPs) in FMS haven't been clearly shown in the diagram. However, from the view of communication network, they actually play the roles of message-package flow transferred through the lower right side in the figure 1. MHS becomes a "loaded channel" or a "dynamical router" that transports WPs guided by the attached message code of package within the FMS "network".

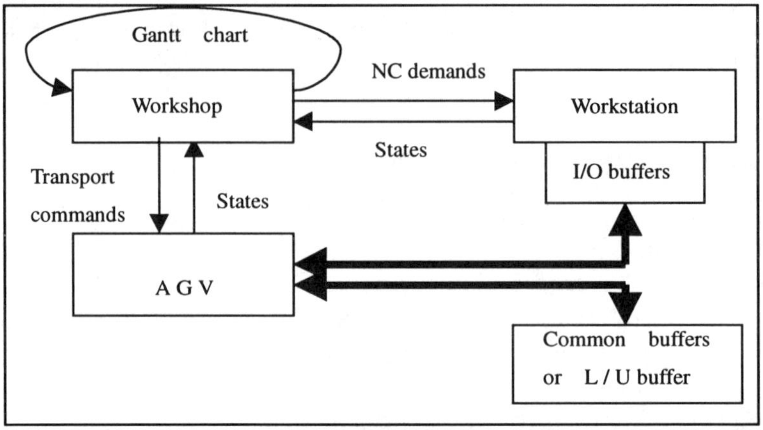

Fig. 1, Information processing within different resources in FMS

The codes attached with each WP may be considered as a "protocol" information that can be understood by MHS. They indicate the WSs where the WP can be picked up/put down and how it can be processed. Thus, during the IS modelling, a WP do need to keep its private information (like an object encapsulates its main features in OOP) such as name, state and special processing routes to show its existence, states and the expected next treatment.

All the WSs keep a rather simple and similar working cycle. Inspiring of their different types, each WS has its own private information codes that can be identified by other objects via its exchange interface. The terminal named "input-stand" on WS is used for accepting WP, and "output-stand" as a peer for sending WP. The machining loop of a WS consists of the following steps:

(1) After checking, take a right WP from the input-stand of WS.
(2) Clamp (fix) WP at the right position and changing cutter in the same time.
(3) Call the NC programming, start machining and finally push the processed WP to the output-stand of WS. Wait to be picked away.

The MHS completes the task of passing WPs (also the attached codes) within different machining loops (similar to message interleaving within DCS). Its main function includes:

(1) Take commands from shop-controller, choose the right WP, set up the picking-up position and releasing position for this WP.
(2) Ride MHS (AGV) along the track, stop AGV at predefined points, load/unload the WP towards different kinds of buffer area.
(3) Keep the WIP (work in process) in a queue for next processing.

The general shop-controller plays several very important roles in the top level, including:

(1) Receive the new batch data, provide several manufacturing schedules in off-line.
(2) Simulate these scheduling programs within the available shop model (discussed in next section), choose one of them based on their performance evaluations.
(3) Control the WP flow at the I/O terminal of FMS, monitor buffer capacity and machining quality to avoid deadlock, collision and other conflicts occurring on-line.
(4) Take the system state-messages within FMS and also its environment, especially the failure signal of tool or machine, decide if a rescheduling is needed in this moment.

SPECIFICATION AND MODELLING OF FMS VIA PETRI NET

Mainly, there are 4 classes of entities existed in a manufacturing system: WS, WP, buffer, and MHS. Several kinds of MHS can be used in the FMS, such as robot, mechanical arm and/or automated guided vehicles (AGV). Here, only an AGV is considered. Besides, there are many styles of buffer in an FMS, for instance, a single common buffer aside the AGV track, the input/output stand of WS or the load/unload area of the shop. Physically, the WP in FMS is not fully free. It should be attached with other entities in the system, existing as WP in WS, WP in buffer or WP in AGV.

In accordance with the system behavior, 4 classes of objects are defined in the FMS. By using the Unified Modelling Language (UML), the corresponding attributes and methods of these objects, and also the relation between these object classes can be explained in a special diagram (that is neglected because of limited space) [Rat 1999].

Let each operation of a WP is defined as a job. Then, for a general "n WPs-m WSs" FMS production problem, the process planning of a given WP type defines a set of jobs handling by different WS resources in accordance with a specified sequence (route). On the other hand, for a definite WS, there is a set of WPs that can be accepted to form a service chain during the same batch task. From the system analytical view, FMS belongs to the discrete event systems (DES) and keeps the feature of asynchronous concurrency. Therefore, control mechanisms based on Petri net model are preferred by most of the authors.

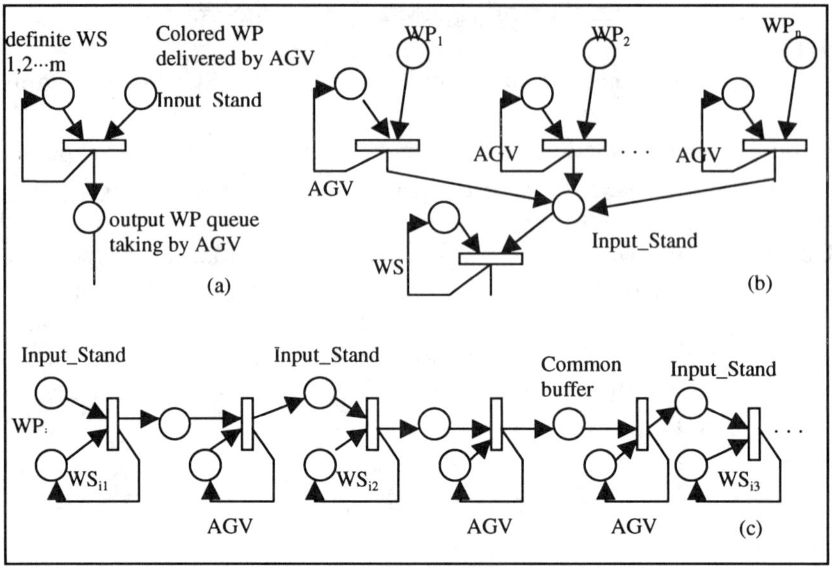

Fig. 2, (a) processing unit, (b) queue waiting for WS service, (c) WP processing sequence

Suppose different codes have been defined for identifying all kinds of WPs and WSs. According to object-oriented Petri net or high level Petri net [Jaf 1992, Wang 1996, Chen 1997], we may identify these differences with attributes of object or their "colored" codes.

A colored WP is going to be processed if it has been put into an input-stand of a "free" WS (ready to work). As soon as a job has been processed, the WP would be put to the output-stand immediately if it is empty. Then, it is waiting to be transferred to other buffer (maybe a common buffer or an input-stand of a WS) by AGV.

As an elementary job-processing unit, the queue of WPs waits for a service to be given by resource WS. The processing sequence of a given type WP during the batch can be shown in figure 2(a), (b), and (c) respectively. At the same time, a Petri net model that denotes AGV connecting all of WS loops and realising inter-loop message passing between them is shown.

Based on these sub-models, a general model of FMS can be summarized, that may combine all the activities and behaviors of the main resources, and provide a complete dynamic view for the coupled processes in whole system.

In accordance with the model framework of a FMS, several successful FMS simulation tools built on the language of OOP (C++, UML) or various Petri Net Tools have been developed nowadays [Wang 1996, Chen 1997, Qu 1999]. The modular structure of software makes the components can be reused for different batch tasks and shops. It is really not difficult to follow the heterogeneous routing traces of all the WPs in a FMS [Wu 1999].

However, the modelling efforts listed above can only create an environment to test the FMS freely. To make an efficient use to prove the potential ability of FMS, one must keep a criterion-guided (simply, benefit-directed) control strategy, that is realized by controller's command-sequence to obtain an optimal simulation result. That means different controllers must be used to supervise both the activities of WP transportation and job processing to ensure the whole system working normally and get a fairly good performance finally.

MONITORING AND CONTROL IN THE INTELLIGENT MANUFACTURING SYSTEM

Some authors have pointed out that the monitoring and control structure of the FMS shop floor is often constructed in a hierarchical way. That means there are many monitors and controllers positioned in different system level supervising respective objective aims. They can be specified from a threshold index, a critical value or an event (Boolean action) in a definite point to the general view considering the states of whole shop.

Among all these tasks, the control blocks in equipment level are relatively mature, such as PLC and continuous variable controllers. However, their monitoring usually require some special techniques/instruments to help high precision measurement and control, such as pattern recognition of a physical entity, shape characteristics extraction sensor, feature classifier and some non-parametric clustering algorithms.

Besides some special hardware instruments, the information structure of the monitoring system may build on a general artificial intelligence (AI) and sensor-fusion platform that combines measuring data, training data, knowledge and pictures of WPs/cutters positioned on the equipment and some respective algorithms. Although it is important for WSs and AGVs in real cases, from the top view of a VM system, their influences are local. They can be designed and tested separately.

The monitoring tasks in the co-ordination and decision level usually need rather simple instruments. However, high level monitoring emphasises on providing enough information for shop controller. Many indices have to be kept for monitoring at every moment/step to assure the system free from blocking. Some important factors must be followed up-to-date, including the number of WIP remaining in the shop, the number of free common buffer, the length of waiting queue for different WSs, the total throughput time of different WPs, etc. Based on such a high level monitoring, a global-shop diagnosis can be made by shop controller that must predict and solve all the faults and mistakes in an early and harmless stage.

Control problems at these levels concentrate on how to create a stable and reliable environment for the effective use of FMS. Firstly, it must make use of the Gantt chart and the records of real-time state monitoring to manage the behaviours of the whole system.

Static scheduling is an off-line task completed by high-level shop controller to arrange the WPs entering/processing sequence for different WSs. According to the schedule, Gantt chart is obtained as a reference for whole operation process to give the detail process step by step graphically. However, Gantt chart can't be precisely executed in practical shop in most cases. There is always a time deviation between the execution records and Gantt chart. Therefore, monitoring high-level states sounds important in giving a diagnosis to make sure if a deviation can be absorbed in its propagation or it makes system to crash few steps later, and checking if all the performances of the system are satisfied or accepted. That means a high-level controller takes care of the performances of whole FMS. In the normal cases, it keeps as small deviation with the Gantt chart to make FMS work in order and in a high efficiency. In an abnormal case, it follows the control strategy to keep the system free from deadlock, at least it may sure that the blocking state can be recovery from an accident very quickly.(even in rescheduling)

Besides making Gantt chart in the off-line case, adjusting input interval between WPs is one of the main activities of shop controller. The problem is when too many WPs allowed entering, the system might be blocked, but less input always causes low production efficiency and also low resources utilities. Since the information structure of FMS belongs to a fairly large and distributed discrete event system, a formal model and powerful numerical algorithm should be considered. The typical problem that need to be solved in high-level is generally with analytical approach or heuristics, that contains:

(1) An algorithm to make static/dynamic scheduling, re-scheduling and AGV transportation scheduling subject to a criterion considering production benefit [Zhao 2001].
(2) An algorithm to manage and control the time length/interval of I/O queues of material flow subject to both production effectiveness and safeness [Ram 1995, Mur 1986].

The high performance properties expected by modern manufacturing systems encourage people to develop more advance techniques/methods for practical uses. Nowadays, many mathematical programming methods (including operations research OR), soft-computing algorithms (including evolutionary, genetic, simulated annealing and fuzzy) and AI/expert systems (AI/ES) have been used to establish a complicated algorithm-base for solving these problems [Baker 1998]. The information structure of high-level problem-solvers is shown in figure 3. For a FMS input controller, its algorithm has to work in parallel with the modules of simulation and monitoring to obtain the results for them in on-line uses.

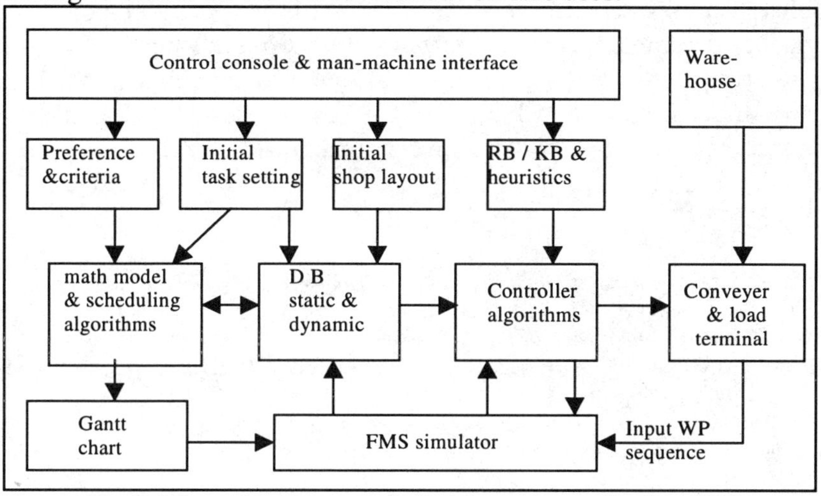

Fig. 3, Information structure of scheduler and controller in a FMS

Although different approaches have been employed for different monitoring and control problems, there are still many resources and information must be shared by those problem-solvers. Some obvious ones are:

(1) Static/dynamic database storing documents of shop resources, batch task and operation records,
(2) Rule-base/knowledge-base and the inference mechanisms that may generate many new conclusions according to the up-to-date system states in dynamic DB,
(3) Data/knowledge and their various reasoning results achieved for solving one special problem in real time would be useful for other problem-solvers' reference.

It means a blackboard architecture [McM 1996] of manufacturing systems is helpful.

CONCLUSION

The paper gave a complete information structure for a modern manufacturing shop. From the structure model of main components to the overall system, a formal model is given for specifying the main requirements of FMS. Several realization approaches and techniques for the virtual FMS, such as its simulation, monitoring and control are discussed.

Corresponding to the practical needs of objective system and the technical features of products in both hardware/software, an experienced designer may decide the number of functional modules which should be integrated together for the given project. It is worth noting that some modules recommended in this paper are still being developed from the practical and commercial view. However, the rapid advances of IT, AI, SE, and also the computer/mechatronic devices and techniques would much improve the abilities in solving most of these problems. An ideal, fully automated and person-less VM shop can be realised quickly.

REFERENCES

[1] Baker, A.D., *A survey of factory control algorithms that can be implemented in a multi-agent heterarchy: dispatching, scheduling and pull,* Journal of Manufacturing Systems, Vol.17, No.4, 1998, pp297-319

[2] Booch, G., Object-Oriented Analysis and Design with Applications, Benjamin/Cummings, 1994

[3] Chen, K. and S.Lu, *A Petri net and entity-relationship diagram based object-oriented design method for manufacturing system control,* Inter. Journal of Computer Integrated Manufacturing, Vol. 10, No. 1, 1997, pp 14-28

[4] Coad, P. and E.Yourdon, Object-Oriented Design with Applications, Benjamin/Cummings, 19 90

[5] Jafari, M., *An architecture for a shop floor controller using colored Petri nets*, Inter. Journal of Flexible Manufacturing Systems, Vol. 4, No. 2, 1992, pp 159-182

[6] McManus, J.W.and W.L.Bynum, *Design and analysis techniques for concurrent blackboard systems,* IEEE Tr. on System, Man and Cybernetics, part A, Vol. 26, No.6, 1996, pp669-680

[7] Murata, T., N.Komoda, K. Matsumoto and K. Haruna, *A Petri net based controller for flexible and maintainable sequence control and its application in factory automation*, IEEE Trans. Industrial Electronics, Vol. 33, No. 1, 1986, pp 1-8

[8] Qu-yang, C., *Developing a Petri net based simulation model for a modified hierarchical shopfloor control framework,* Inter. J. of Production Research, Vol. 37, 1999, pp3139-3167

[9] Raman, Narayan, *Input control in job shops,* IIE Tr. Vol.27, 1995, pp 201-209

[10] Rational Software about Rational ROSE, http ://www.ration.com/UML/Rosel.html 1999

[11] Venkatesh, K. and M.C.Zhou, *Object-oriented design of FMS control software based on object modelling technique diagram and Petri nets.* Journal of Manufacturing Systems, Vol. 17, No. 2, 1998, pp 118-136

[12] Wang, L., *Object-oriented Petri nets for modelling and analysis of automated manufacturing systems,* Computer Integrated Manufacturing Systems, Vol. 26, No. 2, 1996, pp 111-125

[13] Wu, Z., *CEM/T nets, a high level Petri net for FMS modelling,* Inter. Journal of Intelligent Control and Systems, Vol. 3, No. 3, 1999, pp 377-387

[14] Zhao, C. and Z.Wu, *A genetic algorithm approach to flexible scheduling problems in FMS,* will appear in Inter. J. of Flexible Manufacturing Systems, Vol. 13, No. 1, 2001

Managing the Flow of Information on the Factory Floor

Docki Saraswati, Sumiharni Batubara, Reny Mulyadi, Amelia Mulyadi
Industrial Engineering Department, Trisakti University, Jakarta, Indonesia
Em: dsrianto@cbn.net.id

Keywords Computer Integrated Manufacturing, Computer Aided Process Planning, Group Technology, Scheduling.

Abstract This paper presents findings of a study aimed at developing a suitable information system for manufacturing feature interactions in handling manufacturing process planning. The main objective is to provide a Virtual Manufacturing System Planning based on Computer Integrated Manufacturing. It begins with the incoming parts and utilises the proposed parts database that serves as a source for the parts codification module. This module requires Computer Aided Process Planning to provide the routing plans that will identify the machines and tools required. Utilising tools and machines databases will provide the information for parts scheduling, which determines the promised delivery time.

1. INTRODUCTION

Controlling the factory floor has been an elusive goal for manufacturers, especially when customers are more demanding than ever before. Manufacturers should respond quickly to the demands of the customer since the product innovation cycle has changed, from manufacturer- oriented to customer-oriented market [9]. Customers take it for granted that the manufacturer will deliver a product on time. Integrating the systems on the production floor is critical to respond to the rapid and dynamic changes in the product cycle.

The aim of this research is to develop a suitable information system to handle the flow of information on the factory floor. In this regard, information technology plays a dominant role in integrating the whole system.

The proposed flow of information is divided into three parts. First, information on incoming parts, which includes the routing plan using a specific NC machine and tools. Second, it identifies machine availability and determines the use of tools, or vice versa. Finally, it assesses the availability of tooling to determine the relationship between the machines and tools. This allows a more systematic approach to production scheduling, which will determine the use of tooling. The interactivity will also ensure the appropriate delivery time.

On a larger scale, this study will make use of the Internet to create a forum that allows customers access to available information at a given time.

An extensive research will be conducted at an industrial plant located at the Jababeka Industrial Estate in West Java. This plant specialises in spring compression, spring extension and spring torsion.

2. RESEARCH PROBLEM

The current process plan in the plant is performed manually. This approach requires an experienced and knowledgeable planner who has access to data of machine availability, machine capabilities and stock availability. This also means that a planner is involved in retrieval and manipulation of a great deal of information. The result of the process plan is then manifested in the form of printed text, lists and drawing. Moreover, this approach has a great possibility of producing different routing-sheets for similar parts at different times. Therefore, the use of computer has certainly made planning more efficient [5].

This research will explore tools and machines availability. Current planning and controlling of tools for the CNC/NC machines is done manually, using *stock cards*. Using these cards, the coiling production sub-department records all the data flow of tooling. Usually, after each process is completed, the tools remain with the machines. The operator believes this is critical in saving time for when the next process comes. Unfortunately, by doing this, the coiling production sub-department does not have information on the tooling status and this presents a constraint in, or even prevents them from, making a production schedule. The transformation of tooling data into information that benefits production scheduling is therefore required.

3. RESEARCH METHODOLOGY

This paper is presented as part of an on-going large research project whose main objective is to gain a Virtual Manufacturing System Planning based on Computer Integrated Manufacturing (CIM), particularly on the production floor. Based on the information made available, it is possible for managers to make decisions in remote locations. Information Engineering is used as a methodology in applying the structures techniques to the enterprise. Its consists of 4 stages: Information Strategy Planning, Business Area Analysis, System Design and Construction [6].

The framework of research methodology is described in figure 1:

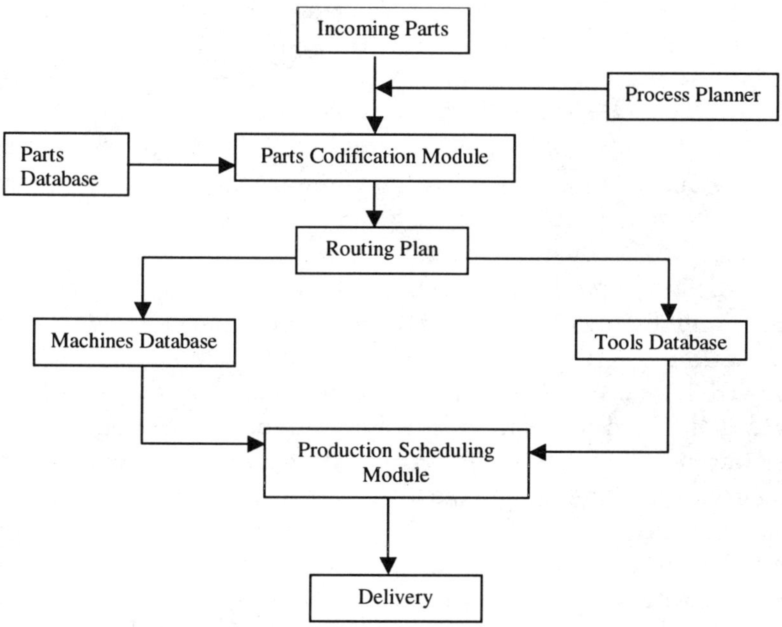

Figure 1 : Framework of research methodology

This research will be divided into three areas:

1. Parts Codification Module
2. Production Scheduling Module
3. Performance System Module

The information system is crucial in fulfilling the aim of this research. By utilising parts, machines and tools databases, the information required by a user becomes accessible. Several papers have been published on the issue, including the concepts of designing an Internet-based system for manufacturing system, managing the flow of information and redesigning it under the web-based system [1,3,9].

4. PARTS CODIFICATION

A process planner is required to identify the incoming parts. For this purpose, integration of technologies that include CAD, CNC or NC is critical.

The incoming parts are classified into parts families, using Group Technology (GT) which is based on their similarities in design or manufacture characteristics and grouping them into part families (Heragu, 94). Depending on how the conceptual scheme is constructed, there are three major types of classification: visual method, part coding analysis and Production Flow Analysis (PFA). This research uses a part coding and classification analysis (PCA) to identify similarities. Part characteristics-based systems form part families according to their design-based features: shapes, sizes and tolerances [7].

Further, the objective of the GT approach is to identify the grouping of parts that share common processing requirements based on their routing sheets. The available information explains the relationship between parts, machines and tools that are required to process them.

A Computer Aided Process Planning (CAPP) system is based on the GT approach, and basically constitutes two approaches, namely variant and generative. In the variant approach, the standard operating plan is identified through GT. If a new plan is required, then the existing standard plan can be retrieved and edited to meet the requirements of the new parts. There are several stages required to establish a variant approach, such as inputting existing GT codes, part process plan database, part drawing database, variant process plan generator and finally standard process plan [2]. In the generative approach, a process plan is made based on information available in the manufacturing database. This system can generate the required and sequenced operations for a new part. All information on manufacturing has to be stored as software. This system utilises decision logic as a tool to generate a process plan. Should the program detect that the information for the new part is not available in the database, then the generator system will proceed to create a new process plan, based on a GT code. This new plan is generated from the interactive input, which is done by a user. This generative approach can be divided into five stages: interactive input, generative process plan generator, operating plan rules, operating plan requirements and part drawing.

The coiling department is a workshop that employs 50 machines consisting of CNC T20 (17 units), HTC (11 units), EH CLS10 (8 units) and EH CLS16 (2 units). Figure 2 shows the product routing in this department.

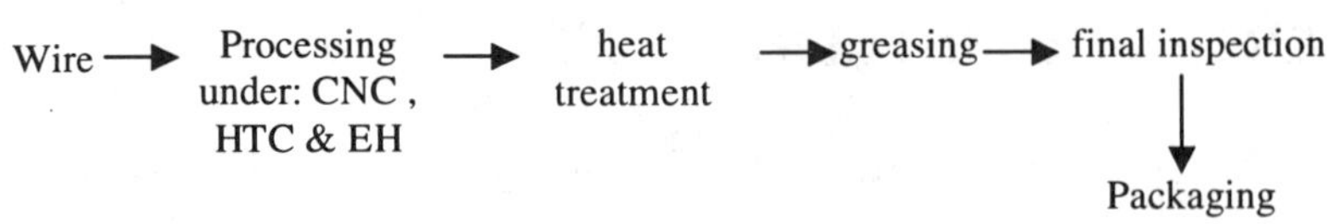

Figure 2: Product routings

There are 3 types of machines: HTC for spring compression, EH for spring extension and CNC for string torsion. Based on parts specification and parts drawing, the codes for spring compression and extension consist of 19 digits, while that for spring torsion is made up of 22 digits.

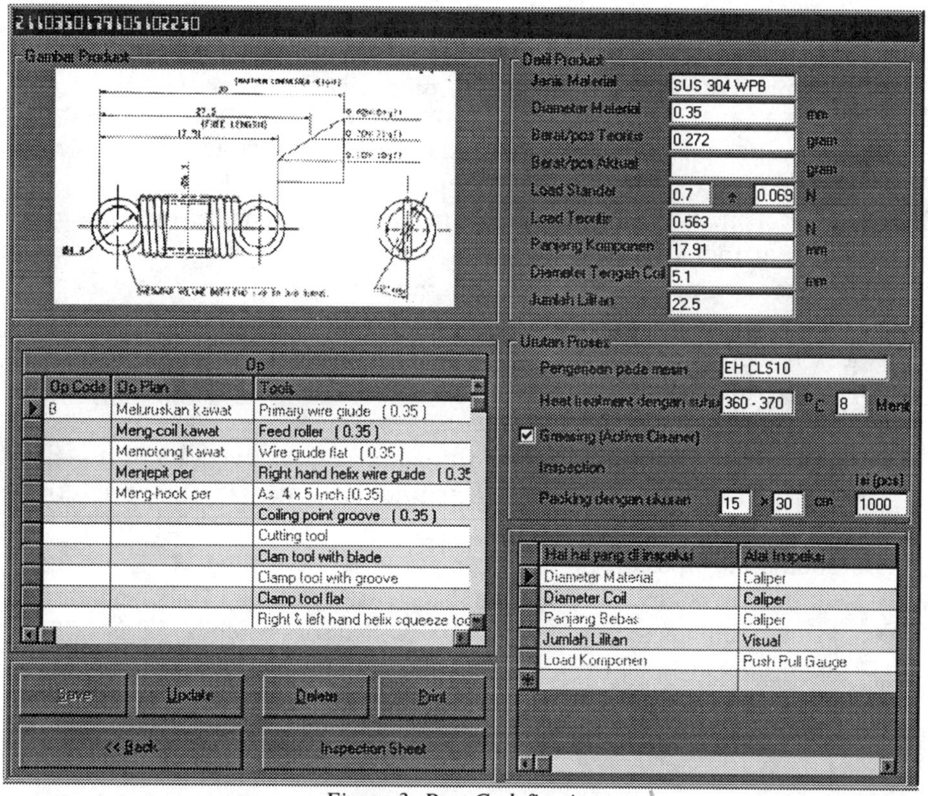

Figure 3: Part Codefication

5. SCHEDULING

Tool requirement for each processing on each machine is associated with part characteristics, such as the diameter of the raw material (wire), the diameter of coil and the number of bending. This means that the variability in type of tools is enormous and can cause problems in controlling the tools.

Therefore, based on the routing plan that is produced under the CAPP system, the PPIC department has to locate the assigned tools using the database. For simplification, the tools have been divided into two categories, main tools and additional tools. They are classified into 5 tool-cutting families, namely adapter tool (9 digits), rolls tool (6 digits), coiling tool (8 digits), bending tool (6 digits) and additional tools (4 digits).

After receiving an order that carries a code number, the PPIC department makes a production scheduling based on the priority sequencing rule for flexible flow shop [8]. There are several possible ways to execute the production schedule and the algorithm is as follows:

1. Check tools inventory, if tools are available, then proceed to step 2, otherwise go to step 3.
2. Check machines availability, if the machine is idle, then go to step 4, otherwise search for the machine that will be available in the shortest time. Unload the tools, then go to step 4
3. Check tools that are attached to the machine, if it requires only one type of tool (e.g. tool a), then go to step 6, otherwise go to step 8.
4. Setting tools to process the new parts, go to step 5
5. Process the new part according to the routing plan.
6. Check if more than one tool is required (number of tools a), set up times for when all tools are ready for the new routing plan. Select the appointed machine. Go to step 7
7. Check the types of tools that are still needed on the machine for the new routing plan, and unload unselected tools. Calculate the total time required. Go to step 9
8. Determine the types of tools that are required. Go to step 11
9. Check if there is any idle machine, otherwise go to step 10. Calculate the total time required to process the part on that idle machine (T2). If T2 < T1, then unload tools from a selected machine. Set up tool required on the machine. Go to step 5, otherwise go to step 10
10. Unload unused tools from the machine. Set up required tools on the machine then go to step 5.
11. Check the number of each type of tools > 1, then go to step 12, otherwise go to step 13.
12. Determine the time when the types of tools are available, then go to step 13
13. Select the suitable machine. Choose the selected machine, then go to step 7

6. IMPLEMENTATION

This model is run under Visual Basic 6.0, using Microsoft access for the databases. Several data dictionaries are included, such as data dictionary for customers, delivery orders, parts knowledge-based, types of tools, tools status, etc. The scenario is implemented in 20 code parts, which is the delivery order part in March 2000 for the CNC 10 T machine. The result is measured by two performances, namely makespan and mean flow time.

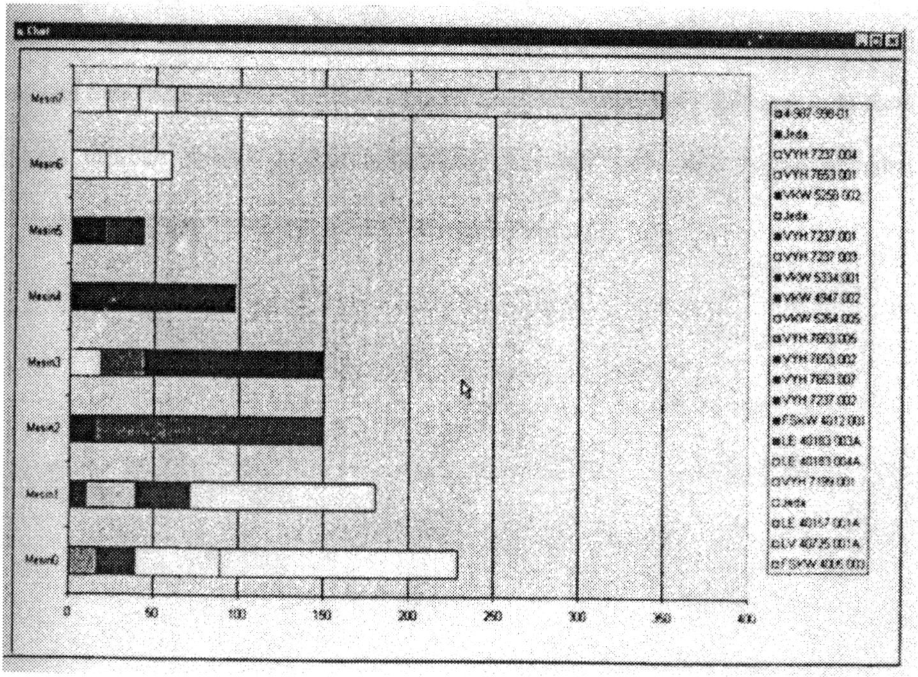

Figure 3: Submenu Form Gantt Chart

7. CONCLUSION

This paper describes the requirements of databases to support parts codification and scheduling based on delivery orders, machines and tools availability. Computer Aided Process Planning is used to provide parts codification. The resulting routing-plan will lead the parts to the machines and tools that are required.

8. ACKNOWLEDGEMENT

This research is conducted under the DCRG (Domestic Collaborative Research Grant) Program – URGE (University Research Graduate Education) Project for the Industrial Engineering Department, Trisakti University, under the supervision of Dr. Kadarsah Suryadi of the Bandung Institute of Technology (ITB).

REFERENCES

[1] Behera, Kirt B., *How to Redesign the Factory for Flow Manufacturing-Integrating Product, Process and People*, APICS International Conference Proceedings, 153-157, New Orleans, LA, October 24-27, 1999.

[2] Feng, Shaw C., Chun (Chuck) Zhang, *A Modular architecture for rapid development of CAPP systems for agile manufacturing*, IIE Transactions, 30, 893-903, 1998.

[3] Gunasekaran, A. *Agile Manufacturing: A Framework for Research and Development*, International Journal Production Economics 62, 87-105, 1999.

[4] Heragu, Sunderesh S., *Group Technology and Cellular Manufacturing*, IEEE Transaction Systems, Man and Cybernetics, vol. 24, no. 2, 203-215, February 1994.

[5] Houtzeel, Alexander, *A CAPP Perspective*, Modern Machine Shop, 72-79, November 1995.

[6] Martin, James, *Information Engineering: A James Martin Trilogy, I. Introduction, II. Planning and Analysis, III. Design and Construction*, PHI, 1989, 1990, 1991.

[7] Offodile, O.Felix, Abraham Mehrez, John Grznar, *Cellular Manufacturing: A Taxonomic Review Framework*, Journal of Manufacturing Systems, vol. 13, no. 3, 196-220, 1994.

[8] Pinedo, Michael, Xiuli Chao, *Operations Scheduling, with applications in Manufacturing and Services*, Irwin-MGH, 1999.

[9] Wang, Z.Y., K.P. Rajurkar, A. Kapoor, *Achitecture for Agile Manufacturing and Its Interface with the Computer Integrated Manufacturing*, Journal of Material Processing Technology 61, 99-103, 1996.

Holonic Architecture for Shop-floor Control

A. Tharumarajah, S. Walsh
CSIRO Manufacturing Science and Technology, Australia
Email: a.tharumarajah@cmst.csiro.au

Keywords Manufacturing, holonic manufacturing systems

Abstract Future manufacturing requirements envisage a highly distributed and flexible arrangement of manufacturing entities to permit easy (re) configuration and adaptability to changing operating and demand situations. In such a system, the manufacturing entities, like machines, parts, tools and transports will function similar to autonomous wholes co-operating with each other to achieve production goals. Based on the concepts of holonic systems, this paper proposes an architecture that enables the construction of such flexible systems in order to satisfy the future requirements. Application of the architecture is illustrated for a machining cell.

1 INTRODUCTION

A holonic system has the essential attributes of autonomy and co-operation of its entities, called holons [7]. These attributes of the entities endow the system the capabilities to become adaptive to changes in the operating environment whilst making it easier to re-arrange or re-configure the organisation. This concept has been successfully used to propose holonic manufacturing systems (HMS) [10]. It is envisaged that HMSs will be capable of rapid response to changes in customer demand and operational requirements, through constant adaptation of their function and organisation. A holonic shop floor would resemble a community of machine, transport, part and other types of holons that interact with each other to plan, schedule and execute their activities.

Architecture to realise a holonic shop floor would include communication among the manufacturing holons and control of their individual and joint activities. It has to also consider the problem solving organisation and the mechanisms of co-ordination for addressing typical production optimisation problems that a shop floor faces. The architecture proposed here is comprehensive in addressing these issues. It builds on a few generic types of holons that provide a high degree of self-similarity , which in turn reduces the demands on integrating new components and benefits the quick and easy reconfiguration. Application of this architecture is illustrated for a machining cell.

The paper is organised as follows. Section two discusses the concepts of holonic systems. Section three proposes a definition of holon types suitable for manufacturing. Building on this definition, the next section describes a holonic architecture for a shop-floor environment. Section six illustrates the application of this architecture to a machining cell. This is followed by conclusion.

2 HOLONIC CONCEPT

The concept of holonic systems portray a system being made up of a set of autonomous entities or holons that can co-operate amongst themselves and self-steer their total behaviour. This self-steering quality is one of its abilities to self-organise to cope with changes in its environment either by adapting its behaviour and/or actively re-configuring the arrangement of its constituent parts. But, how does this quality come about in holonic or other similar systems? The answer lies in the connections, interactions, and feedback loops between the parts (i.e. holons) of the system. These parts act in a manner that is mutually accommodating to maintain self-consistency. That is, in one respect the entities act to preserve themselves, but to do so they have to be accommodating, otherwise their very existence as a viable unity may be threatened. These two typical properties are fundamental and are put forward by Koestler [7] as the attributes of autonomy and co-operation of holons.

Koestler defines the systems made up of holons and exhibiting characteristics of self-organisation as holarchies. A holarchy is not confined to a single level, rather it can be a multi-level structure. At each level, though, the members of the holarchy (the members can be holons or holarchies) would have the twin attributes of being autonomous and co-operative. This way, there is no strict hierarchy as is common in organisations, but there exists an invisible boundary that encapsulates and defines the environment in which a set of holons operate. In order to maintain these characteristics, a holarchy defines the basic rules of co-operation for its member holons and thereby limits their autonomy.

3 A BRIEF REVIEW OF LITERATURE

Both single- and multi-level holonic architectures have been proposed. The former is where there is no definition of a holarchy and only a set of specific types of holons and how they co-ordinate their actions are defined. For instance, the specific types may include functional holons such as scheduling, planning, execution and monitoring [1, 8, 5, 6]. In other cases, the holons represent manufacturing entities such as machines and parts, and have embedded functional capabilities to plan and schedule [3, 4]. Co-ordination among the holons can be an in-built function [10] or a separate holon by itself [3, 4, 9].

Constructing a multi-level structure by simply extending a single-level architecture can be difficult without appropriate definitions of holarchies to represent the representing diverse characteristics of a system. To this end, Fletcher and Deen [2] propose a HMS global architecture. In this architecture a Co-ordination Domain provides the mechanism for the holons to work together with other holons on joint tasks. It is task oriented rather than performance oriented. The difference is that the former is for short-term gains while the latter provides for stability and long-term performance optimisation as intended by the definition of a holarchy. Further, purely task-oriented definition often lacks the representational requirements of manufacturing to model capability-based organisational groupings (e.g. a machining cell) as holarchies. Also, it becomes difficult to specialise the implementation of task-oriented holarchy definitions.

The PROSA architecture [12] overcomes this limitation by proposing and building single- and multi-level holonic architectures on four basic types of holons suitable for manufacturing:

- Resource holon: represents a physical entity (like a machine) with procedural knowledge and functions to make allocation decisions and organise co-operation with other holons.
- Product holon: represents a product with the required processing knowledge to assure correct processing.
- Order holon: represents a customer or make-to-stock order, or generally a task that need to be executed.
- Staff holon: represents specialist knowledge for taking decisions, such as an advisory role.

To define multi-level structures, the basic types are used. For instance, holons of a type are clustered together to form one aggregated holon. Also, in other cases such as an Order holon, the architecture proposes exclusivity in ownership of needed resource holons by an order holon to prevent deadlock situations. The PROSA architecture raises two questions. First is how restrictive a forced assignment of ownership will be on the flexibility of the system. For example, how will an Order holon deal with a situation requiring routing flexibility or to deal with job transfers when disturbances occur. . Secondly is how well it supports temporary membership of holons in a holarchy. Further, a clear definition of holarchy types are notably absent in the proposed architecture and is left to the specifics of implementation.

4 PROPOSED ARCHITECTURE

4.1 Basic entities

The architecture we propose extend the two basic types of entities in a holonic system, namely the holon and the holarchy. To further clarify the definitions of these two types, a holon is differentiated by its atomic nature, i.e. it provides an individual function and cannot be further divided. In contrast, a holarchy provides a societal function in that it functions so as to facilitate the smooth operation of its members.

In order to differentiate between these two types, the holons are aptly named a-Holons to highlight their atomic nature. A holarchy is named as a Co-operation Domain (CD). The definitions of both the a-Holons and CDs are explained below:

4.1.1 a-Holon

The basic features of an a-Holon are:

- It is a communication environment capable of sending and receiving messages to or from other a-Holons or CDs.
- It has information processing and physical processing capabilities where appropriate.
- It can engage in joint decision making activities by co-operating with other a-Holons, such as executing a Contract-net negotiation process.
- It can request and become a member of a CD and where appropriate relinquish its membership.
- An a-Holon, however, has to be a member of at least one CD, its home base.

4.1.2 Co-operation Domain

For practical applications, the invisible boundary of a holarchy can be difficult to construct. Also, while holarchies accommodate only stable associations there may be associations that are temporary. The concept of Co-operation Domain is introduced to deal with these difficulties. It can be thought of as a mechanism for the holons to initiate a holarchy or an association and its membership and manage the interactions. It would define common rules of engagement and, in essence, provides a closure that encapsulates the interactions among its members and thus acts similar to a whole. The management of the interactions can be left to the a-Holons or to a co-ordinator role.

The basic features of a CD are as follows:

- A CD is a communication environment capable of sending and receiving messages from its members or from other outside CDs or holons.
- Initialises and retains an association or membership among a set of holons. The member holons may request entry or exit from the membership thereby allowing temporary membership.
- Permits multi-level structures to be constructed by allowing definition of CDs as members.
- Facilitates a smooth (and optimised) operation of the association: The facilitator role is specialised according to the implementation of a CD.

In addition, a CD has a shared memory space, similar to a memory table, for storing information that is transparent to all the member holons. This information is placed and deleted by the members only. It serves as a way of providing status information of members and tasks being executed. Where a holon has multiple memberships, it is assumed that it has sufficient information to access the particular shared memory appropriate for the purpose.

4.2 Constructing a holonic structure

A holonic structure is created by first defining individual holons and associating them to a base CD. The base CD acts similar to a home. When dynamic associations are created, a holon that is already a member as defined initially can become a member of the new CD. Multiple memberships can also be defined initially, but only one base CD can be defined for a holon.

By implication, a holon uses its CD's services for communicating with other holons as shown in figure 1. Initially, the architecture specifies the CD and the member holons. This information is recorded in a CD register for address look-up. In this implementation, the CD register is common to all CDs. To keep up-to-date on changes, the contents of a register are communicated to all CDs (if operating in different platforms) when membership details change.

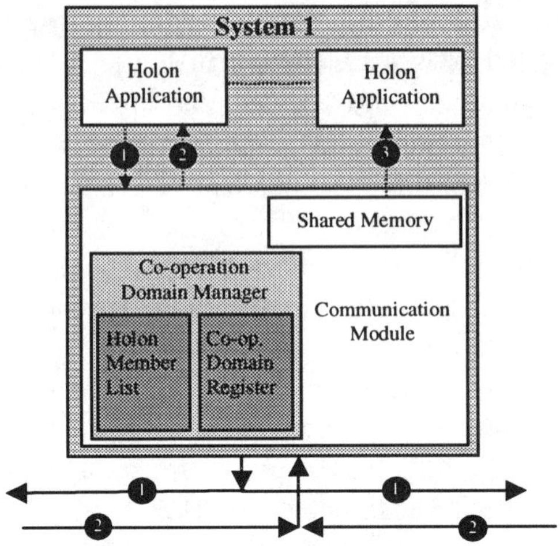

❶ Holon sends message to another Holon (forwarded by the base Co-operation Domain)

❷ Holon recieves message from another Holon (forwarded by the base Co-operation Domain)

❸ Holon inserts/updates/retrieves data object (s) in shared memory

Figure 1: Basic Communication Architecture

5 Demonstration application

The application of the proposed architecture is illustrated for a machining cell shown in figure 2. The cell has four machines capable of producing similar part types. The parts are brought in and taken out of the cell buffer by a gantry transport. Another gantry transport services transportation between the buffer and the machines.

The holonic model of the cell as shown in figure 3 is constructed using the following specialised holons:

- Machine holon: It is derived from an a-Holon type. Other than the basic functions of an a-Holon, a Machine holon has capabilities to process parts using the processing information supplied by a part processing database (PPDB as shown in figure 3), entertain calls for proposal from Order holons and schedule the jobs.
- Order holon: This too is derived from an a-Holon and assumes the definition of a customer order as defined in PROSA [12]. An Order holon has capabilities to call for proposals from the Machine holons and select a suitable machine for the job.
- Cell Broker (CBR): This represents the Co-operation Domain of the machines in the cell. The functions of the broker include managing communication among its members and with other CDs, and job transfer among machines when disruptions or delays are encountered. Also, it

provides rules that limit the bidding freedom of the machines, such as the number of awarded bids a machine can have. Job transfer among machines is made possible by machines periodically interrogating the shared memory managed by the CD. That is, once a machine decides that it cannot fulfil the award conditions, it gives up the job by putting it in the shared memory for others to consider.

- Production Manager (PM): This represents a CD for the Order holons. Its main function is to prioritise the order release into the shop. Other functions such as grouping orders for processing are also contemplated.

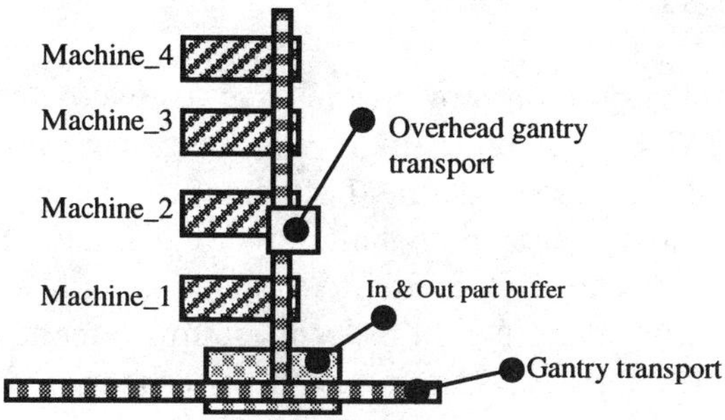

Figure 2: Layout of cell

The holonic structure of the cell is defined top-down by first defining the CBR and then creating and associating the Machine holons as members. As alluded to earlier, an a-Holon is at least associated with one CD. A similar approach is taken for the Orders holons by first defining the PM and associating the Order holons as the customer orders arrive.

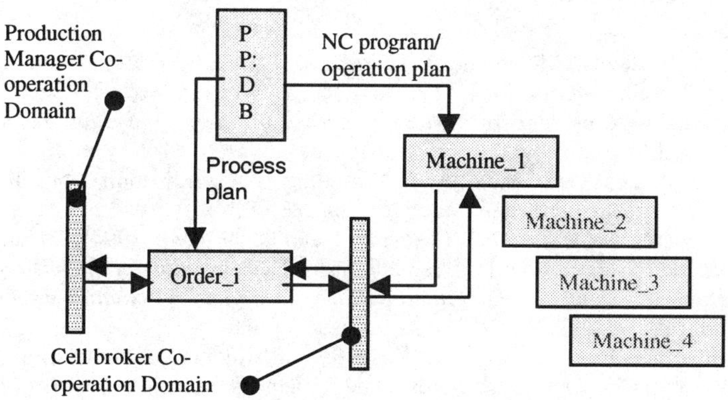

Figure 3: Holonic architecture of cell

The operation of the holonic system is initiated by the behaviour embedded in the holons. For instance, when the Order holons are first created they implement an unprovoked action of calling for bids to process the required jobs to execute the customer order. This, in turn, makes the Machine holons that receive the call for proposal message to respond by submitting bids that are subsequently evaluated and a job awarded to a select Machine holon.

Extensions of this architecture are in defining specialised holons for transports and input/output buffers to complete the holonic structure of the cell.

5. CONCLUSION

The holonic architecture proposed here have many advantages. Using the basic a-Holon and CD types both single and multi-level holonic structures can be constructed. It allows stable holarchies to be constructed while permitting the members of such a holarchy to maintain their individual autonomy properties. The CD is defined so as to allow local rules of operation to prevail in holarchies in order to optimise member performance including bidding for jobs and dealing with disturbances. The architecture lends itself to be built progressively and permits flexibility of organisation by allowing changes in CD membership. The communication structure proposed is general enough to be implemented across different platforms.

Currently, the holonic cell architecture illustrated in this paper is being used to develop a holonic manufacturing system of a multi-cell industrial FMS testbed.

6 REFERENCES

[1] Agre J.R., Elsley G., McFarlane D., Cheng J. and Gunn B., (1994), "Holonic Control of a Water Cooling System for a Steel Rod Mill", *Proceeding of the Forth International Conference on Computer Integrated Manufacturing and Automation Technology,* Oct. 10-12, pp. 134-141.

[2] Fletcher M. and Deen M. (1999), Modelling of Coordination Domains for Holon Interactions, HMS Workshop, Stuttgarrt, June 23-29.

[3] Guo L., Hasegswa T., Luh P.B., Tamura S., and Oblak J.M., (1994), "Holonic Planning and Scheduling for a Robotic Assembly Testbed", *Proceeding of the forth International Conference on Computer Integrated Manufacturing and Automation Technology,* Oct. 10-12, pp. 142-149.

[4] Guo L., Luh P.B., and Kyoya Y., (1998), "Holonic Manufacturing Scheduling: Architecture, Cooperation mechanism and Implementation", *Computers in Industry,* No 37, pp. 213-231

[5] Heikkila T., Jarviluoma M., and Juntunen T., (1997), "Holonic Control for Manufacturing Systems: Functional Design of a Manufacturing Robot Cell", *International Journal of Integrated Computer-Aided Engineering,* Vol. 4, No. 3, pp. 202-218.

[6] Heikkila T., Kollingbaum M., Valckenaers P., and Bluemink G., (1999), "manAge: An Agent Architecture for Manufacturing Control", *Proceedings of the Second International Workshop on Intelligent Manufacturing Systems,* September 1999, Leuven, Belgium, pp. 127-136.

[7] Koestler, A. (1967), *The Ghost in the machine*, Arcana books, London.

[8] McFarlane D., Marett B., Elsley G., and Jarvis D., (1995), "Application of Holonic Methodologies to Problem Diagnosis in a Steel Rod Mill", *IEEE International Conference on Systems, Man and Cybernetic, International Systems for 21st Century,* 22-25 Oct, Vancouver, BC, Canada, Vol. 1, pp. 940-5.

[9] Ng A.H.C., Yeung R.W.H., and Cheung E.H.M., (1996), "HSCS - The Design of a Holonic Shopfloor Control System", *Proceedings 1996 IEEE Conference on Emerging Technologies and Factory Automation. ETFA '96,* 18-21 Nov., Kauai, HI, USA, Vol. 1, pp. 179-85.

[10] Seidel, D. and Mey, M. 1994, IMS - Holonic Manufacturing Systems: Glossary of Terms, In Seidel D. and Mey M. (eds), *IMS - Holonic Manufacturing Systems: Strategies Vol. 1,* March, IFW, University of Hannover, Germany.

[11] Tharumarajah A., and Wells A. J., (1997), A Behaviour-Based Approach to Scheduling in Distributed Manufacturing Systems, Integrated Computer-Aided Engineering, Issue 4(4), pp. 235-249.

[12] Van Brussel H., Wyns J., Valkenaers P., Bongaerts L., and Peeters P. (1998), Reference architecture for holonic manufacturing systems: PROSA, Computers in Industry, v. 37, pp. 255-274.

Software Technology for Design System Integration

P. Bertok
Department of Computer Science, Royal Melbourne Institute of Technology, Australia
Em: pbertok@rmit.edu.au

J.P.T. Mo
CSIRO Manufacturing Science and Technology, Australia

Keywords CAD, Application integration

Abstract Even fairly recent software can become easily outdated due to the very fast development in software technology. It is very common that the business logic of the application is still up to scratch, but the user interface, or the development platform has become obsolete. Re-use of legacy components of legacy systems is preferred to redevelopment. An often-used solution is integrating the existing software into a new system via re-engineering. This paper proposes an integration platform based approach for integrating legacy and new components into CAD systems. It analyses different integration patterns and elaborates on their usability. The approach described was successfully applied to the development of the system Virtual CAD.

1 INTRODUCTION

New software techniques and technologies are introduced at a very fast pace; user interfaces, development platforms and other components are superseded before products using them reach the end of their lifetime. Hence, existing systems can become outdated due to external reasons, while their business logic is still performing well. It becomes necessary to reengineer the "old" system, but that may have some difficulties. Modifying the software to reflect the changes in the new environment is usually not feasible, as the documentation may be incomplete, the developers of the original system are already not around to help etc. and the effort needed may be difficult to justify.

Replacing whole applications just because of changes in the environment is also not feasible in many cases. Applications, or at least their most important features may have to be retained for other reasons as well, e.g. because too many users or other applications depend on it. Hence, new methods have become necessary, which reduce the effort of reengineering.

The emergence of application service providers (ASPs), who offer sophisticated applications to organizations, created additional demand for the integration of applications, as shown in [1].

This paper gives a brief overview of reengineering methods, and presents an example of reengineering and integrating different components into a new system called Virtual-CAD. The integration described here consists of two major steps: (1) connecting legacy systems to the new system, and (2) adding new services or utilities, such as access control, to the new system.

2. REENGINEERING AND SYSTEM INTEGRATION

There are two basic approaches to reengineering legacy systems. One approach requires a detailed understanding of the legacy system, all its internal architecture and operations. This is called *white-box* reengineering. The other method requires only a minimum understanding of the legacy system's interfaces, as it attempts to access legacy system functions via these interfaces, without any detailed understanding of the operation. This is also known as *black box* reengineering. The white-box approach provides an opportunity to make major changes to the legacy system, which can be beneficial in the long term. The work involved is substantial, and possibly includes some degree of reverse engineering the old system before making the changes. The black-box method, on the other hand, avoids any changes to the internal workings of the legacy system. The work required is minor compared to the white box approach, and interruptions to the operation of the system are also reduced.

2.1 Wrapping

The black-box method is usually implemented by using encapsulation. The legacy system is accessed via external interfaces only, the behaviour of the system need not be fully known. The essence of the legacy system is maintained in its original, untouched form. The black-box method with encapsulation is often referred to as *wrapping*, and distributed object technology, such as CORBA, is very often used for implementation.

2.2 Wrapping methods

There are three major wrapping strategies, as discussed in [2]. These are the Ad-Hoc, Legacy-First and Component-First methods. The Ad-Hoc method, as the name suggests, is a one-off solution, individually tailored to a particular application. The work is carried out without a standard methodology and without any general characteristics that can be ported from one case to another. This method is only a temporary fix: the result is a unique product that can later display symptoms similar to the original problem and substantial changes will become necessary. Its application can really be justified only if it can be performed with minimum effort.

The Legacy-First approach attempts to preserve the legacy system's functionality in the new environment. The aim is to preserve the system or its functionality in the new environment, and no effort is made to update, modify or improve the legacy system. This solution can extend the life of the legacy system in its current form, which is very often necessary with widely used, popular software. In many cases the user may not even notice any difference between using the old and new systems. In practice, this method builds a shell around the legacy system, and interfaces the system functions with the new environment. The new system is inherently constrained by the legacy system, even though the shell itself can manipulate the data if required, and there can be a limited opportunity to make some improvements. All development efforts are concentrated on building this shell, which means the software development process can easily be kept under control.

The Component-First approach puts the requirements of the new system first, while using the related functions of the legacy system. During specification of the new system the legacy system's services are examined, modifications are considered and a revised set of services is defined. In this process the reusable functions are identified and the necessary changes are elaborated. The new system can have additional components, e.g. can have a new type of user interface, and eventually the user may not notices that the business logic and other internals of the system remained the same. The Component-First method still aims at minimising the changeover effort though, as individual components of the legacy system are the building blocks of the new system. But the approach gives more control over the system by allowing a reassessment of functionality.

3. SOFTWARE DESIGN PATTERNS AND REENGINEERING

Using design patterns for recurring problems in software development is becoming increasingly popular. While each task is a unique case with specific problems, solutions to similar problems offer great help and can significantly reduce the development effort. Each pattern describes a problem scenario together with the description of the solution, and this is presented in a general form not relating to any particular application. There are many handbooks documenting dozens of patterns, e.g. [3], [4], and having a particular problem a related pattern can easily be searched for. In the context of reengineering, common approaches such as black box, Component-First, etc. present recurring scenarios and using patterns is an obvious choice. As each case has its own individual characteristics, patterns need to be adapted, and there can be more than one applicable pattern. Finding the most suitable one needs a good understanding of the problem at hand as well as a good knowledge of design patterns.

In the following sections several black-box reengineering wrapper patterns are analysed, and a pattern for component access control is also described.

3.1 Wrapping Patterns

3.1.1 Proxy Pattern

In the Proxy pattern an object is represented by another, proxy object [3]. The proxy object provides access to the represented object, but hides all the implementation details. This is a very frequently used pattern, remote proxies represent objects in remote address spaces, virtual proxies store partial information only and create/load expensive parts of objects on demand, smart proxies perform additional functions before or after accessing the represented object. The proxy can operate without knowing the type of the real object: accessing the represented object via an abstract interface allows uniform treatment of objects.

In the Proxy pattern one object is represented by another object, i.e. it implements a one-to-one mapping. This can be useful e.g. in Legacy-First type wrapping applications.

3.1.2 Adapter Pattern

In many cases a specific service component needs to be accessed by a client who has an incompatible interface. An adapter can convert one interface into another, and can also perform functions not provided by the server [3].

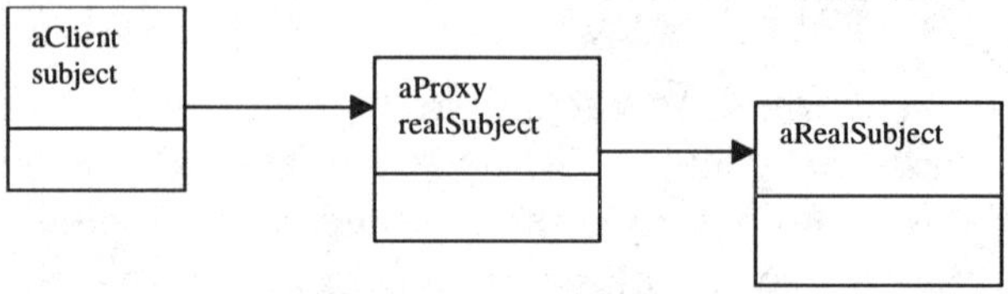

Figure 1 The Proxy Pattern

An adapter can implement a one-to-one or a one-to-many mapping, i.e. an object can represent one or more adaptee objects. This feature can be useful for Legacy-First or Component-First based wrapping.

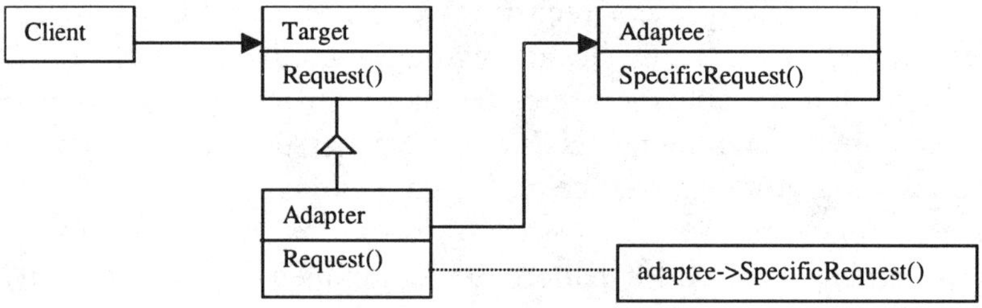

Figure 2 Adapter Pattern

3.1.3 Façade Pattern

To provide a common interface to a group of classes, components or subsystems the Façade pattern can be used. Classes encapsulated by a Façade can also be accessed individually. The main advantage of a Façade is that classes' and other components' collaborations and dependencies can be simplified and decoupled from clients by the Façade.

The Façade pattern provides one-to-many mapping.

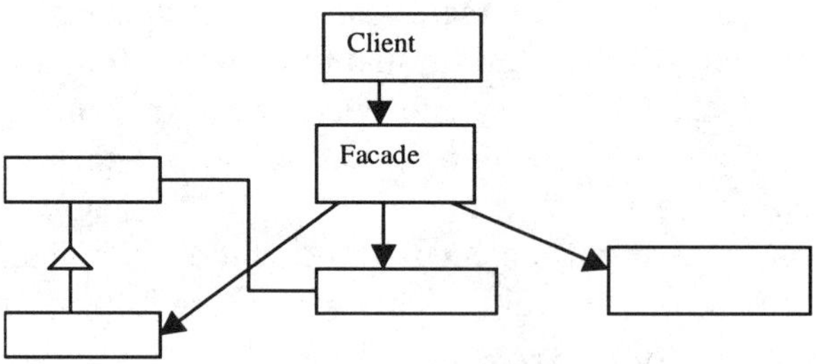

Figure 3 Facade Pattern

3.1.4 Wrapper Façade Pattern

The Wrapper Façade is a version of the Façade pattern, but it is more focused towards reengineering. It implements a simpler structure and can encapsulate data structures and functions to provide conveniently usable and maintainable interfaces [5]. It can also use dynamic assignment of methods, but this requires a great deal of expertise at implementation.

The Wrapper Façade provides one-to-many mapping. In the Virtual-CAD system this pattern was used to interface legacy systems, as it offered a simply manageable mapping between legacy system functions and new requirements.

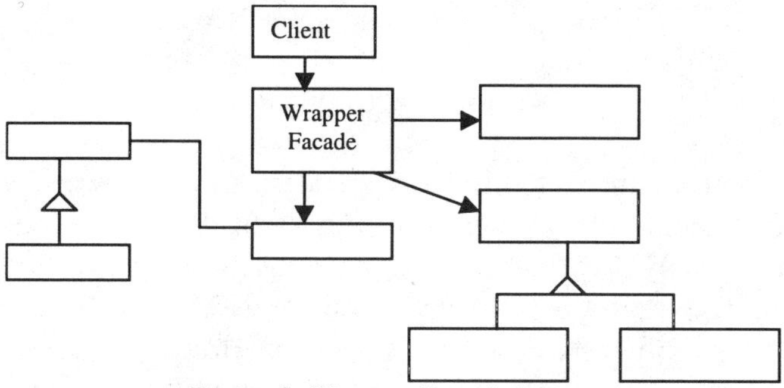

Figure 4 Wrapper Facade Pattern

3.2 A Pattern for controlling access to design components

Users of a system want to access different data in the system, and each user may have different access rights. Access control is provided across the whole system and is encapsulated in the system. It is the job of the system administrator to define access policies of individual objects.

While access rights are connected to users, a user can play several roles during interacting with a system. Hence, access rights are better represented by being connected to roles than by being connected to users. User roles can be modelled with interface objects, and in this model an actor represents a particular role performed by a user [6]. A system can offer different ways or uses to the user, and the set of all these uses form the total functionality of the system. A particular usage of the system can be described in terms of use-cases, i.e. where each case can contain one or more uses [7] [8] [9]. An example of use cases of a design system is given in Figure 5, each oval representing a use case. Before starting to use the system the user is authenticated; this is represented by the *Login* use case. Then the user may want to view a design component; this will involve the *ReadData* use case. Finishing the session will involve the *Logout* use case.

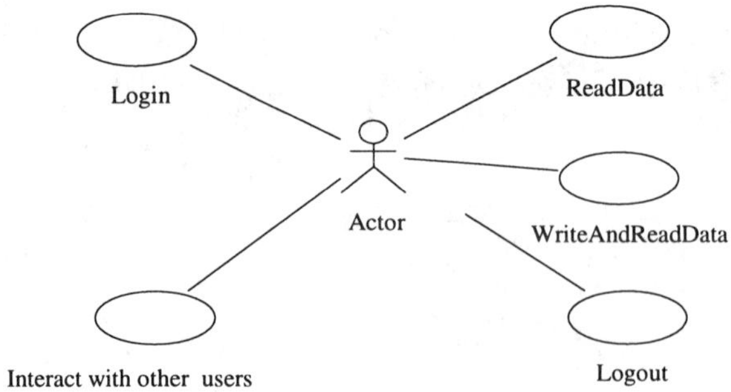

Figure 5 Use Case Model

Access control is reflected in software architecture as well, as illustrated in the object-oriented class diagram in Figure 6. When a user logs into the system, the *SecurityManager* will validate the user's name and password. Then, during a session a new object can be created the following way. The *FactoryManager* creates a *Factory* object and returns a reference of the generated object to the user, who can then use it to access the object. Security related information remains within the system: access rights are assigned by the *FactoryManager* at creation, and they are not passed on to the user.

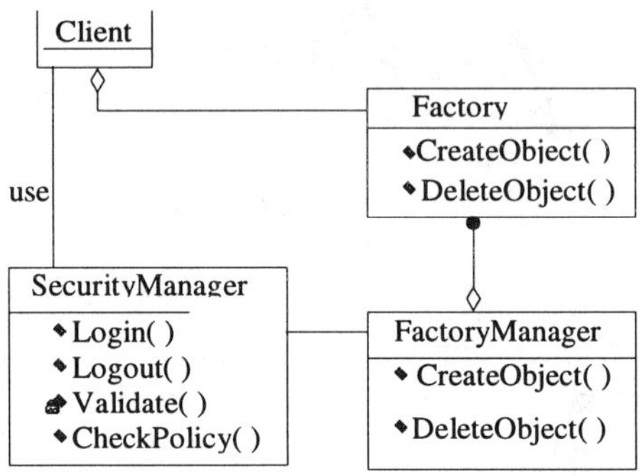

Figure 6 Object-Oriented Diagram of Access Control

Once created, objects can be accessed only via a proxy that enforces access rights and prevents unauthorized access.

4. CONCLUSION

Design systems may need to incorporate different components, some of which may be legacy systems with established usage. Reengineering of some sort is a common way of bringing legacy systems in line with new requirements. Wrapping is a straightforward reengineering method that can be implemented with reasonable effort. Using design patterns is a common way of software development, and several patterns can be used for wrapping.

It was found that

- Legacy-First wrapping can be implemented by using the Proxy pattern. If some enhancements are needed, the Adapter pattern can be more suitable.
- Component-First wrapping may need a more complex pattern, like Façade, that allows new functionality be built on top of the legacy system. The pattern can hide the legacy system and satisfy new requirements while still using the old system's components.

Patterns can also be used to implement additional functionality over integrated components, and an example of enforcing security and access rights illustrated the usage of component access control.

Design system integration with the help of patterns as described here was successfully tested and applied to the development of the Virtual CAD system described in [10].

5. REFERENCES

[1] Computerised Facility Integration, http://www.gocfi.com

[2] Morin T. (1999). Migrating Legacy Systems to CORBA, http://developer.netscape.com:80/viewsource/morin_corba/morin_corba.html, 1999

[3] Gamma E. Helm R. Johnson R. Vlissides J. (1995). Design Patterns, Addison-Wesley, 1995

[4] Rising L. (2000). The Pattern Almanach, Addison-Wesley, 2000

[5] Schmidt D. (1999). Wrapper Façade, C++ Report, February 1999

[6] Jacobson I. (1995) Object-Oriented Software Engineering – A Use Case Driven Approach, Addison-Wesley, 1995

[7] Sigfried S. (1996) Understanding Object-Oriented Software Engineering, IEEE Computer Society Press, 1996

[8] Vadaparty, K. (2000) Use Cases - Basics, Journal of Object-Oriented Programming, Feb. 2000

[9] Vadaparty, K. (2000) Refining the Use-Case Model, Journal of Object-Oriented Programming, May 2000

[10] P. Bertok, J.P.T. Mo, Woodman S. (2000) Collaboration and Application Integration: Distributed Design with Virtual CAD, in this volume

Agility Through Design - The Holonic Multi-cell Control System (HoMuCS) Architecture

J. Schnell, G. Langer, C. Sørensen
Department of Manufacturing Engineering – DTU, Denmark
Em: js@ipt.dtu.dk, gilad@ipt.dtu.dk, cs@ipt.dtu.dk

Keywords Holonic manufacturing systems, capabilities, machine integration

Abstract In a Holonic system the issues concerned with interfacing to the physical equipment are of utmost importance. The two performance criteria for agile systems defined as structural and operational performance have to be optimised. This necessitates a clear model of the capabilities of the physical equipment encapsulated within the Holon. This is achieved through a generic model representing the states and capabilities of the physical equipments, which the Holon uses for structural and operational changes. This model is implemented in the HoMuCS architecture, which is a specific implementation of the Holonic Manufacturing System theory, with the VMD class. The paper briefly describes the Holonic Manufacturing Systems (HMS) and the HoMuCS architecture. It then supplements with a more in depth description of the characteristics and architecture of this VMD and its capability model. The use of a VMD class is exemplified through a case study.

1 INTRODUCTION

With the extensive growth and extent of information technology into the manufacturing domain and the growing need of manufacturers to increase the flexibility of their manufacturing facilities, new paradigms for manufacturing are becoming more attractive. One of these new paradigms is the Holonic Manufacturing System theory proposed by the HMS consortium [1]. It presents a solution for manufacturing systems that has the ability to accommodate the increasingly dynamic characteristics of the manufacturing environment. This involves many novel features that need to be considered one of them being the manufacturing system and its control, commonly known as Shop Floor Control (SFC).

Current research at the Department has resulted in the Holonic Multi-cell Control System concept that consists of a methodology and system-architecture [2]. HoMuCS is still under development and this paper describes some of the on going research in the integration of the physical manufacturing equipment.

2 HOLONIC MANUFACTURING SYSTEMS

The Holonic Manufacturing System (HMS) concept has evolved as a result of research performed during the last decade. It is inspired by the work of the Hungarian sociologist Athur Köstler. In his book 'The Ghost in the Machine' [3], he proposes a method to describe human behavior and social structures. He introduces the term 'Holon' to describe the basic element of the social structure e.g. the individual. The term 'holon' is made up from the Greek word 'holos' meaning whole and 'on' as in electron that is 'part of' a whole - the atom. The structures that the holons create is called holarchies and are neither just hierarchic nor heterarchic but can be anything in between.

2.1 HOMUCS

HoMuCS is acronym for 'Holonic Multi-cell Control System' and names the research in future manufacturing systems carried out at Department of Manufacturing Engineering at the Technical University of Denmark. HoMuCS proposes an architecture and methodology that can be used to engineer manufacturing execution - or shop floor control systems based on the HMS paradigm.

2.1.1 The architecture

The HoMuCS engineering concept consists of four elements as shown in Figure 1. The central element here is the architecture, which is the set of customizable components for the Holons. These components are abstract meaning that they have to be customized, as part of the engineering process in order to obtain the holons and other structural elements needed for implementation. These are supported by functional requirements defined by the functional models. Each Holon in the architecture describes a generic component of a shop floor control system. The complete description of the architecture and corresponding models are given in [2].

At the moment the system architecture is implemented using the Java programming language and can be viewed in more detail on the HoMuCS website (http://www.homucs.org). The fundamental objective of the architecture is to provide design simplicity and quality of the engineering of industrial HoMuCS solutions.

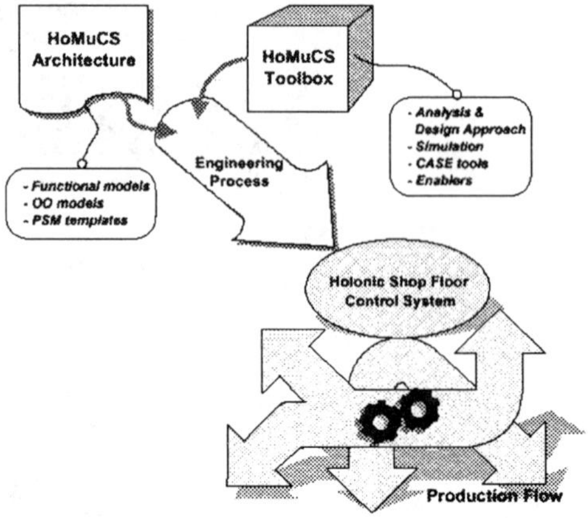

Figure 1: The HoMuCS engineering concept.

2.1.2 The HoMuCS Methodology

Implementation of the architecture is performed according to the HoMuCS methodology [2]. The basic idea is that the architecture is used as a set of guidelines and customizable building blocks that have to be specified and extended for implementation of a HoMuCS. The architecture's functional models are used during the analysis phase of the methodology as functional guidelines for a system implementation. Thus during the analysis phase they are extended to include the needed description for implementation of specific Holon functionality that is not included in their generic form given by the architecture.

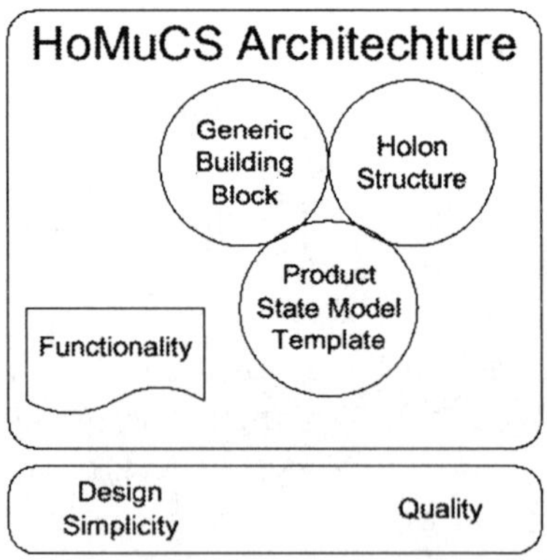

Figure 2: Overview of the HoMuCS architecture

The methodology aids the development of the HoMuCS by directing the assembly of specialization of the abstract components of the architecture. In this way the holonic characteristics of a HoMuCS are secured.

2.2 The structure of the Holon

The HoMuCS architecture defines a resource holon as an abstraction of some production equipment ranging from single machines and plc's to cells, shops and factories. In the context of this paper the main interest is the structure of this types of Holons.

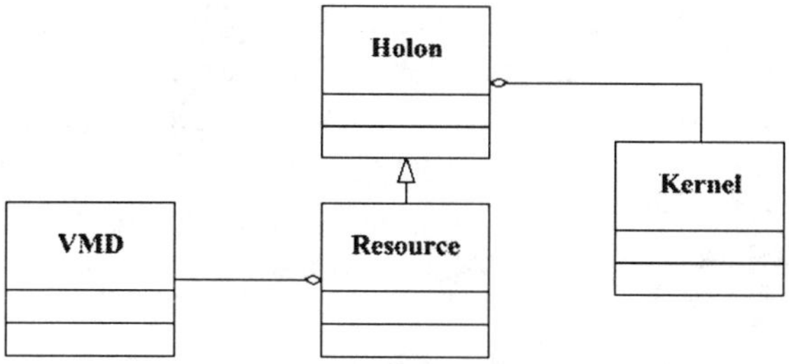

Figure 3 UML structure of the Holon class

A Resource Holon aggregates two main objects, the Kernel and the VMD, Figure 3. Figure 3The Kernel contains the holon's decision logic, which is used to interact using both long and short-term interaction. The Kernel links the holon to the other holons in the holarchy and gives the holon its co-operative behavior. The VMD is the integration of the physical part of the holon. It is the integration of the physical manufacturing equipment that is the main focus of this paper.

3 Integration with manufacturing equipment

One important property about holonic manufacturing systems is that the concept, unlike most traditional manufacturing control systems, does not separate the manufacturing system from the manufacturing control system. The implication of this is that it is essential that the actual state and capabilities of the equipment be known to the holon at any time. Thus a Resource Holon has to be able to specify what tasks it is capable of performing. Such information is an essential requirement during the dynamic process of resource allocation in a HoMuCS.

The physical properties of a machine or machine tool determine the capabilities of that resource. Furthermore the required processing function is a synthesis of two or more basic capabilities. For example when a part has to be moved between locations by a robot. The robot consists of a mechanism and a gripper, where the gripper holds the part that has to be moved and the robot mechanism moves the gripper and the part in space. Here the gripper determines the size of the object that could be moved and the robot mechanism determines the weight and the distance that artifacts can be moved. Thus in a HoMuCS a Robot Holon requires this information to evaluate whether it is capable of executing the task, which it assesses based on the state and capabilities of the physical robot that it encapsulates.

Previous research has shown the benefits of creating a generic interface to the physical equipment [4]. Such an interface, implemented as a software component, gives access to the physical functions of the manufacturing device. Moreover it solves the problem regarding the access to the state and capabilities of the physical equipment. In the HoMuCS architecture this component is defined as the VMD. The name VMD is inspired by the Manufacturing Message Specification, MMS (ISO 9506) [5], and is an acronym for Virtual Manufacturing Device.

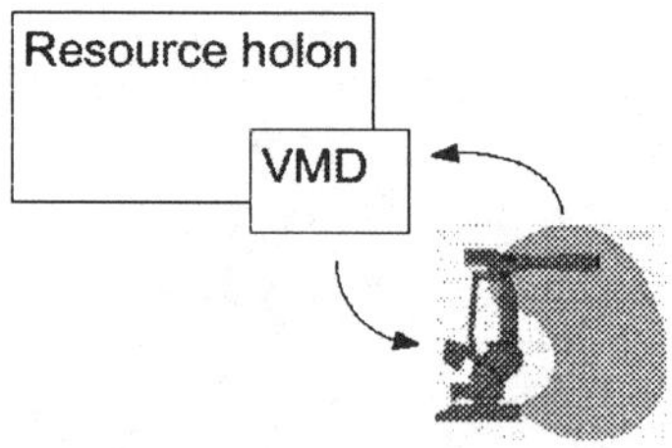

Figure 4 The VMD connects the software and hardware systems

Apart from enabling access to various functions of the equipment, the VMD has a mechanism to evaluate and communicate the capabilities of the equipment. In other words the VMD provides the holon with a generic view and interpretation of the physical equipment that is an integrated part of it. In order do so it has to be able to acquire all the information from the equipment such as states, functions, limitations, etc. and present them as a model of capabilities to the Holon. The holon in turn uses this in its execution logic.

The process of modelling and presenting these capabilities is the main area of focus of this specific research. Currently the capabilities are communicated as text strings that the kernel recognizes and maps them to processing and handling tasks in the manufacturing system.

The VMD performs as a transparent layer between the physical system and the software part of the system, Figure 4. Furthermore it inherently allows for development of a HoMuCS without the need to access the real manufacturing system. This is done by implementing a VMD that emulates the functionality of the physical equipment. This enables the development of SFC systems without direct access to the physical machines and the transfer of the developed system onto the real system by merely changing the VMD implementations.

Currently a prototype implementation of HoMuCS for a steel plate mill is developed that is used to further develop and test the HoMuCS architecture. A brief presentation of this work that focuses on the details regarding integration of the production equipment is given in the following.

4 THE OSS CASE

Odense Steel Shipyard (OSS) produces the world largest container ships. The company is one of the few Scandinavian shipyards, which have survived and prospered, in the strong competition of the last decade. As competition is becoming even harder, OSS continuously thrives to improve the efficiency of their processes. One of these improvements is the implementation of a steel plate milling-cell, an alternative process to using flame cutting for the chamfering process of steel plates. A HoMuCS prototype of this cell has been developed and in the following the results and experiences from this work will be described.

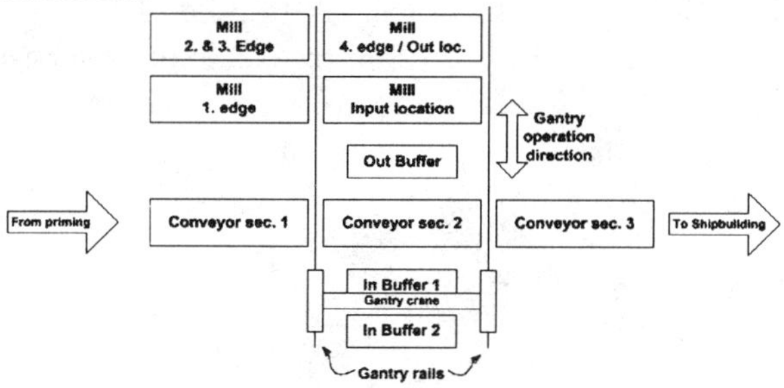

Figure 5 Cell layout

The prototype has a number of holons, three conveyor holons, the crane holon, three buffer holons and the mill holon. Moreover a holon representing the priming station prior to this cell and one representing the shipbuilding after this cell are modeled as wrappers connecting between the holonic cell and the rest of the shipyard.

In this system steel plates are transferred on the conveyors from the priming cell and into the shipbuilding. Some of these plates are assigned for the skin of the ship's hull and need to have their edges chamfered by the milling machine. These plates are lifted off the conveyor and processed by the mill.

In the HoMuCS the steel plate is represented as a plate holon that has to be lifted off the conveyor. To fulfill this requirement a resource holon, with the necessary capabilities that enables it to hold the plate and move it from the conveyor to another location is acquired for that task. The crane's kernel knows through its VMD that it possesses these capabilities and is assigned for the task. From the conveyor the plate is routed to the mill that performs the chamfering. However if the mill is not ready to process that plate, the plate is rejected and has to find an alternative location where it can be placed until the mill is ready to process it. The plate has to find a resource that has the capability of storing it for some time. Buffers have this capability and the plate is moved to one of these and placed there until the mill is ready to process the plate.

The order of the plates is prioritized by the due date. An algorithm is used to minimize the chances that plates with lower priority are placed on top of plates with higher priority. The system allows to introduce more buffers can creating locations that have the capability of storing the steel plates. Since the real manufacturing cell was not available during the development of the HoMuCS an emulation model of the cell was built using a standard simulation software package. Emulation is a representation of an object or system with respect to behavior and operation. For example a processing machine on a shop floor is emulated by representing its behavior when initiated externally by a control system or a human operator. The simulation software used allowed for multiple socket-connections and thus a connection was created to every element/holon in the cell. Figure 6 shows the model as it looks in the simulation software.

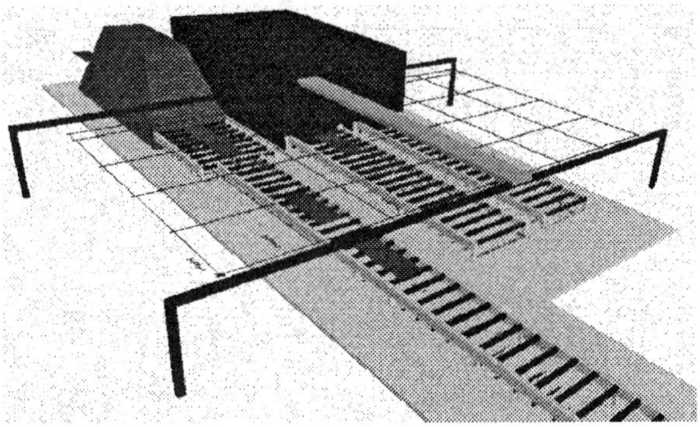

Figure 6 The emulation model

5 CONCLUSION

Development of a generic model for representing the capabilities of the Holonic Resources is an essential part of the architecture since it defines the set of tasks that resources are capable of executing. This is vital for Holonic resources since it allows for holon to make their own plans and execute them – or in other word to be autonomous. The use of the VMD, as shown in this paper, is a potentially good way of solving the interface problem in a holonic architecture. Not only does it give a generic interface but serves as the base for the implementation of the capabilities model. As the paper points out the actual capability model for a resource still needs much more research. Moreover the mechanism and process where the VMD translates the information from the physical equipment to the generic capability that a Resource Holon can use is still very conceptual and is currently the object of intense research.

Using a generic interface to the physical manufacturing equipment facilitates the integration of the machines that the HoMuCS system is dependent on. These generic interfaces to the physical devices make it easy to reconfigure the device itself and even takeaway the machine and replace it with another. This plug-ability also enables the developer to develop and test the system in a virtual environment before taking the system into the production. These properties are secured by the use of the VMD. The case study was part of the ongoing research effort in the field of HMS, which is dedicated to developing the HoMuCS concept. The author will continue the research around the development of a capability model for the HoMuCS architecture. This will specifically involve a method that enables the Holon to combine capabilities that are for example needed for execution of complex production tasks.

6 REFERENCES

[1] Van Brussel, H., Valckanaers, P., Holonic Manufacturing Systems Technical Overview, http://hms.ifw.uni-hannover.de/public/hms_tech.html, PMA Katholieke Universiteit Leuven, Belgium, Dec. 1995.

[2] Langer, G., 1999, HoMuCS – A methodology and architecture for Holonic Multi-cell Control Systems, ISBN 87-90855-00-0

[3] Koestler, A., The Ghost in the Machine, Arkana Books, London 1989, ISBN 0140191925

[4] Schnell, et al., 1999, Development of a robot holon using an open modular controller, Proceedings of 1999 IEEE International Conference on Control Applications, Hawaii, USA

[5] SISCO, 1995, Overview and Introduction to the Manufacturing Message Specification (MMS), Revision 2, Systems Integration Specialists Company, Inc., USA.

Implementation of a Layer Structured Control System on the 'Glue Logic'

Masayuki Takata
The University of Electro-Communications, Japan
takata@cc.uec.ac.jp

Eiji Arai
Osaka University, Japan
aria@mapse.eng.osaka-u.ac.jp

Keywords Manufacturing work-cell control system, Infrastructural software system, Subsumption architecture, Control architecture

Abstract This paper describes an infrastructural system designed for applications ill the factory automation, which is named *Glue Logic*, and a real-time control system with layer structure build oil this system. The Glue Logic provides an environment which supports the programming, controlling and monitoring the production system, and also supports efficient manufacturing programming with high program modularity and re-usability. Furthermore, this system includes event notification message sending and condition monitoring features by means of active database technique. The authors proposed an architecture paradigm for scalable intelligent control systems, and implemented a model train control system on the Glue Logic as a sample, which has layered architecture derived from the *subsumption architecture*.

1 INTRODUCTION

The Glue Logic is an infrastructural system designed to make building manufacturing work-cell control systems easy and flexible [1] [2]. This system binds multiple application software modules, referred as *agents*, developed separately, and coordinates agents by means of inter-process massage passing. As the Glue Logic supports event notification and condition monitoring features based on active database scheme, users can easily build event-driven applications.

All the data and agents in a system can be represented by symbolic names defined in the Glue Logic, and are accessed without any knowledge on implementation of other agents. Glue Logic compliant agents are easy to re-use, and the users can build large libraries of application agents. This makes the life cycle and the reliability of such agents are extended, and their development cost is greatly reduced.

This paper describes the Glue Logic designed as the infrastructural system for factory automation applications, and the scalable intelligent control architecture paradigm to make up control systems, of which intelligence can be added in steps. Section 2 illustrates description of the Glue Logic. Section 3 discusses the programming paradigm of the layered structured control architecture proposed by the authors, a modification of the *subsumption architecture*. In Section 4, a sample scalable intelligent control system implemented in this paradigm is discussed. Lastly, in section 5, the architecture of the coming multi-agent real-time control systems are discussed, and the Glue Logic is evaluated.

2 THE GLUE LOGIC

2.1 ARCHITECTURE

The Glue Logic has been developed to support application programming by means of data sharing, event notification and condition monitoring. As the system uses inter-process communication internally over the network, the Glue Logic can play the roles of the infrastructure of the distributed manufacturing work-cell control systems. Furthermore, the Glue Logic is effective not only in the distributed work-cell environment, but also in the single work-cell system, to keep application agents simple and highly independent from others.

The Clue Logic provides a framework shown in Figure 1. In this figure, the shaded part shows the service of the Glue Logic, playing the role of the MES. The boxes above represent application processes referred as agents, which utilize the function of the Glue Logic.

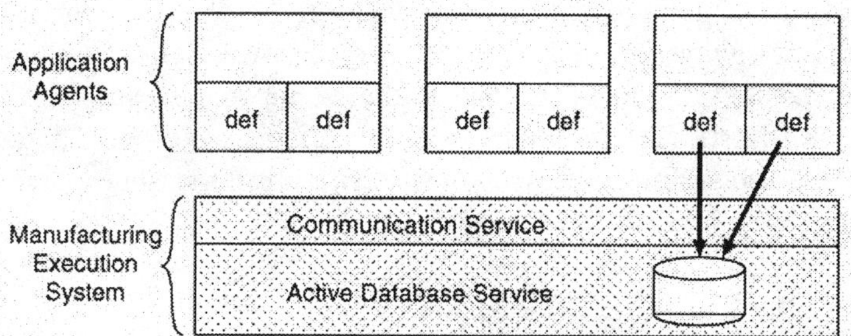

Figure 1: Framework provided by the Glue Logic.

In this figure, *def* represents the formal definition of the data to be imported from or exported to other agents. These information is centralized at the active database part of the Glue Logic.

The Glue Logic relays all inter-process communication among its agents, and manages all shared data. Because of this, the Glue Logic can send change notification messages, when the values of the shared data are altered. As the counterpart of the communication can be virtualized by relaying all the inter-agent communication, each agent can be independent from adding, deleting and altering other agents.

2.2 IMPLEMENTATION

The design of the Glue Logic is based on the client-server model of transaction processing, as shown in Figure 2, though there is no need for general users to know about its implementation.

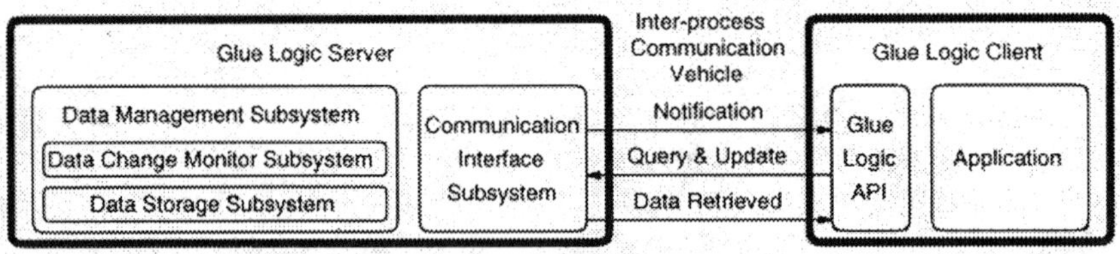

Figure 2: Configuration of the Glue Logic

The server process of the Glue Logic is a specific process running on a specific processor. All agents' processes communicate only with this server process over network, and there is no redundancy in this system.

The server process consists of two major parts: the communication interface subsystem and the data management subsystem. The former exchanges information with agents running in both the same work-cell controller and remote work-cell controllers over the network.

The data management subsystem consists of also two parts: the data change monitor subsystem and the data storage subsystem. The data storage subsystem manages the association of the *name* and the *value*. The data change monitor subsystem monitors the changes in the data storage subsystem and sends out the data change notification messages, and executes depending data evaluation.

2.3 BEHAVIOR OF THE GLUE LOGIC

The atomic element of the Glue Logic is the tuple of a *name* and its *value*. The *name* resembles variable identifier, and can have a value. The name is a sequence of some identifiers, separated by a period, such as abc.ijk.xyz. Using this format, users can denote data structure by the sequence of identifiers. The agent programmers can implement arbitrary data structures. In the elements of one structure, their names contain same identifier sequence in its leading part. The trailing part of their name differs from each other. The leading common part is called a *stem* and the trailing part is called a *variant.*

Each name may have some *attributes*. The attributes denote optional characteristics of corresponding names, and the Glue Logic changes its behavior according to the values of attributes. As the value of the name, application programs may specify one of followings; integer, floating point real, character string, expression and link. As the name itself is not typed, users may bind any types of data in turns. If an agent accesses the name bounded to an expression type value, the expression is evaluated and the result is returned as the value. Using the link, users can point another name. This is equivalent to the *symbolic link* used in the UNIX file systems.

2.4 ACATIVE DATABASE FEATURES

In order to eliminate data polling and to decrease the network load, the feature of data change notification is introduced. The agents, which want to receive a change notification of a certain name's value, can register one's agent ID the name. The agent ID list of the notification destination is kept as the value of *Inform To* attribute.

On the time when the Glue Logic server receives data update request, it searches for the agents registered as the notification destination, and then notify to all the registered agents. In this way, the application programs are freed from polling to find status change.

Some agents using the Glue Logic may need to know the value of name being a certain constant value, or the values of names satisfy a certain condition. The Glue Logic can be set to send a notification message only if a certain condition is met.

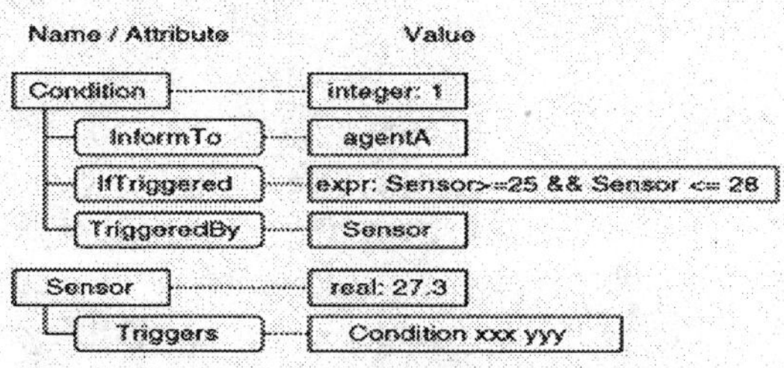

Figure 3: Data used by the condition monitoring.

As shown in Figure 3, a name in the Glue Logic can have dependence lists as the values of *Triggers* and *TriggeredBy* attributes. If one or more elements of the list in the some name's *TriggeredBy* attribute is updated, the value of the name itself is updated to have the result of an expression, which is also registered as the value of *IfTriggered* attribute. If this new value differs from the former, the data change notification is sent to its notification destinations.

With this mechanism, user can monitor whether sensor read-out stays within a control limit, or can implement multi-way branching by comparing value of a name to some given constants. These checking may be requested by each agents separately, users can add other sensors to be watched or branching target without modifying existing agents.

3 TOWARDS SCALABLE INTELLIGENT CONTROL ARCHITECTURE

To obtain scalable intelligent control systems, the authors started from making a programming paradigm for such control systems. The word *scalable* means that the architecture of the control systems permits the step-wise addition of much more abstracted intelligence.

The *scalable intelligent control architecture* designed by authors consists of multiple layers, and is derived from the *subsumption architecture* proposed by Brooks [3].

3.1 Subsumption Architecture

The subsumption architecture consists of multiple layers, and was designed for mobile robot control hardware systems. Each layer consists of multiple finite-state machines and are connected by low-speed serial communication links.

The layers in the subsumption architecture are divided by task-achieving behavior, not by sequence of data flow which is frequently found. Layers operate simultaneously, where lower layers continue to realize low level feed-back operations (i.e. less abstracted operations), while higher layers controls the behavior of the system by offsetting the control output of next lower layers (i.e. more abstracted, supervisory operations).

With the Brooks' subsumption architecture, it is very easy to add-on much more intelligent controlling and planning, by integrating much more higher level layer which monitors status of lower layer and emits directive information.

3.2 Scalable Intelligent Control Architecture

The authors have modified the subsumption architecture, in order to fit it to multiple software agent control systems.

The *scalable intelligent control architecture* is also one of layer structured architectures. Each layer consists of one or more software agents, and their communications are strongly supported by the Glue Logic. Lower layers process plant operational or tactical tasks, while higher layers process strategic or user-interface tasks. And, lower layers send messages to next higher to show the current *abstracted* status of the controlled system, while higher layers send messages to next lower to show the current sub-goal or the final goal. The scalable intelligent control architecture is illustrated in Figure 4.

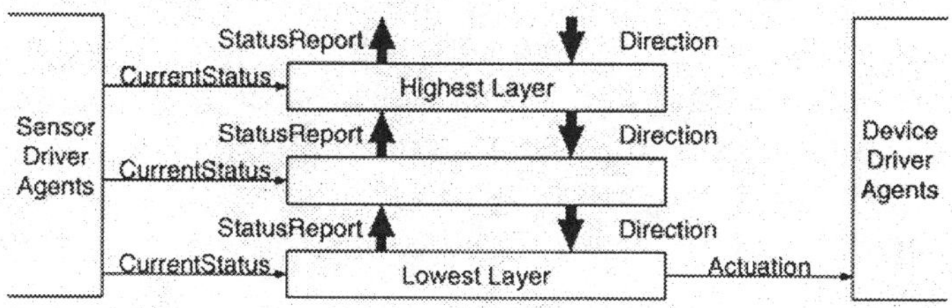

Figure 4: The Scalable Intelligent Control Architecture.

4 IMPLEMENTATION OF THE SUBSUMPTION ARCHITECTURE WITH THE GLUE LOGIC

Using the Glue Logic, a real-time N gauge model train control system has been developed, in order to evaluate the Glue Logic itself and visualize an AGV traffic control system. In this system, a few trains are controlled within a single train layout including many points and crossings, and their speed is controlled without crashing. All trains are given their destination sections, and their routing and their speed are planned and controlled adaptively at run-time.

This system consists of about 8 kinds, approximately 20 agents in three layers, including user-interface agents.

4.1 The Model Train Layout

The train layout is divided into 48 sections, as shown in Figure 5. This train layout consists three loops and two of them are overlapping in zones E and G, and those loops are interchangeable with two X crossing points in zones B and D. There are four escape lines to permit over-taking and passing on-coming trains in zones A, C, F and H.

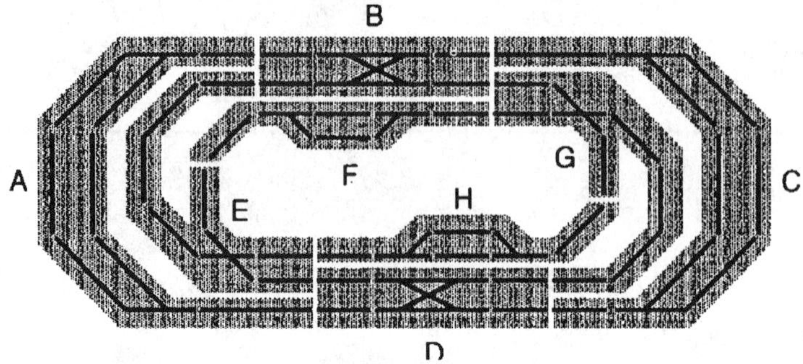

Figure 5: The Railway Layout

The power driving train is provided to each sections independently. The power is controlled to realize three speeds by means of pulse width modulation. Some sections contain one points each, and those points are controlled electrically by point control agents.

4.2 Attributes of the Trains and Sections

Each trains has *destination section* and *maximum speed* as their attributes. Also they have *maximum reserved section number* as their attributes, which limits the number of sections to be exclusively reserved before the train itself enters into them.

Each section has information of its neighboring sections. There no agent which has entire route map. In the case of the section reserved, the train ID, its destination and its maximum speed are also kept in the section data structure.

4.3 Running Control Layer

In this control system, each train should reserve the section before it enters, and it tries to reserve multiple sections if possible. This section reservation scheme is the base of system integrity, and the most important functionality of this lowest layer. In accordance with the number of reserved successive sections, the running speed of the train is designated.

If another train is ahead of a train, the train fails to reserve the section in which the other train is. As the number of reserved section decreases, the running speed of the following train is limited, or the following one is stopped until the reserved section is released.

4.4 Routing Control Layer

This layer designates route by supplying the check points (i.e. subgoals) to the destination. As the running control layer accept only one destination, this layer should pass an appropriate sub-goal as the train's current position changes.

This layer can accept a sequence of sub-goals or some combinational behaviors, such as loop *forever on outer loop*.

4.5 User-Interface Layer

This layer is the highest one because the next higher layer is the human operator. Generally, as the source of the intension is a human, the uppermost control layer consists of agents working as command receptors, status reporters and operation loggers.

In this model train control system, the user-interface layer accepts sequences of destinations for each trains, and other control parameters for them. There are also status display agents, which shows the status of the track at that moment, the status of each trains and some vital internal parameters.

4.6 Result of the Implementation

With this control system, the authors implemented the trailing behavior of multiple trains, the waiting for other trains at a crossing points and the safe merging at the junction points.

At this time, this system controls up to 4 trains, which raise about 10 events per one second, and the control system operated with the control frequency of approximately 10 Hz.

The authors are going to add on much more higher layer on the current control system, in order to implement much more intelligent behavior, such as scheduling with state forecasting using knowledge-base system agents. Even in that case, lower layers require no modification.

5 CONCLUSION

In this paper, the outline of the Glue Logic and the scalable intelligent control architecture are described. The authors believe in the effectiveness of the concept of infrastructural system for agent based processing and the feasibility of the scalable intelligent control architecture, which binds multiple tasks to be executed in the manufacturing work-cells, defined as the agents.

To develop flexible manufacturing system control software, it should take less cost, less time and more reliability, as manufacturer may have to develop a new manufacturing control software for producing only one instance. The library which consists of widely used agents is strongly required for this requirement, and the concept of the subsumption architecture is the most efficient way to refine manufacturing system stepwisely.

The authors would like to emphasize that the smart mechanism of the Glue Logic is the very thing to make the programming system powerful and easy to be programmed, especially in the execution environment which deals with and coordinates large and complicated application agents.

6 REFERENCES

[1] Takata, M., Arai, E. (1997): "The Glue Logic: An Information Infrastructure System for Distributed Manufacturing Control," *Proc. of the Intl, Conf on Manufacturing Milestones toward the 21st Century*, The Japan Society of Mechanical Engineers, pp.549 – 554, Tokyo. Japan, July 23-25, 1997.

[2] Takata, M., Arai, E. (1996): "The Glue Logic: All Integrated Programming/Execution Environment, for Distributed Manufacturing Work-Cell Control System." *Information Infrastructure Systems for Manufacturing (Ed. Goossenaerts, J., Kimura, F., Wortmann.. H.), Proc. of DIISM '96*, pp.181 - 192, Chapman & Hall, 1997.

[3] Brooks, R. A. (1986): "A Robust Layered Control System For A Mobile Robot," IEEE *Journal of Robotics and Automation*, Vol. RA-2, No. 1, 1986